MW01178445

Statistical and Computational Techniques in Manufacturing

J. Paulo Davim (Ed.)

Statistical and Computational Techniques in Manufacturing

Editor
J. Paulo Davim
University of Aveiro
Campus Santiago
Department of Mechanical Engineering
Aveiro
Portugal

ISBN 978-3-642-25858-9 e-ISBN 978-3-642-25859-6
DOI 10.1007/978-3-642-25859-6
Springer Heidelberg New York Dordrecht London

Library of Congress Control Number: 2011946096

Printed on acid-free paper

Springer is part of Springer Science+Business Media (www.springer.com)

Preface

In recent years, has been increased interest in developing statistical and computational techniques for applied in manufacturing engineering. Today, due to the great complexity of manufacturing engineering and the high number of parameters used conventional approaches are no longer sufficient. Therefore, in manufacturing, statistical and computational techniques have achieved several applications, namely, modelling and simulation manufacturing processes, optimisation manufacturing parameters, monitoring and control, computer-aided process planning, etc.

The chapter 1 of the book provides design of experiment methods in manufacturing (basics and practical applications). Chapter 2 is dedicated to stream-of-variation based quality assurance for multi-station machining processes (modelling and planning). Chapter 3 described finite element modelling of chip formation in orthogonal machining. Chapter 4 contains information on GA-fuzzy approaches (application to modelling of manufacturing process) and chapter 5 is dedicated of single and multi-objective optimization methodologies in CNC machining. Chapter 6 described numerical simulation and prediction of wrinkling defects in sheet metal forming. Finally, chapter 7 is dedicated of manufacturing seamless reservoirs by tube forming (finite element modelling and experimentation).

The present book can be used as a research book for final undergraduate engineering course or as a topic on manufacturing at the postgraduate level. Also, this book can serve as a useful reference for academics, manufacturing and computational sciences researchers, manufacturing, industrial and mechanical engineers, professional in manufacturing and related industries. The interest of scientific in this book is evident for many important centers of the research, laboratories and universities as well as industry. Therefore, it is hoped this book will inspire and enthuse others to undertake research in this field of statistical and computational techniques in manufacting.

The Editor acknowledges Springer for this opportunity and for their enthusiastic and professional support. Finally, I would like to thank all the chapter authors for their availability for this work.

Aveiro, Portugal
February 2012

J. Paulo Davim

Contents

Contributors

Viktor P. Astakhov
(Chap.1)
General Motors Business
Unit of PSMi, 1255 Beach Ct., Saline
MI 48176, USA
Email: astvik@gmail.com

J. V. Abellan-Nebot
(Chap.2)-Corresponding author
Department of Industrial Systems
Engineering and Design,
Universitat Jaume I, Av. Vicent Sos
Baynat s/n. Castelló, C.P. 12071. Spain
Email: abellan@esid.uji.es

J. Liu
(Chap.2)
Department of Systems and Industrial
Engineering
The University of Arizona. Tucson,
AZ 85704. USA
Email: jianliu@sie.arizona.edu

F. Romero Subiron
(Chap.2)
Department of Industrial Systems
Engineering and Design,
Universitat Jaume I, Av. Vicent Sos
Baynat s/n. Castelló, C.P. 12071. Spain
Email: fromero@esid.uji.es

Amrita Priyadrashini
(Chap.3)
Department of Mechanical Engineering,
Indian Institute of Technology,
Kharagpur, Kharagpur- 721 302,
West Bengal, India
Email: amrita@mech.iitkgp.ernet.in

Surjya K. Pal
(Chap.3)–Corresponding author
Department of Mechanical Engineering
Indian Institute of Technology,
Kharagpur- 721 302, West Bengal, India
Email: skpal@mech.iitkgp.ernet.in

Arun K. Samantaray
(Chap.3)
Department of Mechanical Engineering,
Indian Institute of Technology,
Kharagpur- 721 302, West Bengal, India
Email: samantaray@mech.iitkgp.ernet.in

Arup Kumar Nandi
(Chap.4)
Central Mechanical Engineering
Research Institute (CSIR-CMERI)
Durgapur-713209, West Bengal, India
Email: nandi@cmeri.res.in

Nikolaos Fountas
(Chap.5)
Department of Mechanical Engineering
Technology Educators,
School of Pedagogical and
Technological Education (ASPETE),
GR-14121, N. Heraklion Attikis, Greece
Email: n.fountas@webmail.aspete.gr

Agis Krimpenis
(Chap.5)
Department of Mechanical Engineering
Technology Educators,
School of Pedagogical and
Technological Education (ASPETE),
GR-14121, N. Heraklion Attikis, Greece
Email: a.krimpenis@webmail.aspete.gr

Nikolaos M. Vaxevanidis
(Chap.5) – Corresponding author
Department of Mechanical Engineering
Technology Educators,
School of Pedagogical and
Technological Education (ASPETE),
GR-14121, N. Heraklion Attikis, Greece
Email: vaxev@aspete.gr

J. Paulo Davim
(Chap.5)
Department of Mechanical Engineering,
University of Aveiro,
Campus Santiago, 3810-193,
Aveiro, Portugal
Email: pdavim@ua.pt

M. P. Henriques
(Chap.6)
GRIDS Research Group,
Centre for Automation and Mechanical
Technology (TEMA),
Department of Mechanical Engineering,
University of Aveiro,
3810-193 Aveiro, Portugal
Email: marisahenriques@ua.pt

T. J. Grilo
(Chap.6)
GRIDS Research Group,
Centre for Automation and
Mechanical Technology (TEMA),
Department of Mechanical Engineering,
University of Aveiro,
3810-193 Aveiro, Portugal
Email: a38296@ua.pt

R. J. Alves de Sousa
(Chap.6)
GRIDS Research Group,
Centre for Automation and Mechanical
Technology (TEMA),
Department of Mechanical Engineering,
University of Aveiro,
3810-193 Aveiro, Portugal
Email: rsousa@ua.pt

R. A. F. Valente
(Chap.6) – Corresponding author
GRIDS Research Group,
Centre for Automation and
Mechanical Technology (TEMA),
Department of Mechanical Engineering,
University of Aveiro,
3810-193 Aveiro,
Portugal
Email: robertt@ua.pt

Luís M. Alves
(Chap.7)
IDMEC, Instituto Superior Técnico,
Universidade Técnica de Lisboa,
Av. Rovisco Pais s/n,
1049-001 Lisboa,
Portugal
Email: luisalves@ist.utl.pt

Pedro Santana
(Chap.7)
OMNIDEA, Aerospace Technology
and Energy Systems,
Tv. António Gedeão,
9, 3510-017 Viseu,
Portugal
Email: pedro.santana@omnidea.net

Nuno Fernandes
(Chap.7)
OMNIDEA, Aerospace Technology
and Energy Systems,
Tv. António Gedeão,
9, 3510-017 Viseu,
Portugal
Email:nuno.fernandes@omnidea.net

Paulo A. F. Martins
(Chap.7) – Corresponding author
IDMEC, Instituto Superior Técnico,
Universidade Técnica de Lisboa,
Av. Rovisco Pais s/n,
1049-001 Lisboa,
Portugal
Email: pmartins@ist.utl.pt

1

Design of Experiment Methods in Manufacturing: Basics and Practical Applications

Viktor P. Astakhov

General Motors Business Unit of PSMi, 1255 Beach Ct., Saline MI 48176, USA
astvik@gmail.com

This chapter discuses fundamentals and important practical aspects of design of experiments (DOE) in manufacturing. Starting with the basic terminology involved in DOE, it walks a potential reader through major types of DOE suitable for manufacturing testing. Particular attention is paid to the pre-process decisions and test preparation. The most common DOE as the full and fractional factorial including response surface analysis as well as special DOE as the Taguchi DOE sieve DOE, split-plot DOE, and group method of data handling (GMDH) are discussed. The adequate examples are given for some special DOE. The chapter ends with detailed description of the strategy and principal steps in DOE.

1.1 Introduction

1.1.1 Design of Experiment as a Formal Statistical Method

Genomics researchers, financial analysts, and social scientists hunt for patterns in vast data warehouses using increasingly powerful statistical tools. These tools are based on emerging concepts such as knowledge discovery, data mining, and information visualization. They also employ specialized methods such as neural networks, decisions trees, principal components analysis, and a hundred others. Computers have made it possible to conduct complex statistical analyses that would have been prohibitive to carry out in the past. However, the dangers of using complex computer statistical software grow when user comprehension and control are diminished. Therefore, it seems useful to reflect on the underlying philosophy and appropriateness of the diverse methods that have been proposed.

This could lead to a better understanding of when to use given tools and methods, as well as contribute to the invention of new discovery tools and refinement of existing ones. This chapter concentrates on one of the most powerful statistical method known as design of experiments (hereafter, DOE) or experimental design.

DOE is an statistical formal methodology allowing an experimentalist to establish statistical correlation between a set of input variables with a chosen outcome of the system/process under study under certain uncertainties, called uncontrolled inputs. The visualization of this definition is shown in Figure 1.1 where (x_1, x_2,...x_n) are n input variables selected for the analysis; (y_1, y_2,...y_m)) are m possible system/process outputs from which one should be selected for the analysis; and (z_1, z_2,...z_p) are p uncontrollable (the experimentalist has no influence) inputs (often referred to as noise).

The system/process is designated in Figure 1.1 as black box[1], i.e. it is a device, system or object which can be viewed solely in terms of its input, output, and transfer (correlation) characteristics without any knowledge of its internal workings, that is, its implementation is "opaque" (black). As a result, any model established using DOE is not a mathematical model (although it is expressed using mathematical symbols) but formal statistical or correlation model which does not have physical sense as a mathematical model derived using equations of mathematical physics with corresponding boundary conditions. Therefore, no attempt should normally be made to derive some physical conclusions from the model obtained as statistical correlation can be established between an input and the output which are not physically related.

For example, one can establish a 100% correlation between the rate of grass growth in his front yard and the level of water in a pond located in the neighborhood by carrying our perfectly correct statistical analysis. However, these two are not physically related. This physically misleading conclusion is obtained only because the amount of rainfall that affects both the rate of grass growth and the level of water was not considered in the statistical analysis. Therefore, the experimentalist should pay prime attention to physical meaning of what he is doing at all stages of DOE in order to avoid physically meaningless but statistically correct results. In the author's opinion, this is the first thing that a potential user of DOE should learn about this method.

---

[1] The modern term "black box" seems to have entered the English language around 1945. The process of network synthesis from the transfer functions of black boxes can be traced to Wilhelm Cauer who published his ideas in their most developed form in 1941. Although Cauer did not himself use the term, others who followed him certainly did describe the method as black-box analysis.

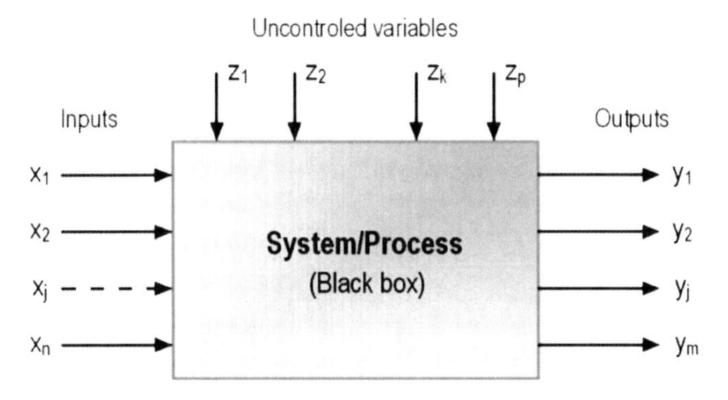

Fig. 1.1. Visualization of definition of DOE

1.1.2 Short History

For centuries, the common experimentation method was conducted using OFAT (one-factor-at-a-time) experimentation. OFAT experimentation reached its zenith with the work of Thomas Edison's "trial and error" methods [1]. In OFAT, a single variable is varied at a time, keeping all other variables in the experiment fixed. The first factor is fixed as a "good" value, the next factor is examined, and on and on to the last factor. Because each experimental run considers only one factor, many runs are needed to obtain sufficient information about the set of conditions contributing to the problem. This consumes a hefty amount of time and money, along with running a high risk of error. Another limitation is that when factors change, they generally change together, so it is impossible to understand the best solution by pointing to a single, isolated factor. Traditional OFAT experimentation frequently reduces itself to no methodology whatsoever—just trial and error and a reliance upon guesswork, experience, and intuition of its users. As a result, OFAT experiments often unreliable, inefficient, time consuming and may yield false optimum condition for the process [2].

Time has changed with growth of manufacturing. Industrial researches no longer could afford to experiment in a trial-and-error manner, changing one factor at a time, the way Edison did in developing the light bulb. A need was felt to develop a better methodology to run tests where many factors and their combination affect the outcome.

A far more effective method is to apply a systematic approach to experimentation, one that considers all factors simultaneously. That approach is called design of experiments (DOE). A methodology for designing experiments was proposed by Ronald A. Fisher, in his innovative book *The Design of Experiments* [3] the first addition of which was published in 1935 [4]. He was described by Anders Hald, a Danish statistician who made contributions to the history of statistics, as "a genius who almost single-handedly created the foundations for modern statistical science." [5].

Although Fisher is regarded as the founder of the modern methods of design and analysis of experiments, it would be wrong, however, to image that there had been no development of such a technique before Fisher. One of the very first scientific papers to relate to the topic of design of experiments was published in the statistical journal Biometrika in 1917 [6]. The author, who used the famous pseudonym "Student", was also responsible for developing "Student's t-test" in 1908. His true identity was Willian Sealey Gosset.

In the 1920s, Fisher developed the methods into a coherent philosophy for experimentation and he is now widely regarded as the originator of the approach. He made a huge contribution to statistics in general, and to design of experiments in particular, from his post at the Rothamsted Experimental Station in Harpenden, UK. Fisher initiated the principles of DOE and elaborated on his studies of "analysis of variance". He also concurrently furthered his studies of the statistics of small samples.

Perhaps even more important, Fisher began his systematic approach of the analysis of real data as the springboard for the development of new statistical methods. He began to pay particular attention to the labor involved in the necessary computations performed by hand, and developed methods that were as practical as they were founded in rigor. In 1925, this work culminated in the publication of his first book, *Statistical Methods for Research Workers* [7]. This went into many editions and translations in later years, and became a standard reference work for scientists in many disciplines.

Although DOE was first used in an agricultural context, the method has been applied successfully in the military and in industry since the 1940s [8]. Besse Day, working at the U.S. Naval Experimentation Laboratory, used DOE to solve problems such as finding the cause of bad welds at a naval shipyard during World War II. W. Edwards Deming taught statistical methods, including experimental design, to Japanese scientists and engineers in the early 1950s at a time when "Made in Japan" meant poor quality. Genichi Taguchi, the most well known of this group of Japanese scientists, is famous for his quality improvement methods using modified fractural DOE presently well-known as Taguchi method [9]. One of the companies where Taguchi first applied his methods was Toyota. Since the late 1970s, U.S. industry has become interested again in quality improvement initiatives, now known as "Total Quality" and "Six Sigma" programs [10, 11]. DOE is considered an advanced method in the Six Sigma programs, which were pioneered at Motorola and General Electric and nowadays becoming common in the automotive industry [12].

Classical full and fractional factorial designs [13] were already in use at the beginning of the 20th century, while more complex designs, such as D-optimal [14], came with the arrival of modern computers in the 1970s. It was also around that time that more advanced regression analysis and optimization techniques were developed [15].

Today, corporations across the world are adopting various computer-enhanced DOEs as a cost-effective way to solve serious problems afflicting their operations. In the author's opinion, the highest level of DOE is the so-called group of inductive learning algorithms for complex system modeling [16, 17]. The most powerful method in this group is the Group Method of Data Handling (GMDH). It is the inductive sorting-out method, which has advantages in the cases of rather complex objects, having no definite theory, particularly for the objects with fuzzy characteristics. Although the algorithms of this method are well developed, they are not a part of standard DOE computer programs.

1.1.3 What Is This Chapter All about?

A great number of papers, manuals, and books have been written on the subject in general and as related to manufacturing in particular. Moreover, a number of special (for example, Satistica), specialized (for example, Minitab) and common (for example, MS Excel) computer programs are available to assist one to carry out DOE with endless examples available in the Web. Everything seems to be known, the terminology, procedures, and analyses are well developed. Therefore, a logical question why this chapter is needed should be answered.

The simple answer is that this chapter is written from the experimentalist side of the fence rather than the statistical side used in vast majority of publication on DOE. As the saying goes "The grass is always greener on the other side of the fence," i.e. "statisticians" often do not see many of real-world problems in preparing proper tests and collecting relevant data to be analyzed using DOE. Moreover, as mentioned above, the dangers of using complex computer statistical easy-to-use software with colorful user interfaces and looks-nice graphical representation of the results grow while user comprehension, and thus control are diminished.

This chapter does not cover basic statistics so that a general knowledge of statistics including probability concept, regression, and correlation analysis, statistical distributions, statistical data analysis including survey sampling has to be refreshed prior to working with this chapter. Rather it presents the overview of various DOE to be used in manufacturing commenting its suitability for particular cases.

1.2 Basic Terminology

Terms commonly used in DOE are [1, 2, 8, 18]:

- *Response*: A response is the dependent variable that corresponds to the outcome or resulting effects of interest in the experiment. One or more response variables may be studied simultaneously.
- *Factors*: A factor is a variable contribute to the response.
- *Levels*: The levels are the chosen conditions of the factor under study.
- *Orthogonality* can be thought of as factors independence. It in an experiment results in the factor effects being uncorrelated and therefore more easily interpreted. The factors in an orthogonal experiment design are varied independently of each other. The main results of data collected using this design can often be summarized by taking differences of averages and can be shown graphically by using simple plots of suitably chosen sets of averages. In these days of powerful computers and software, orthogonality is no longer a necessity, but it is still a desirable property because of the ease of visualizing, and thus explaining results.
- *Blocks/Blocking*: A block is a homogenous portion of the experimental environment or materials that bears certain variation effects on the response(s). A block may be a batch of material supplied by a vendor or products manufactured in a shift on a production floor. Blocking is a method for increasing precision by removing the effect of known nuisance factors. An example of a known nuisance factor is batch-to-batch variability. In a blocked design, both the baseline and new procedures are applied to samples of material from one batch, then to samples from another batch, and so on. The difference between the new and baseline procedures is not influenced by the batch-to-batch differences. Blocking is a restriction of complete randomization, since both procedures are always applied to each batch. Blocking increases precision since the batch-to-batch variability is removed from the "experimental error."
- *Randomization* is a method that protects against an unknown bias distorting the results of the experiment. An example of a bias is instrument drift in an experiment comparing a baseline procedure to a new procedure. If all the tests using the baseline procedure are conducted first and then all the tests using the new procedure are conducted, the observed difference between the procedures might be entirely due to instrument drift. To guard against erroneous conclusions, the testing sequence of the baseline and new procedures should be in random order such as B, N, N, B, N, B, and so on. The instrument drift or any unknown bias should "average out."
- *Replica/replication* is a result/process of a complete repetition of the same experimental conditions, beginning with the initial setup. Replication increases the signal-to-noise ratio when the noise originates from uncontrollable nuisance variables common in real-world manufacturing. Replication increases the sample size and is a method for increasing the precision of the experiment. However, the 'size' of experiment (cost, time) increases at the same rate, more difficult to make a great number of test at the same

conditions. A special design called a *Split Plot* can be used if some of the factors are hard to vary.

- *Treatments*: A treatment is the condition or a factor associated with a specific level in a specific experiment.
- *Experimental units*: Experimental units are the objects or entities that are used for application of treatments and measurements of resulting effects.
- *Factorial experimentation* is a method in which the effects due to each factor and to combinations of factors are estimated. Factorial designs are geometrically constructed and vary all the factors simultaneously and orthogonally. Factorial designs collect data at the vertices of a cube in k-dimensions (k is the number of factors being studied). If data are collected from all of the vertices, the design is a full factorial, requiring 2^k runs provided that each factor is run at two levels.
- *Fractional factorial* experimentation includes a group of experimental designs consisting of a carefully chosen subset (fraction) of the experimental runs of a full factorial design. The subset is chosen so as to exploit the sparsity-of-effects principle to expose information about the most important features of the problem studied, while using a fraction of the effort of a full factorial design in terms of experimental runs and resources. As the number of factors increases, the fractions become smaller and smaller (1/2, 1/4, 1/8, 1/16, ...). Fractional factorial designs collect data from a specific subset of all possible vertices and require 2^{k-q} runs, with 2^{-q} being the fractional size of the design. If there are only three factors in the experiment, the geometry of the experimental design for a full factorial experiment requires eight runs, and a one-half fractional factorial experiment (an inscribed tetrahedron) requires four runs.

To visualize DOE terminology, process, and its principal stages, consider the following diagram of a cake-baking process shown in Figure 1.2. There are five aspects of the process that are analyzed by a designed experiment: factors, levels of the factors, experimental plan, experiments, and response.

1.3 Response

The response must satisfy certain requirements. First, the response should be the *effective* output in terms of reaching the final aim of the study. Second, the response should be *easily measurable*, preferably quantitatively. Third, the response should be a *single-valued function* of the chosen parameters. Unfortunately, these three important requirements are rarely mentioned in the literature on DOE.

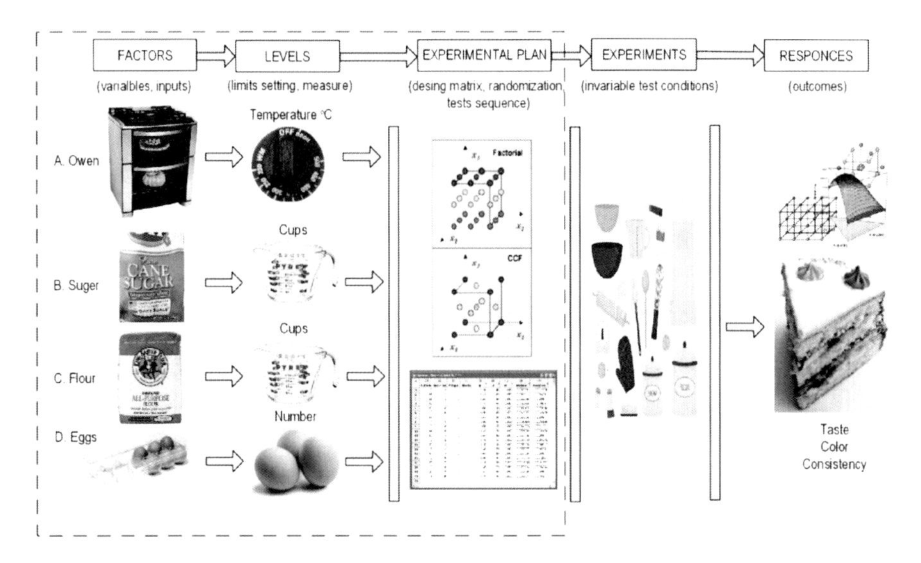

Fig. 1.2. Visualization of DOE terminology, process and its principal stages

1.4 Levels of the Factors

Selections of factors levels is one of the most important yet least formalized, and thus discussed stage. Each factor selected for the DOE study has a certain global range of variation. Within this range, a local sub-range to be used in DOE is to be defined. Practically it means that *the upper* and *lower limits* of each included factors should be set. To do this, one should use all available information such as experience, results of the previous studies, expert opinions, etc.

In manufacturing, a selection of the upper limit of a factor is a subject of equipment, physical, and statistical limitations. *The equipment limitation* means that the upper limit of the factor cannot be set higher than that allowed by the equipment used in the tests. In the example considered in Figure 1.2, the oven temperature cannot be selected higher than that the maximum temperature available in the oven. In machining, if the test machine has maximum spindle rpm of 5000 and drills of 3 mm are to be tested that the upper limit of the cutting speed cannot be more than 47.1 m/min. *The physical limitation* means that the upper limit of the factor cannot be set higher than that where undesirable physical transformation may occur. In the example considered in Figure 1.2, the oven temperature cannot be selected higher than that under which the edges of the cake will be burned. In machining, the feed should be exceed the so-called breaking feed (the uncut chip thickness or chip load) [19] under which the extensive chipping or breakage of the cutting insert occur. *The statistical limitation* is rarely discussed in the literature on DOE in manufacturing. This limitation sets the upper limit to the maximum of a factor for which the automodelity of its variance is still valid. This assures much less potential problems with the row variances in the further statistical analysis of the test results.

There is another limitation on the selection of the upper and lower limits of the factors chosen for DOE. This limitation requires that *the factor combinations should be compatible*, i.e. all the required combinations of the factors should be physically realizable on the setup used in the study. For example, if a combination of cutting speed and feed results in drill breakage, then this combination cannot be included in the test. Often, chatter occurs at high cutting regimes that limits the combinations of the regime parameters. In the example considered in Figure 1.2, it could happen that for a certain number of eggs and the low limit of the oven temperature, the consistency of the baked cake cannot be achieved no matter what is the baking time is. Although it sounds simple, factors compatibility is not always obvious at the stage of the selection of their limits. If some factors incompatibility is found in the tests, the time and resources spent on this test are wasted and the whole DOE should be re-planned from scratch.

Mathematically, the defined combination of the selected factors can be thought of as a point in the multi-dimensional factorial space. The coordinates of this point are called *the basic (zero) levels of the factors*, and the point itself is termed as *the zero point* [1, 20].

The interval of factor variation is the number which, when added to the zero level, gives the upper limit and, when subtracted from the zero level, gives the lower limit. The numerical value of this interval is chosen as the unit of a new scale of each factor. To simplify the notations of the experimental conditions and procedure of data analysis, this new scale is selected so that the upper limit corresponds to +1, lower to −1 and the basic level to 0. For the factors having continuous domains, a simple transformation formula is used

$$x_i = \frac{\tilde{x}_i - \tilde{x}_{i-0}}{\Delta \tilde{x}_i} \tag{1.1}$$

where x_i is a new value of the factor i, $\tilde{x}_i$ is its true (or real) value of the factor (the upper or lower limit), $\tilde{x}_{i-0}$ is the true (real) value of the zero level of the factor, $\Delta \tilde{x}_i$ is the interval of factor variation (in true (real) units) and i is the number of the factor.

In the example considered in Figure 1.2, assume that the maximum oven temperature in the test is selected to be $\tilde{x}_{A-\max} = 200^o C$ and the minimum temperature is $\tilde{x}_{A-\min} = 160^o C$. Obviously, $\tilde{x}_{A-0} = 180^o C$ so that $\Delta \tilde{x}_A = 20^o C$. Than the maximum of factor A in the scale set by Eq. (1.1) is calculated as $x_{A(\max)} = (200-180)/20 = +1$ while its minimum is calculated as $x_{A(\min)} = (160-180)/20 = -1$. The other factors in the test shown in Figure 1.1 are set as follows: $\tilde{x}_{B-\min} = 1\text{cup}$, $\tilde{x}_{B-\max} = 2\text{cup}$; $\tilde{x}_{C-\min} = 2\text{cup}$, $\tilde{x}_{C-\max} = 3\text{cup}$; $\tilde{x}_{D-\min} = 1\text{egg}$, $\tilde{x}_{D-\min} = 3\text{eggs}$.

Equation (1.1) is used in practically any type of DOE for direct and reverse transformations of the factors. The latter is needed to convert the correlation model obtained as a result of using DOE in the real scale of factors. Unfortunately, this is routinely forgotten step in the representation of DOE results.

1.5 Experimental Plan – Factorial Experiments

A very useful class of designed experiments is a range of factorial experiments, in which the treatments are combinations of levels of several factors. These are used in many applications of experimental design, but especially in technological experiments, where the factors might be, for example, time, concentration, pressure, temperature, etc. As mentioned above, it is very common to use a small number of levels for each of the factors, often just two levels, in which case a design with k treatment factors has 2^k treatments and is called a 2^k factorial design. As an example, in a experiment, if there were 10 systematic parameters then a full 210 factorial might have each systematic parameter set at $\pm 1\sigma$; of course in this case it would be usual as well to have one or more runs at the central 'mean value' or 'best guess' of all the systematics.

Factorial designs, including fractional factorials, have increased precision over other types of designs because they have built-in internal replication. Factor effects are essentially the difference between the average of all runs at the two levels for a factor, such as "high" and "low." Because each factor is varied with respect to all of the factors, information on all factors is collected by each run. In fact, every data point is used in the analysis many times as well as in the estimation of every effect and interaction. Additional efficiency of the two-level factorial design comes from the fact that it spans the factor space, that is, puts half of the design points at each end of the range, which is the most powerful way of determining whether a factor has a significant effect.

1.5.1 Full Factorial Design

Referring to Figure 1.2, Table 1.1 gives the settings for a 2^4 factorial experiment; usually the order of the runs would be randomized, but the structure of the experiment is easier to see in the un-randomized form. The real variables are added to the table (called the design matrix which defines all the runs and levels in the experiment) for clarity – normally the design matrix includes only the code variables for simplicity.

Table 1.1. A 2^4 factorial design of 16 runs, with the response labeled according to conventional notation for the factor levels. Real variables are added for clarity.

run	Coded variables				Real variables				response
	A	B	C	D	Temp	Sugar	Flour	Egg	
1	−1	−1	−1	−1	160	1	2	1	$y_{(1)}$
2	−1	−1	−1	+1	160	1	2	3	y_d
3	−1	−1	+1	−1	160	1	3	1	y_c
4	−1	−1	+1	+1	160	1	3	3	y_{cd}
5	−1	+1	−1	−1	160	2	2	1	y_b
6	−1	+1	−1	+1	160	2	2	3	y_{bd}
7	−1	+1	+1	−1	160	2	3	1	y_{bc}
8	−1	+1	+1	+1	160	2	3	3	y_{bcd}
9	+1	−1	−1	−1	200	1	2	1	y_a
10	+1	−1	−1	+1	200	1	2	3	y_{ad}
11	+1	−1	+1	−1	200	1	3	1	y_{ac}
12	+1	−1	+1	+1	200	1	3	3	y_{acd}
13	+1	+1	−1	−1	200	2	2	1	y_{ab}
14	+1	+1	−1	+1	200	2	2	3	y_{abd}
15	+1	+1	+1	−1	200	2	3	1	y_{abc}
16	+1	+1	+1	+1	200	2	3	3	y_{abcd}

The run called '1', for example, has all four factors set to their low level, whereas the run called '2', has factors A, B, and C set to their low level and D set to its high level. Note that the estimated effect in going from the low level of A, say, to the high level of A, is based on comparing the averages of 8 observations taken at the low level with 8 observations taken at the high level. Each of these averages has a variance equal to 1/8 the variance in a single observation, or in other words to get the same information from an OFAT design one would need 8 runs with A at -1σ and 8 runs with A at $+1\sigma$, all other factors held constant. Repeating this for each factor would require 64 runs, instead of 16. The balance of the 24 design ensures that one can estimate the effects for each of the four factors in turn from the average of 8 observations at the high level compared to 8 observations at the low level: for example the main effect of D is estimated by

$$\left(\begin{array}{l} y_{abcd} - y_{abc} + y_{abd} - y_{ab} +_{acd} - y_{ac} + y_{ad} - y_a + y_{bcd} - \\ y_{bc} + y_{bd} - y_b + y_{cd} - y_c + y_d - y_{(1)} \end{array} \right) \Big/ 8 \tag{1.2}$$

and similar estimates can be constructed for the effects of B and C.

Note that by constructing these four estimates, four linear combinations of 16 observations were used. One linear combination, the simple average, is needed to set the overall level of response, leaving 11 linear combinations not yet used to estimate anything. These combinations are in fact used to estimate the interactions of various factors, and the full set of combinations is given by the set of signs in Table 1.2.

Table 1.2. The 2^4 factorial showing all of the interaction effects

run	A	B	C	D	AB	AC	AD	BC	BD	CD	ABC	ABD	ACD	BCD	ABCD
1	−1	−1	−1	−1	+1	+1	+1	+1	+1	+1	−1	−1	−1	−1	+1
2	−1	−1	−1	+1	+1	+1	−1	+1	−1	−1	−1	+1	+1	+1	−1
3	−1	−1	+1	−1	+1	−1	+1	−1	+1	−1	+1	−1	+1	+1	−1
4	−1	−1	+1	+1	+1	−1	−1	−1	−1	+1	+1	+1	−1	−1	+1
5	−1	+1	−1	−1	−1	+1	+1	−1	−1	+1	+1	+1	−1	+1	−1
6	−1	+1	−1	+1	−1	+1	−1	−1	+1	−1	+1	−1	+1	−1	+1
7	−1	+1	+1	−1	−1	−1	+1	+1	−1	−1	−1	+1	+1	−1	+1
8	−1	+1	+1	+1	−1	−1	−1	+1	+1	+1	−1	−1	−1	+1	−1
9	+1	−1	−1	−1	−1	−1	−1	+1	+1	+1	+1	+1	+1	−1	−1
10	+1	−1	−1	+1	−1	−1	+1	+1	−1	−1	+1	−1	−1	+1	+1
11	+1	−1	+1	−1	−1	+1	−1	−1	+1	−1	−1	+1	−1	+1	+1
12	+1	−1	+1	+1	−1	+1	+1	−1	−1	+1	−1	−1	+1	−1	−1
13	+1	+1	−1	−1	+1	−1	−1	−1	−1	+1	−1	−1	+1	+1	+1
14	+1	+1	−1	+1	+1	−1	+1	−1	+1	−1	−1	+1	−1	−1	−1
15	+1	+1	+1	−1	+1	+1	−1	−1	−1	−1	−1	−1	−1	−1	−1
16	+1	+1	+1	+1	+1	+1	+1	+1	+1	+1	+1	+1	+1	+1	+1

For example, the interaction of factors A and B is estimated by the contrast given by the fourth column of Table 1.2

$$\left\{ \begin{array}{l} y_{abcd} + y_{abc} + y_{abd} + y_{ab} - y_{bcd} - y_{bc} - y_{bd} - y_b - \\ \left(y_{acd} + y_{ac} + y_{ad} + y_a - y_{cd} - y_c - y_d - y_{(1)} \right) \end{array} \right\} \Big/ 8 \tag{1.3}$$

which takes the difference of the difference between responses with A at high level and A at low level with the difference between responses with B at high level and B at low level. The column of signs in Table 1.2 for the interaction effect AB was obtained simply by multiplying the A column by the B column, and all the other columns are similarly constructed.

This illustrates two advantages of designed experiments: the analysis is very simple, based on linear contrasts of observations, and as well as efficiently estimating average effects of each factor, it is possible to estimate interaction effects with the same precision. Interaction effects can never be measured with OFAT designs, because two or more factors are never changed simultaneously.

The analysis, by focusing on averages, implicitly assumes that the responses are best compared by their mean and variance, which is typical of observations that follow a Gaussian distribution. However the models can be extended to more general settings [1].

The discussed factorial DOE allows accurate estimation of all factors involved and their interactions. However, the cost and time need for such a test increase with the number of factors considered. Normally, any manufacturing test includes a great number of independent variables. In the testing of drills, for example, there are a number of tool geometry variables (the number of cutting edges, rake angles, flank angles, cutting edge angles, inclination angles, side cutting edge back taper angle, etc.) and design variables (web diameter, cutting fluid hole shape, cross-sectional area and location, profile angle of the chip removal flute, length of the cutting tip, the shank length and diameter, etc.) that affect drill performance.

Table 1.3 shows the number of runs needed for the so-called full factorial DOE considered above. In laboratory conditions, at least three repetitions for the same run are needed while if the test is run in the shop floor conditions then at least 5 repetitions should be carried out in randomized manner. The poorer the test conditions, the greater the number of uncontrollable variables, the greater number of repetitions at the same point of the design matrix is needed to pass statistical analysis of the obtained experimental data.

Therefore, there is always a dilemma. On one hand, to keep the costs and time spent on the test at reasonable level it is desirable to include into consideration only a limited number of essential factors carefully selected by the experts. On the other hand, if even one essential factor is missed, the final statistical model may not be adequate to the process under study. Unfortunately, there is no way to justify the decisions made at the preprocess stage about the number of essential variables prior the tests. If a mistake is made at this stage, it may show up only at the final stage of DOE when the corresponding statistical criteria are examined. Obviously, it is too late then to correct the test results by adding the missed factor. The theory of DOE offers a few ways to deal with such a problem [20, 21].

The first relies on the collective experience of the experimentalist(s) and the research team in the determination of significant factors. The problem with such an approach is that one or more factors could be significant or not, depending on the particular test objectives and conditions. For example, the backtaper angle in drills is not a significant factor in drilling cast irons, but it becomes highly significant in machining titanium alloys.

A second way is to use screening DOE. This method appears to be more promising in terms of its objectivity. A screening DOE is used when a great number of factors are to be investigated using a relatively small number of tests. These kinds of tests are conducted to identify the significant factors for further full factorial analysis. The most common type of screening DOE is the so-called the fractional factorial DOE.

Table 1.3. Two-level designs: minimum number of runs as a function of number of factors for full factorial DOE

Number of factors	Number of runs	Number of repetitions	
		3	5
1	2	6	10
2	$4 = 2^2$	12	20
3	$8 = 2^3$	24	40
4	$16 = 2^4$	48	80
5	$32 = 2^5$	96	160
6	$64 = 2^6$	192	320
7	$128 = 2^7$	384	640
8	$256 = 2^8$	768	1280
9	$512 = 2^9$	1536	2560
10	$1024 = 2^{10}$	3072	5120

Very often, especially in manufacturing settings, the factors correspond to underlying quantitative variables, and the levels, denoted ±1, are codes for particular numerical values: temperatures at 80 and 100 degrees, for example. In such cases the choice of factor levels involves both subject matter expertise, and at least in the early stages, considerable guesswork. As well, the goal of the experiment might not be to compare responses at different factor settings, but to find the combination of factor settings that leads to minimum or maximum response.

Factorial designs adapted to these conditions are called *response surface designs* (RSM). The basic idea is that the response y is a continuous function of some input variables x_1, x_2, and so on, and factorial designs are used sequentially to explore the shape of the response surface. Sequential experimentation in the relevant range of x–space usually begins with a screening design, to quickly assess which of several factors have the largest effect on the response. Then second stage is a factorial design at new values of the underlying variables, chosen in the direction of increasing (or decreasing) response. Near the maximum additional factor levels are added, to model curvature in the response surface.

The first goal for RMS is to find the optimum response. When there is more than one response then it is important to find the compromise optimum that does not optimize only one response. When there are constraints on the design data, then the experimental design has to meet requirements of the constraints. The second goal is to understand how the response changes in a given direction by adjusting the design variables. In general, the response surface can be visualized graphically. The graph is helpful to see the shape of a response surface; hills, valleys, and ridge lines.

A plot showing how the response changes with respect to changes in the factor levels is a response surface, a typical example of which is shown in Figure 1.3a as a three-dimensional perspective plot. In this graph, each value of x_1 and x_2 generates a y–value. This three-dimensional graph shows the response surface from the side and it is called a response surface plot.

Sometimes, it is less complicated to view the response surface in two-dimensional graphs. The contour plots can show contour lines of x_1 and x_2 pairs that have the same response value y. An example of contour plot is as shown in Figure 1.3b.

In order to understand the surface of a response, graphs are helpful tools. But, when there are more than two independent variables, graphs are difficult or almost impossible to use to illustrate the response surface, since it is beyond 3-dimension.

RSM methods are utilized in the optimization phase. Due to the high volume of experiments, this phase focuses on a few highly influential variables, usually 3 to 5. The typical tools used for RSM are central composite design (CCD) and the Box-Behnken design (BBD) [1]. With the aid of software, the results of these complex designs are exhibited pictorially in 3D as 'mountains' or 'valleys' to illustrate performance peaks.

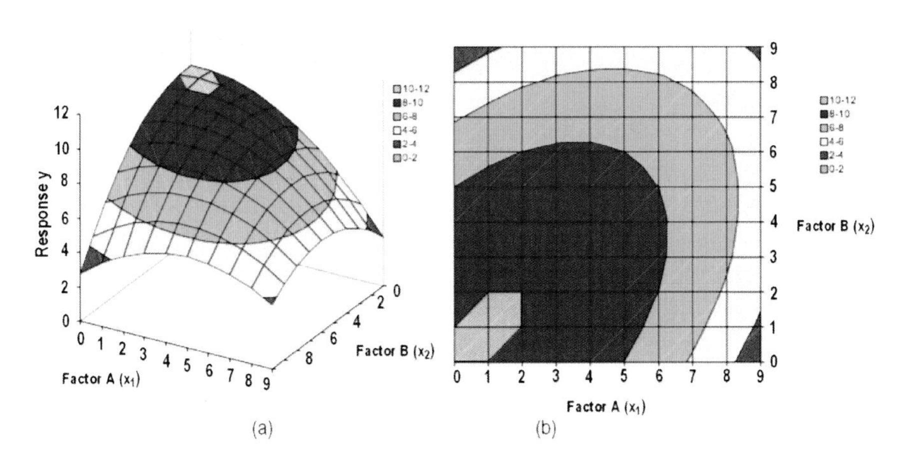

Fig. 1.3. Graphical representation of the outcome of RSM: (a) response surface plot, (b) control plot

A class of two-level factorial designs which allow reduction of the number of runs while still allowing estimation of important effects is the family of two-level fractional factorial designs. The construction of two-level fractional factorial designs is based on a structure of two-level full factorial designs.

In practice, interactions involving three or more factors are often negligible. One can take advantage of the fact that some of the effects estimable in the full factorial are negligible to reduce the number of runs required for the experiment. In the above-considered example, one may assume that the 3-way and 4-way interactions are negligible. This information can then be used either to study additional variables without increasing the number of runs or to reduce the number of runs in the present study.

For example, a fifth factor E, for example butter (measured in cups being 0.5 cup – low level, 1.5 cup – the upper level) can be introduced to the 2^4 factorial study (Table 1.2). A full factorial design with 5 variables requires $2^5 = 32$ runs. Instead of adding a fifth column into the design matrix represented by Table 1.2 which would require twice as many runs, the new fifth variable is assigned to the column for ABCD interaction to obtain its setting. That is, in the first run E would be set to its high level, in the second run to its low level, in the third run to its low level, and so on, following the pattern of +1 and −1 in the last column of Table 1.2. The resulting contrast $\left(y_{(1)} - y_a - y_b + y_{ab} \pm ...\right)/8$ is estimating the main effect of factor E (i.e. the difference between responses on the high level of E to the low level of E), but it is also estimating the ABCD interaction: these two effects are completely aliased. The working assumption is that the ABCD interaction is likely to be very small, so any observed effect can be attributed to E. The main effects of A, B, C, and D are estimated as before, and one now has information on 5 factors from a 16 run design. However all the main effects are aliased with 4 factor interactions: for example A is aliased with BCDE, B with ACDE, and so on. Further, all 2 factor interactions are aliased with 3 factor interactions. Again, the working assumption is typically that any observed effect is more likely to be due to a 2 factor interaction than a 3 factor interaction.

Table 1.4 gives the new design matrix. This design is called a half-fraction of 32-run full factorial design. Because it has five variables, with one of them defined in terms of other 4, it is a 2^{5-1} fractional factorial design. Note that the number of runs is equal to $2^{5-1} = 2^4 = 16$, half the number required for a full 5-factor factorial test.

Table 1.5 shows the complete aliasing pattern for the 2^{5-1} fractional factorial design from setting E = ABCD. I is used to represent the identity element and consists of a column of 16 plus signs. Such a column is normally is a part of any design matrix and used to estimate the response average in the regression analysis of the data.

Table 1.4. Design matrix for 2^{5-1} fractional factorial DOE

run	Coded variables					Real variables				
	A	B	C	D	E	Temp	Sugar	Flour	Egg	Butter
1	−1	−1	−1	−1	+1	160	1	2	1	1.5
2	−1	−1	−1	+1	−1	160	1	2	3	0.5
3	−1	−1	+1	−1	−1	160	1	3	1	0.5
4	−1	−1	+1	+1	+1	160	1	3	3	1.5
5	−1	+1	−1	−1	−1	160	2	2	1	0.5
6	−1	+1	−1	+1	+1	160	2	2	3	1.5
7	−1	+1	+1	−1	+1	160	2	3	1	0.5
8	−1	+1	+1	+1	−1	160	2	3	3	0.5
9	+1	−1	−1	−1	−1	200	1	2	1	0.5
10	+1	−1	−1	+1	+1	200	1	2	3	1.5
11	+1	−1	+1	−1	+1	200	1	3	1	1.5
12	+1	−1	+1	+1	−1	200	1	3	3	0.5
13	+1	+1	−1	−1	+1	200	2	2	1	1.5
14	+1	+1	−1	+1	−1	200	2	2	3	0.5
15	+1	+1	+1	−1	−1	200	2	3	1	0.5
16	+1	+1	+1	+1	+1	200	2	3	3	1.5

Table 1.5. Aliasing pattern for 2^{5-1} fractional factorial DOE

I = ABCDE
A = BCDE
B = ACDE
C = ABDE
D = ABCE
E = ABCD
AB = CDE
AC = BDE
AE = BCD
BC = ADE
BD = ACE
BE = ACD
CD = ABE
CE = ABD
DE = ABC

It follows from Table 1.5 that each main effect and 2-factor interaction is aliased with a 3 or 4-way interaction which was assumed to be negligible. Thus, ignoring effects expected to be unimportant, one can still obtain estimates of all main effects and 2-way interactions. As a result, fractional factorial DOE is often used at the first stages of the study as a screening test to distinguish obtain significant factors, and thus to reduce the number of factors used the full factorial tests.

The process of fractionalization can be continued; one might for example assign a new factor F, say, to the ABC interaction (which is aliased with DE), giving a 26–2 design, sometimes called a 1/4 fraction of a 26. This allows one to assess the main effect of 6 factors in just 16 runs, instead of 64 runs, although now some 2 factor interactions will be aliased with each other.

There are very many variations on this idea; one is the notion of a screening design, in which only main effects can be estimated, and everything else is aliased. The goal is to quickly assess which of the factors is likely to be important, as a step in further experimentation involving these factors and their interactions. Table 1.6 shows an 8 run screening design for 7 factors. The basic design is the 23 factorial in factors A, B, and C shown in the first 3 columns; then 4 new factors have been assigned to the columns that would normally correspond to the interactions BC, AC, AB, and ABC.

Table 1.6. A screening design for 7 factors in 8 runs, built from a 2^3 factorial design

	A	B	C	D	E	F	G
1	−1	−1	−1	+1	+1	+1	−1
2	−1	−1	+1	−1	−1	+1	+1
3	−1	+1	−1	−1	+1	−1	+1
4	−1	+1	+1	+1	−1	−1	−1
5	+1	−1	−1	+1	−1	−1	+1
6	+1	−1	+1	−1	+1	−1	−1
7	+1	+1	−1	−1	−1	+1	−1
8	+1	+1	+1	+1	+1	+1	+1

Any modern DOE software (for example, Statistica, Design-Ease® and Design-Expert®, Minitab, etc.) can generate design matrix for fractional factorial DOE and provide a table of interactions similar to that shown in Table 1.5.

As mentioned above, the objective of DOE is to find the correlation between the response and the factors included. All factors included in the experiment are varied simultaneously. The influence of the unknown or non-included factors is minimized by properly *randomizing* the experiment. Mathematical methods are used not only at the final stage of the study, when the evaluation and analysis of the experimental data are conducted, but also though all stages of DOE, i.e. from the formalization of *priory* information till the decision making stage. This allows answering of important questions: "What is the minimum number of tests that should be conducted? Which parameters should be taken into consideration? Which method(s) is (are) better to use for the experimental data evaluation and analysis?" [20, 21]. Therefore, one of the important stages in DOE is the selection of the mathematical (correlation) model. An often quoted insight of George Box is, "All models are wrong. Some are useful"[22]. The trick is to have the simplest model that captures the main features of the process.

Mathematically, the problem of DOE can be formulated as follows: define the estimation E of the response surface which can be represented by a function

$$E\{y\} = \phi(x_1, x_2, ..., x_k) \tag{1.4}$$

where y is the process response (for example, cutting temperature, tool life, surface finish, cutting force, etc.), $x_i, i = 1,2,...k$ are the factors varied in the test (for example, the tool cutting edge angle, cutting speed, feed, etc.).

The mathematical model represented by Eq.(1.4) is used to determine the gradient, i.e., the direction in which the response changes faster than in any other. This model represents the response surface, which is assumed to be continuous, two times differentiable, and having only one extreme within the chosen limits of the factors.

In general, a particular kind of the mathematical model is initially unknown due to insufficient knowledge on the considered phenomenon. Thus, a certain approximation for this model is needed. Experience shows [21] that a power series or polynomial (Taylor series approximations to the unknown true functional form of the response variable) can be selected as an approximation

$$y = \beta_0 + \sum_{i=1}^{p} \sum_{\substack{j=1 \\ i \neq j}}^{p} \beta_{ij} x_i x_j + \sum_{i=1}^{p} \sum_{j=1}^{p} \sum_{\substack{k=1 \\ i \neq j \neq k}}^{p} \beta_{ijk} x_i x_j x_k + ... \tag{1.5}$$

where β_0 is the overall mean response, β_i is the main effect of the factor ($i = 1,2, ..., p$), β_{ij} is the two-way interaction effect between the *i*th and *j*th factors, and β_{ijk} is the three-way interaction effect between the *i*th, *j*th, and *k*th factors.

A general recommendation for setting the factor ranges is to set the levels far enough apart so that one would expect to see a difference in the response. The use of only two levels seems to imply that the effects must be linear, but the assumption of *monotonicity* (or nearly so) on the response variable is sufficient. At least three levels of the factors would be required to detect curvature.

Interaction is present when the effect of a factor on the response variable depends on the setting level of another factor. Two factors are said to have interaction when the effect of one factor varies under the different levels of another factor. Because this concept is very important in DOE, consider an example to clarify the concept. Suppose that Factor A is the silicon content is a heat-treatable aluminum. Suppose that factor B is the heat treating temperature. In this case the response values may represent the hardness. Table 1.7 shows the response depending upon the factors' setting. A plot of the data (Figure 1.4a) shows how response is a function of factors A and B. Since the lines are almost parallel, it can be assumed that little interaction between the variables occurs.

Consider another set of data as shown in Table 1.8 In this instance, the response is the relative wear rate, factor A is the shaft speed, and factor B may be the type of lubricant. Again, a plot of the data (Figure 1.4b) shows how response depends on the factors. It can be seen that B_2 is a better lubricant at low speed but not at high speed. The crossing of the lines indicates an interactive effect between the factors. In this case, factors A and B are interrelated, i.e. not independent.

Table 1.7. A factorial experiment with no factors interaction

Factor	B_1	B_2
A_1	20	30
A_2	40	52

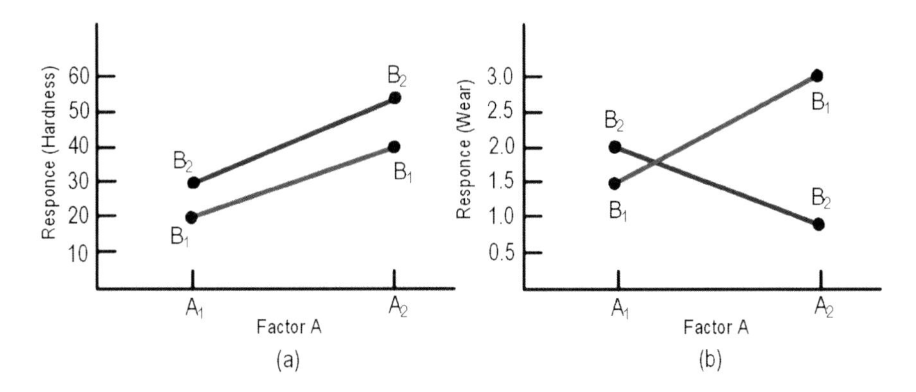

Fig. 1.4. Concept of interaction: (a) no interaction of factors A and B, (b) interaction of factors A and B

Table 1.8. A factorial experiment with factors interaction

Factor	B_1	B_2
A_1	1.5	2.0
A_2	3.0	0.8

The β_{ij} terms in Eq.(1.5) account for the two-way interactions. Two-way interactions can be thought of as the corrections to a model of simple additivity of the factor effects, the model with only the β_i terms in Eq.(1.5). The use of the simple additive model assumes that the factors act separately and independently on the response variable, which is not always a very reasonable assumption.

The accuracy of such an approximation would depend upon the order (power) of the series. To reduce the number of tests at the first stage of experimental study, a polynomial of the first order or a linear model is sufficiently suitable. Such a model is successfully used to calculate the gradient of the response, thus, to reach the stationary region. When the stationary region is reached then a polynomial containing terms of the second, and sometimes, the third order may be employed.

Experience shows [23, 24] that a model containing linear terms and interactions of the first order can be used successfully in metal cutting. Such a model can be represented as

$$y = \beta_0 + \sum_i \beta_i x_i + \sum_{ij} \beta_{ij} x_i x_j \tag{1.6}$$

The coefficients of Eq.(1.6) are to be determined from the tests. Using the experimental results, one can determine the regression coefficients b_1, b_i, and b_{ij}, which are estimates for the theoretical regression coefficients β_1, β_i, and β_{ij}. Thus, the regression equation constructed using the test results has the following form

$$E\{y\} = \hat{y} = b_0 + \sum_i b_i x_i + \sum_{ij} b_{ij} x_i x_j \tag{1.7}$$

where $\hat{y}$ is the estimate for $E\{y\}$.

The problem of mathematical model selection for the object under investigation requires the formulation of clear objective(s) of the study. This problem occurs in any study, but the mathematical model selection in DOE requires the quantitative formulation of the objective(s) called the response, which is the result of the process under study or its output. The process under study may be characterized by several important output parameters but only one of them should be selected as the response.

As mentioned above, the response must satisfy certain requirements. First, the response should be the *effective* output in terms of reaching the final aim of the study. Second, the response should be *easily measurable*, preferably quantitatively. Third, the response should be a *single-valued function* of the chosen parameters. Unfortunately, these three important requirements are rarely mentioned in the literature on DOE.

Most statistical software (for example, Statistica, Design-Ease® and Design-Expert®, Minitab, etc.) can estimate the significant coefficients of the chosen model and verify the model adequacy using standard statistical procedures using ANOVA (analysis of variance methods) [1]. Moreover, they can deduce from the specification of the model which effects are aliased and in some packages and plus it is relatively easy to produce a graphical display of the estimated effects that allows one to assess which effects are nonzero, at least in part so that the 'nearly zero' effects can be pooled to estimate the error.

1.6 Resolution Level

Experimental designs can be categorized by their resolution level. A design with a higher resolution level can fit higher-order terms in Eq.(1.5) than a design with a lower resolution level. If a high enough resolution level design is not used, only the linear combination of several terms can be estimated, not the terms separately. The word "resolution" was borrowed from the term used in optics. Resolution levels are usually denoted by Roman numerals, with III, IV, and V being the most commonly used. To resolve all of the two-way interactions, the resolution level must be at least V [8]. Four resolution levels and their meanings are given in Table 1.9.

Table 1.9. Resolution levels and their meaning

Resolution level	Meaning
II	Main effects are linearly combined with each other ($\beta_i + \beta_j$).
III	Main effects are linearly combined with two-way interactions ($\beta_i + \beta_{jk}$).
IV	Main effects are linearly combined with three-way interactions ($\beta_i + \beta_{jkl}$) and two-way interactions with each other ($\beta_{ij} + \beta_{kl}$).
V	Main effects and two-way interactions are not linearly combined except with higher-order interactions ($\beta_i + \beta_{jklm}$) and ($\beta_{ij} + \beta_{klm}$).

1.7 More Specialized Designs

1.7.1 Orthogonal Array and Taguchi Method

The 8 run screening design illustrated in Table 1.6 is a 2^{7-4} fractional factorial, but is also an example of an orthogonal array, which is by definition an array of symbols, in this case ±1, in which every symbol appears an equal number of times in each column, and any pair of symbols appears an equal number of times in any pair of columns. An orthogonal array of size n × (n – 1) with two symbols in each column specifies an n-run screening design for n – 1 factors. The designs with symbols ±1 are called Plackett-Burman designs and Hadamard matrices defining them have been shown to exist for all multiples of four up to 424.

More generally, an n × k array with m_i symbols in the *i*th column is an orthogonal array of strength *r* if all possible combinations of symbols appear equally often in any r columns. The symbols correspond to levels of a factor. Table 1.10 gives an orthogonal array of 18 runs, for 6 factors with three levels each.

Table 1.10. An orthogonal array for 6 factors each at 3 levels

run	A	B	C	D	E	F
1	−1	−1	−1	−1	−1	−1
2	−1	0	0	0	0	0
3	−1	+1	+1	+1	+1	+1
4	0	−1	−1	0	0	+1
5	0	0	0	+1	+1	−1
6	0	+1	+1	−1	−1	0
7	+1	−1	0	−1	+1	0
8	+1	+1	−1	+1	0	−1
9	+1	+1	−1	+1	0	+1
10	−1	−1	+1	+1	0	0
11	−1	0	−1	−1	+1	+1
12	−1	+1	0	0	−1	−1
13	0	−1	0	+1	−1	+1
14	0	0	+1	−1	0	−1
15	0	+1	−1	0	+1	0
16	+1	−1	+1	0	+1	−1
17	+1	0	−1	+1	−1	0
18	+1	+1	0	−1	0	+1

Orthogonal arrays are particularly popular in applications of so-called Taguchi methods in technological experiments and manufacturing. An extensive discussion is given in [9-11]. For simplest experiments, the Taguchi experiments are same as the fractional factorial experiments in the classical DOE. Even for common experiments used in the industry for problem solving and design improvements, the main attractions of the Taguchi approach are standardized methods for experiment designs and analyses of results. To use the Taguchi approach for modest experimental studies, one does not need to be an expert in statistical science. This allows working engineers on the design and production floor to confidently apply the technique.

While there is not much difference between the two types of DOE for simpler experiment designs, for mixed level factor designs and building robustness in products and processes, the Taguchi approach offers some revolutionary concepts that were not known even to the expert experimenters. These include standard *method for array modifications*, experiment designs to include *noise factors in the outer array*, *signal-to-noise ratios* for analysis of results, *loss function* to quantify design improvements in terms of dollars, treatment of systems with *dynamic characteristics*, etc.

Although the Taguchi method was developed as a powerful statistical method for shop floor quality improvement, a way too many researchers have been using this method as a research and even optimization method in manufacturing, and thus in metal cutting studies (for example, [25-29] that, in the author's opinion unacceptable because the Taguchi methods suffer the same problems as any fractional factorial DOE.

Unfortunately, it became popular to consider the used of only a fraction of the number of test combinations needed for a full factorial design. That interest spread because many practitioners do not take the time to find out, or for other reasons never realized the "price" paid when one uses fractional factorial DOEs including the Taguchi method: (1) Certain interaction effects lose their contrast so knowledge of their existence is gone, (2) Significant main effect and important interactions have aliases – other 'confounding' interaction names. Thus wrong answers can, and often do come from the time, money, and effort of the experiment.

Books on DOE written by 'statistical' specialists add confusion to the matter claiming that interactions (three-factor or higher order) would be too difficult to explain; nor could they be important. The ideal gas law (1834 by Emil Clapeyron), known from high-school physics as

$$PV = nRT \tag{1.8}$$

(where P is the pressure of the confined gas, V is the volume of the confined gas, n is the number of moles of gas, R is gas constant, T is the temperature) plots as a simple graph. It depicts a three-factor interaction affecting y (response) as pressure, or as volume. The authors of these statistical books/papers may have forgotten their course in physics?

The problem is that the ability of the Taguchi method is greatly overstated by its promoters who described Taguchi orthogonal tables as Japan's 'secret super weapon' that is the real reason for developing an international reputation for quality. Claim: a large number of variables could now be handled with practical efficiency in a single DOE. As later details became available, many professionals realized that these arrays were fractional factorials, and that Taguchi went to greater extremes that other statisticians in the degree of fractionating. According to the Taguchi method, the design is often filled with as many single factors for which it has room. The design becomes "saturated" so no degrees of freedom are left for its proper statistical analysis. The growing interest in the Taguchi method in the research and optimization studies in manufacturing attests to the fact that manufacturing researches either are not aware of the above-mentioned 'price" paid for apparent simplicity, or know of no other way to handle more and more variables at one time.

1.7.2 Sieve DOE

Plackett and Burman [30] developed a special class of fractional factorial experiments that includes interactions. When this kind of DOE (referred to as the Plackett–Burman DOE) is conducted properly using a completely randomized sequence, its distinctive feature is high resolution. Despite a number of disadvantages (for example, mixed estimation of regression coefficients), this method utilizes high-contrast diagrams for the factors included in the test as well as for their interactions of any order. This advantage of the Plackett–Burman DOE is very useful in screening tests.

This section presents a simple methodology of screening DOE to be used in manufacturing tests [31]. The method, referred to as the sieve DOE, has its foundation in the Plackett–Burman design ideas, an oversaturated design matrix and the method of random balance. The proposed sieve DOE allows the experimentalist to include as many factors as needed at the first phase of the experimental study and then to sieve out the non-essential factors and interactions by conducting a relatively small number of tests. It is understood that no statistical model can be produced in this stage. Instead, this method allows the experimentalist to determine the most essential factors and their interactions to be used at the second stage of DOE (full factorial or RSM DOE).

The proposed sieve DOE includes the method of random balance. This method utilizes oversaturated design plans where the number of tests is fewer than the number of factors and thus has a negative number of degrees of freedom [32]. It is postulated that if the effects (factors and their interactions) taken into consideration are arranged as a decaying sequence (in the order of their impact on the variance of the response), this will approximate a ranged exponential-decay series. Using a limited number of tests, the experimentalist determines the coefficients of this series and then, using the regression analysis, estimates the significant effects and any of their interactions that have a high contrast in the noise field formed by the insignificant effects.

The initial linear mathematical model, that includes k number of factors (effects), has the following form

$$y = a_0 + a_1 x_1 + \ldots + a_k x_k + a_{12} x_1 x_2 + \ldots + a_{k-1,k} x_{k-1} x_k + \delta \tag{1.9}$$

where a_0 is the absolute term often called the main effect, a_i ($i = 1,k$) are the coefficients of linear terms, a_{ij} ($i = 1,...,k-1$; $j = i+1,...,k$, $i \neq j$) are the coefficients of interaction terms and δ is the residual error of the model.

The complete model represented by Eq. (1.9) can be rearranged as a split of a linear form considering that some x_i designate the iterations terms as

$$y = a_0 + a_1x_1 + ... + a_{k-l,k}x_{k-l} + b_1z_1 + b_2z_2 + ...b_lz_l + \delta = \\ a_0 + a_1x_1 + ... + a_{k-l,k}x_{k-l} + \Delta \tag{1.10}$$

where

$$\Delta = b_1z_1 + b_2z_2 + ...b_lz_l + \delta \tag{1.11}$$

and

$$\sigma^2\{\Delta\} = b_1^2\sigma^2\{z_1\} + b_2^2\sigma^2\{z_2\}... + b_l^2\sigma^2\{z_l\} + \sigma^2\{\delta\} \tag{1.12}$$

In the construction of the split model represented by Eq. (1.10), *(k - l)* significant effects were distinguished and *l* effects were assigned to the noise field. Naturally, the residual variance $\sigma^2\{\Delta\}$ is greater than the tests variance $\sigma^2\{\delta\}$ so that the regression coefficients in Eq.(1.10) will be estimated with greater errors and, moreover, the estimates of the coefficients of this model are mixed. Therefore, the sensitivity of the random balance method is low so that the resultant model has a little significance, and thus should not be used as a valid statistical model. However, this method is characterized by the great contrast of essential effects, which could be distinguished easily on the noisy fields formed by other effects. The latter makes this method the simplest yet highly reliable screening method that can be used at the first stage of testing to distinguish the significant factors and interaction to be used in the full factoring DOE including RSM.

To demonstrate the simplicity of the discussed test, an example of determination of critical factors and their interaction is considered for grinding parameters of a gundrill [31]. The terminal end of a gundrill is shown in Figure 1.5. It is formed with the approach cutting edge angles φ_1 and φ_2 of the outer 1 and inner 2 cutting edges, respectively. These cutting edges meet at the drill point *P*. The location of *P* (defined by distance m_d in Figure 1.5) is critical and it can be varied for optimum performance depending on the work material and the finished hole specifications. The geometry of the terminal end largely determines the shape of the chips and the effectiveness of the cutting fluid, the lubrication of the tool, and removal of the chips. The flank surface 3 having normal flank angle α_{n1} is applied to the outer cutting edge 1 and the flank surface 4 having normal flank angle equal to α_{n2} is applied to the inner cutting edge 2. The rake angle γ is formed by the chip removal flute face 5. To assure drill free penetration, i.e. to prevent the interference of the drill's flanks with the bottom of the hole being drilled, the auxiliary flank 6 known as shoulder dub-off is ground at certain angle φ_4. Their location of the shoulder dub-off is defined by distance b. The detailed description of the gundrill geometry and importance of the parameters selected for the sieve test are presented by the author earlier [33].

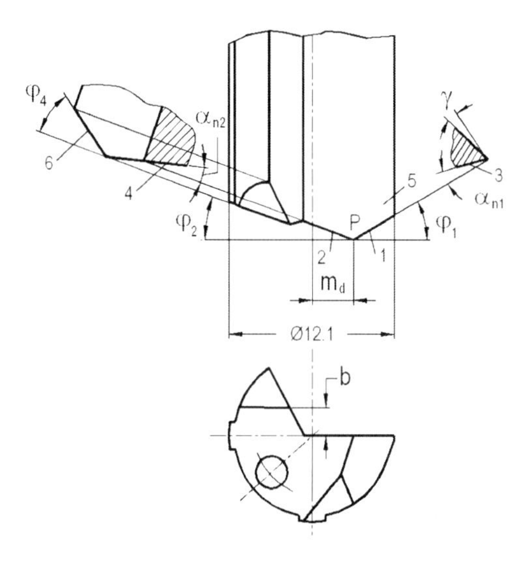

Fig. 1.5. Grinding parameters of the terminal end of a gundrill selected for sieve DOE

Eight factors have been selected for this sieve DOE and their intervals of variation are shown in Table 1.11. The design matrix was constructed as follows. All the selected factors were separated into two groups. The first one contained factors x_1, x_2, x_3, x_4, form a half-replica 2^{4-1} with the defining relation $I = x_1x_2x_3x_4$. In this half-replica, the factors' effects and the effects of their interactions are not mixed. The second half-replica was constructed using the same criteria. A design matrix was constructed using the first half-replica of the complete matrix and adding to each row of this replica a randomly selected row from the second half-replica. Three more rows were added to this matrix to assure proper mixing and these rows were randomly selected from the first and second half-replicas. Table 1.12 shows the constructed design matrix.

As soon as the design matrix is completed, its suitability should be examined using two simple rules. First, a design matrix is suitable if it does not contain two identical columns having the same or alternate signs. Second, a design matrix should not contain columns whose scalar products with any other column result in a column of the same ("+" or "–") signs. The design matrix shown in Table 1.12 was found suitable as it meets the requirements set by these rules. In this table, the responses $\overline{y}_i$ $i = 1...11$ are the average tool life calculated over three independent tests replicas obtained under the indicated test conditions.

Analysis of the result of sieve DOE begins with the construction of a correlation (scatter) diagram shown in Figure 1.6. Its structure is self-evident. Each factor is represented by a vertical bar having on its left side values (as dots) of the response obtained when this factor was positive (the upper value) while the values of the response corresponding to lower lever of the considered factor (i.e. when this factor is negative) are represented by dots on the right side of the bar. As such, the scale makes sense only along the vertical axis.

Table 1.11. The levels of te factors selected for the sieve DOE

Factors	Approach angle φ_1 (°)	Approach angle φ_2 (°)	Flank angle α_{n1} (°)	Drill point offset m_d (mm)	Flank angle α_{n2} (°)	Shoulder dub-off angle φ_4 (°)	Rake angle γ (°)	Shoulder dub-off location b (mm)
Code designation	x_1	x_2	x_3	x_4	x_5	x_6	x_7	x_8
Upper level (+)	45	25	25	3.0	12	45	5	4
Lower level (-)	25	10	8	1.5	7	20	0	1

Table 1.12. Design matrix

Run	Factors								Average tool life (min)	Corrections	
	x_1	x_2	x_3	x_4	x_5	x_6	x_7	x_8	$\overline{y}$	$\overline{y}_{c1}$	$\overline{y}_{c2}$
1	+1	+1	−1	−1	+1	−1	+1	−1	7	18.75	11.11
2	+1	+1	+1	+1	−1	+1	+1	−1	16	11.50	11.50
3	−1	+1	−1	−1	+1	+1	−1	+1	11	11.00	11.00
4	−1	−1	+1	+1	+1	−1	−1	+1	38	21.75	16.61
5	+1	−1	−1	+1	−1	−1	+1	−1	18	13.50	13.50
6	+1	−1	+1	−1	+1	+1	−1	−1	10	21.75	14.61
7	−1	−1	−1	−1	−1	+1	−1	+1	14	14.00	14.00
8	−1	+1	−1	+1	−1	−1	+1	+1	42	25.75	16.61
9	+1	−1	−1	−1	−1	+1	−1	+1	9	20.75	20.75
10	−1	+1	+1	+1	−1	+1	−1	−1	32	15.75	15.75
11	+1	−1	+1	−1	−1	−1	+1	−1	6	17.75	10.61

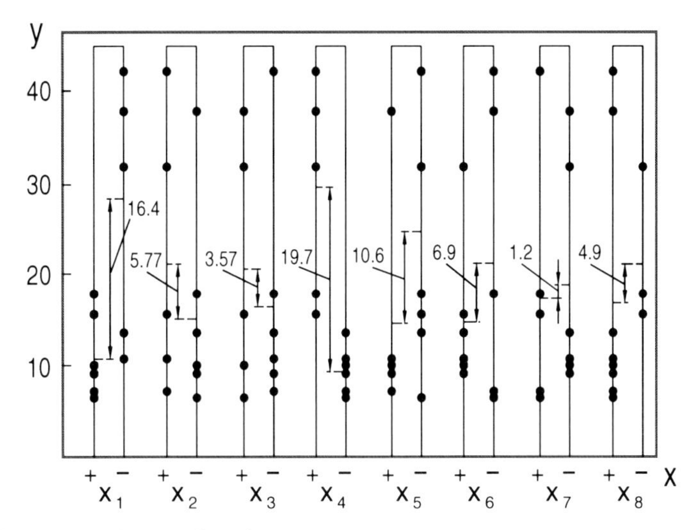

Fig. 1.6. Correlation diagram (first sieve)

Each factor included in the experiment is estimated independently. The simplest way to do this is to calculate the distance between means on the left and right side of each bar. These distances are shown on the correlation diagram in Figure 1.6. As can be seen, these are the greatest for factors x_1 and x_2, and thus these two factors are selected for the analysis. The effects of factors are calculated using special correlation tables. A correlation table (Table 1.13) was constructed to analyze the considered two factors. Using the correlation table, the effect of each selected factor can be estimated as

$$X_i = \frac{\overline{y}_1 + \overline{y}_3 + ... + \overline{y}_n}{m} - \frac{\overline{y}_2 + \overline{y}_4 + ... + \overline{y}_{n-1}}{m} \quad (1.13)$$

where m is the number of $\overline{y}$ in Table 1.13 for the considered factor assigned to the same sign ("+" or "−"). It follows from Table 1.13 that $m = 2$.

The effects of the selected factors were estimated using data in Table 1.13 and Eq.(1.13) as

$$X_1 = \frac{\overline{y}_{1-1} + \overline{y}_{1-3}}{2} - \frac{\overline{y}_{1-2} + \overline{y}_{1-4}}{2} = \frac{17 + 9.3}{2} - \frac{37.3 + 12.5}{2} = -11.75 \quad (1.14)$$

$$X_4 = \frac{\overline{y}_{1-1} + \overline{y}_{1-2}}{2} - \frac{\overline{y}_{1-3} + \overline{y}_{1-4}}{2} = \frac{17 + 37.3}{2} - \frac{9.3 + 12.5}{2} = 16.25 \quad (1.15)$$

Table 1.13. Correlation table (first sieve)

Estimated factor	$+x_1$	$-x_1$	Estimated factor	$+x_1$	$-x_1$
$+x_4$	16 18 $\sum y_{1-1} = 34$ $\overline{y}_{1-1} = 17$	38 42 32 $\sum y_{1-2} = 112$ $\overline{y}_{1-2} = 37.3$	x_4	7 10 9 11 $\sum y_{1-3} = 37$ $\overline{y}_{1-3} = 9.3$	11 14 $\sum y_{1-4} = 25$ $\overline{y}_{1-4} = 12.5$

The significance of the selected factors is examined using the Student's t-criterion, calculated as

$$t = \frac{\left(\overline{y}_{1-1} + \overline{y}_{1-3} + \ldots + \overline{y}_{1-n}\right) - \left(\overline{y}_{1-2} + \overline{y}_{1-4} + \ldots + \overline{y}_{1-(n-1)}\right)}{\sqrt{\sum_i \frac{s_i^2}{n_i}}} \tag{1.16}$$

where s_i is the standard deviation of i-th cell of the correlation table defined as

$$s_i = \sqrt{\frac{\sum_i y_i^2}{n_i - 1} - \frac{\left(\sum_i y_i\right)^2}{n_i\left(n_i - 1\right)}} \tag{1.17}$$

where n_i is the number of terms in the considered cell.

The Student's criteria for the selected factors are calculated using the results presented in the auxiliary table (Table 1.14) as follows

$$t_{X_1} = \frac{\left(\overline{y}_{1-1} + \overline{y}_{1-3}\right) - \left(\overline{y}_{1-2} + \overline{y}_{1-4}\right)}{\sqrt{\sum_i \frac{s_i^2}{n_i}}} = \frac{(17 + 9.3) - (37.3 + 12.5)}{3.52} = -6.68 \tag{1.18}$$

$$t_{X_4} = \frac{\left(\overline{y}_{1-1} + \overline{y}_{1-2}\right) - \left(\overline{y}_{1-3} + \overline{y}_{1-4}\right)}{\sqrt{\sum_i \frac{s_i^2}{n_i}}} = \frac{(17 + 37.3) - (9.3 + 12.5)}{3.52} = 9.23 \tag{1.19}$$

A factor is considered to be significant if $t_{X_i} > t_{cr}$ where the critical value, t_{cr} for the Student's criterion in found in a statistical table for the following number of degrees of freedom

$$f_r = \sum_i n_i - k = 11 - 4 = 7 \tag{1.20}$$

where k is the number of cells in the correlation table.

For the considered case, $t_{99.9}$ = 5.959 (Table 5.7 in [34]) so that the considered factors are significant with a 99.9% confidence level.

Table 1.14. Calculating t-criterion

Cell #	$\sum y_{1-i}$	$\left(\sum y_{1-i}\right)^2$	$\sum y_{1-i}^2$	n_i	s^2	s^2 / n_i
1	34	1156	580	2	2.00	1.00
2	112	12544	4232	3	25.33	8.44
3	37	1369	351	4	2.92	0.73
4	25	625	317	2	4.50	2.25

The discussed procedure is the first stage in the proposed sieve DOE and thus is referred to as the first sieve. This first sieve allows the detection of the strongest factors, i.e. those factors that have the strongest influence on the response. After these strong linear effects are detected, the size of "the screen" to be used in the consecutive sieves is reduced to distinguish less strong effects and their interactions. This is accomplished by correction of the experimental results presented in column $\overline{y}_1$ of Table 1.12. Such a correction is carried out by adding the effects (with the reverse signs) of the selected factors (Eqs. (1.14) and (1.15)) to column $\overline{y}_1$ of Table 1.12, namely, by adding 11.75 to all results at level "$+x_1$" and –16.25 to all results at level "$+x_4$". The corrected results are shown in column $\overline{y}_{c1}$ of Table 1.12. Using the data of this table, one can construct a new correlation diagram shown in Figure 1.7, where, for simplicity, only a few interactions are shown although all possible interactions have been analyzed. Using the approach described above, a correlation table (second sieve) was constructed (Table 1.15) and the interaction x_4x_8 was found to be significant. Its effect is X_{48} = 7.14.

After the second sieve, column $\overline{y}_{c1}$ was corrected by adding the effect of X_{48} with the opposite sign, i.e. –7.14 to all results at level $+x_{48}$. The results are shown in column $\overline{y}_{c2}$ of Table 1.12. Normally, the sieve of the experimental results continues while all the remaining effects and their interactions become insignificant, at say a 5% level of significance if the responses were measured with high accuracy and a 10% level if not. In the considered case, the sieve was ended after the third stage because the analysis of these results showed that there are no more significant factors or interactions left. Figure 1.8 shows the scatter diagram of the discussed test. As can be seen in this figure, the scatter of the analyzed data reduces significantly after each sieve so normally three sieves are sufficient to complete the analysis.

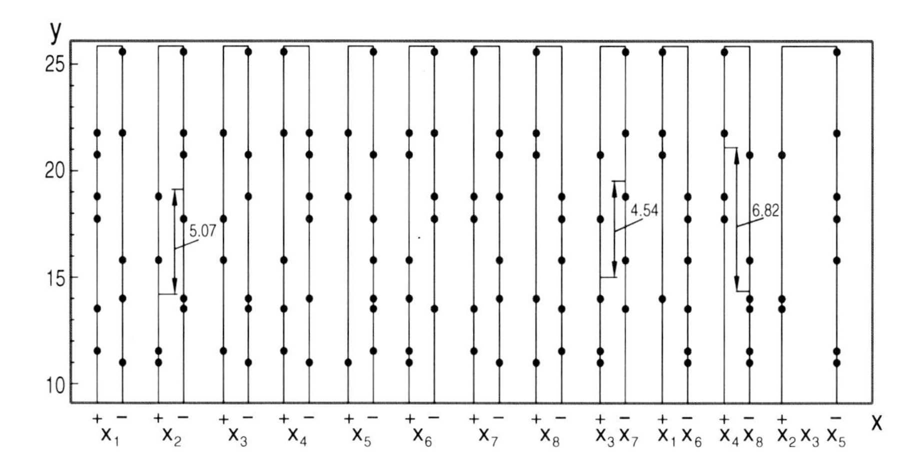

Fig. 1.7. Correlation diagram (second sieve)

Table 1.15. Correlation table (second sieve)

Estimated factor	$+x_3x_7$	$-x_3x_7$	$+x_4x_8$	$-x_4x_8$
$+x_2$	11.5 11 $\sum y_{2-1} = 22.5$ $\bar{y}_{2-1} = 11.25$	18.75 25.75 15.75 $\sum y_{2-2} = 60.25$ $\bar{y}_{2-2} = 20.08$	18.75 25.75 $\sum y_{2-3} = 44.5$ $\bar{y}_{2-3} = 22.25$	11.5 11 15.75 $\sum y_{2-4} = 38.25$ $\bar{y}_{2-4} = 12.75$
Estimated factor	$+x_3x_7$	$-x_3x_7$	$+x_4x_8$	$-x_4x_8$
$-x_2$	14 20.75 17.75 $\sum y_{2-5} = 52.5$ $\bar{y}_{2-5} = 17.5$	21.75 13.5 21.75 $\sum y_{2-6} = 57$ $\bar{y}_{2-6} = 19$	21.75 21.75 17.75 $\sum y_{2-7} = 61.25$ $\bar{y}_{2-6} = 20.42$	13.5 14 20.75 $\sum y_{2-8} = 48.25$ $\bar{y}_{2-5} = 16.08$

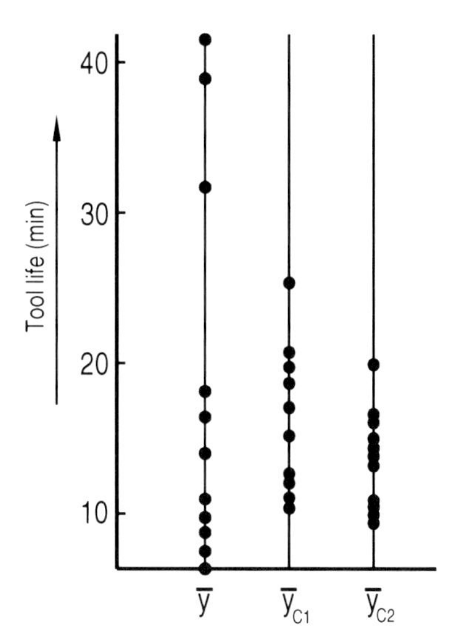

Fig. 1.8. Scatter diagram

The results of the proposed test are summarized in Table 1.16. Figure 1.9 shows the significance of the distinguished effects in terms of their influence on tool life. As seen, two linear effects and one interaction having the strongest effects were distinguished. The negative sign of x_1 shows that tool life decreases when this parameter increases. The strongest influence on tool life has the drill point offset m_d.

Table 1.16. Summary of the sieve test

Stage of sieve	Distinguished factor	Effect	*t*-criterion
Original data	x_1	−11.5	6.68
	x_4	16.25	9.23
First sieve	x_{48}	7.14	4.6
Second sieve	-	-	-

While the distinguished linear effects are known to have strong influence on tool life, the distinguished interaction x_4x_8 has never before been considered in any known studies on gun drilling. Using this factor and results of the complete DOE, a new pioneering geometry of gun drills has been developed (for example US Patent 7147411).

The proposed sieve DOE allows experimentalists to include into consideration as many factors as needed. Conducting a relatively simple sieve test, the significant factors and their interactions can be distinguished objectively and then be used in the subsequent full DOE. Such an approach allows one to reduce the total number of tests dramatically without losing any significant factor or factor interaction. Moreover, interactions of any order can be easily analyzed. The proposed correlation diagrams make such an analysis simple and self-evident.

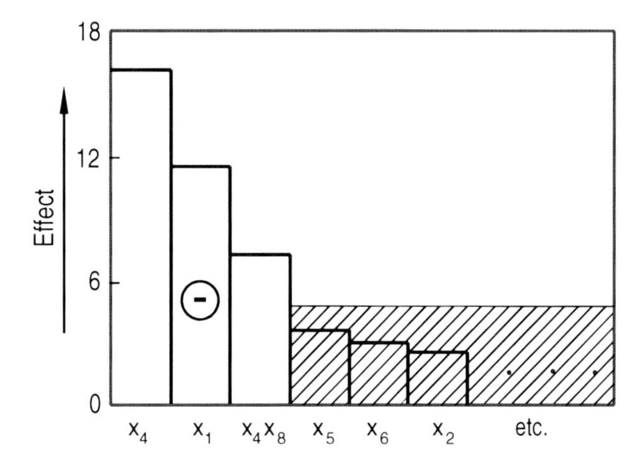

Fig. 1.9. Significance of the effects distinguished by the sieve DOE (Pareto analysis)

1.7.3 Split-Plot DOE

All industrial experiments are split-plot experiments. This provocative remark has been attributed to the famous industrial statistician, Cuthbert Daniel, by Box et al. [35] in their well-known text on the design of experiments. Split-plot experiments were invented by Fisher [36] and their importance in industrial experimentation has been long recognized [37]. It is also well known that many industrial experiments are fielded as split-plot experiments and yet erroneously analyzed as if they were completely randomized designs. This is frequently the case when hard-to-change factors exist and economic constraints preclude the use of complete randomization. Recent work, most notably by Lucas and his coworkers [38], Ganju and Lucas [39], Ju and Lucas [40],Webb et al. [41] has demonstrated that many experiments previously thought to be completely randomized experiments also exhibit split-plot structure. This surprising result adds credence to the Daniel's proclamation and has motivated a great deal of pioneering work in the design and analysis of split plot experiments [42].

In simple terms, a split-plot experiment is a blocked experiment, where the blocks themselves serve as experimental units for a subset of the factors. Thus, there are two levels of experimental units. The blocks are referred to as whole plots, while the experimental units within blocks are called split plots, split units, or subplots. Corresponding to the two levels of experimental units are two levels of randomization. One randomization is conducted to determine the assignment of block-level treatments to whole plots. Then, as always in a blocked experiment, a randomization of treatments to split-plot experimental units occurs within each block or whole plot.

Split-plot designs were originally developed by Fisher [36] for use in agricultural experiments. As a simple illustration, consider a study of the effects of two irrigation methods (factor A) and two fertilizers (factor B) on yield of a crop, using four available fields as experimental units. In this investigation, it is not possible to apply different irrigation methods (factor A) in areas smaller than a field, although different fertilizer types (factor B) could be applied in relatively small areas. For example, if we subdivide each whole plot (field) into two split plots, each of the two fertilizer types can be applied once within each whole plot, as shown in Figure 1.10. In this split-plot design, a first randomization assigns the two irrigation types to the four fields (whole plots); then within each field, a separate randomization is conducted to assign the two fertilizer types to the two split plots within each field.

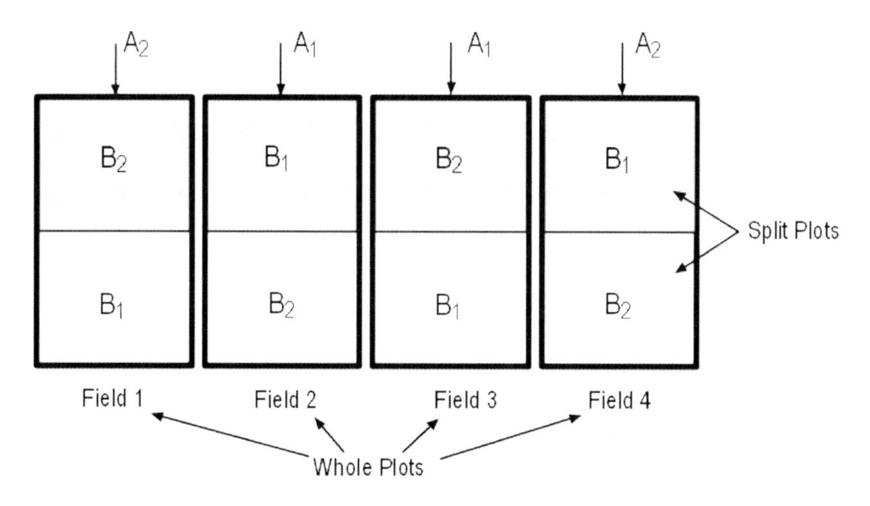

Fig. 1.10. Split plot agricultural layout. (Factor A is the whole-plot factor and factor B is the split-plot factor)

In industrial experiments, factors are often differentiated with respect to the ease with which they can be changed from experimental run to experimental run. This may be due to the fact that a particular treatment is expensive or time-consuming to change, or it may be due to the fact that the experiment is to be run in large batches and the batches can be subdivided later for additional treatments. Box et.al. [35] described a prototypical split-plot experiment with one easy-to-change factor and one hard-to-change factor. The experiment was designed to study the corrosion resistance of steel bars treated with four coatings, C_1, C_2, C_3,and C_4, at three furnace temperatures, 360°C, 370°C, and 380°C. Furnace temperature is the hard-to-change factor because of the time it takes to reset the furnace and reach a new equilibrium temperature. Once the equilibrium temperature is reached, four steel bars with randomly assigned coatings C_1, C_2, C_3, and C_4 are randomly positioned in the furnace and heated. The layout of the experiment as performed is given in Table 1.17. Notice that each whole-plot treatment (temperature) is replicated twice and that there is just one complete replicate of the split-plot treatments (coatings) within each whole plot. Thus, DOE matrix has six whole plots and four subplots within each whole plot.

Table 1.17. Split-plot design and data for studying the corrosion resistance of steel bars

Whole-plot	Temperature (°C)	Coatings (randomized order)			
1.	360	C_2	C_3	C_1	C_4
		73	83	67	89
2.	370	C_1	C_3	C_4	C_2
		65	87	86	91
3.	380	C_3	C_1	C_2	C_4
		147	155	127	212
4.	380	C_4	C_3	C_2	C_1
		153	90	100	108
5.	370	C_4	C_1	C_3	C_2
		150	140	121	142
6.	360	C_1	C_4	C_2	C_3
		33	54	8	46

The analysis of a split-plot experiment is more complex than that for a completely randomized experiment due to the presence of both split-plot and whole-plot random errors. In the Box et al. corrosion resistance example [35], a whole-plot effect is introduced with each setting or re-setting of the furnace. This may be due, e.g., to operator error in setting the temperature, to calibration error in the temperature controls, or to changes in ambient conditions. Split-plot errors might arise due to lack of repeatability of the measurement system, to variation in the distribution of heat within the furnace, to variation in the thickness of the coatings from steel bar to steel bar, and so on.

In the author's experience, ADX Interface for Design of Experiments (http://support.sas.com/documentation/cdl/en/adxgs/60376/HTML/default/overview_sect1.htm) by SAS company is most suitable for the split-plot DOE. Following its manual [43], consider split-plot DOE for tablet production.

The tablet production process has several stages, which include batch mixing, where ingredients are combined and mixed with water, and pellet production, where the batch is processed into pellets that are compressed to form tablets. It is more convenient to mix batches and randomize the treatments within each batch than to mix a new batch for each run. Thus, this experiment calls for a standard two-stage split-plot design.

The moisture content and mixing speed for the batch constitute whole-plot factors, while the factors that control the variety of ways that pellets can be produced from a single batch are subplot factors. The responses of interest are measured on the final tablets. Table 1.18 shows all the variables involved and the stage with which they are associated. The goal of the experiment is to determine which effects are significant. The researcher is interested in both whole-plot factors and subplot factors.

Table 1.18. Factors and their responses in the tablet formulation experiment

Variable	Name	Level	Description	
		Low	High	
Whole Plot	FLUID	90	115	Moisture content (%)
	MIX	15	45	Mixing time (min)
Split Plot	EXTRUDER	36	132	Extruder speed (rpm)
	SCREEN	0.6	1.0	Screen size (mm)
	RESID	2	5	Residence time (min)
	DISK	450	900	Disk speed (rpm)
Response	MPS			Mean particle size (micron)

I. Task List

1. Create the split-plot design.
2. Explore the response data with scatter plots and box plots.
3. Estimate all main effects and as many two-factor interactions as possible between the factors.

II. Start the program

A split-plot design can be created by selecting **File** ⟶ **Create New Design** ⟶ **Split-plot** from the ADX desktop. The main design window will appear as shown in Figure 1.11

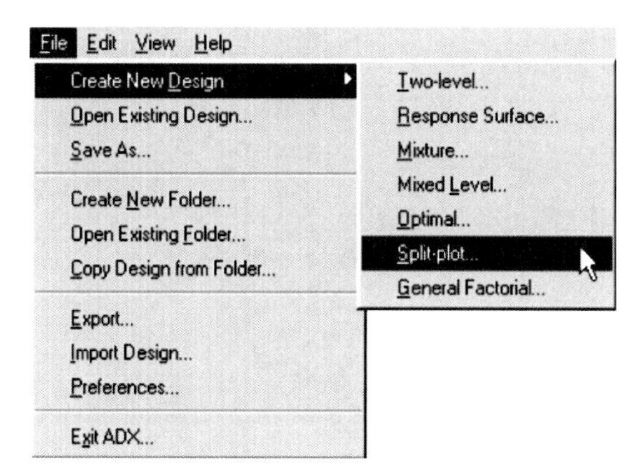

Fig. 1.11. Selecting the split-plot design

III. Defining Variables

To define the variables in this experiment, one should do the following:

1. Click **Define Variables**. The Define Variables window will appear, and the **Whole-Plot Factor** tab will already be selected (Figure 1.12).
2. Define the whole-plot factors.
    2.1. Create two new factors by clicking **Add** and selecting **2**.
    2.2. Enter the factor names given in the whole-plot section of Table 1.12.
3. Define the subplot factors.
    3.1. Click the **Sub-plot Factor** tab (Figure 1.13).
    3.2. Create four new factors by clicking **Add** and selecting **4**.
    3.3. Enter the factor names given in the subplot section of Table 1.12
4. Click the **Block** tab (Figure 1.14). ADX will assign a unique block level to each whole plot when it generates the design, so you do not need to specify the number of block levels. Change the block name to **BATCH**, since each whole plot is a batch of material.
5. Enter the response information.
    5.1. Click the **Response** tab (Figure 1.15).
    5.2. Change the name of the default response to **MPS** and its label to **Mean Particle Size (microns)**.
    5.3. Click **OK** to accept the variable definitions.

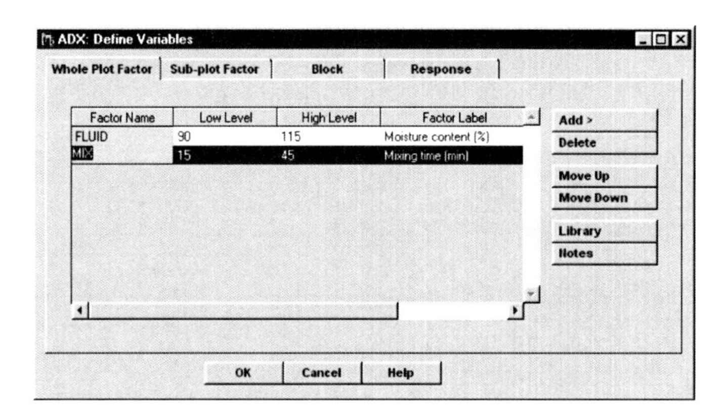

Fig. 1.12. Defining variables, the whole plot

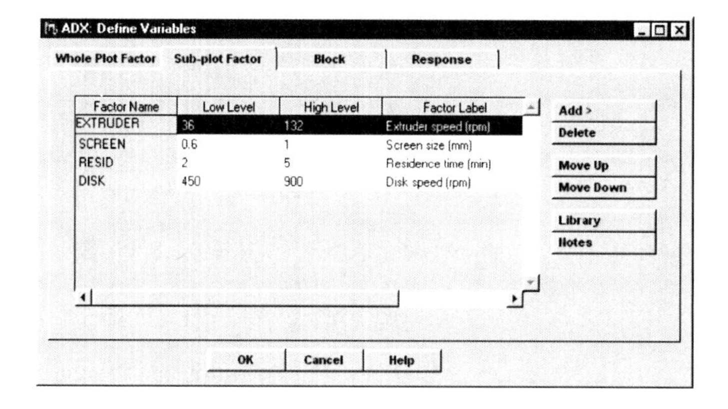

Fig. 1.13. Defining variables, the sub-plot

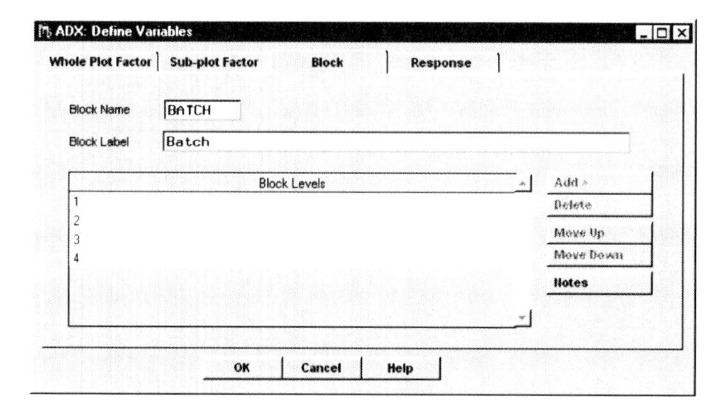

Fig. 1.14. Assigning block levels

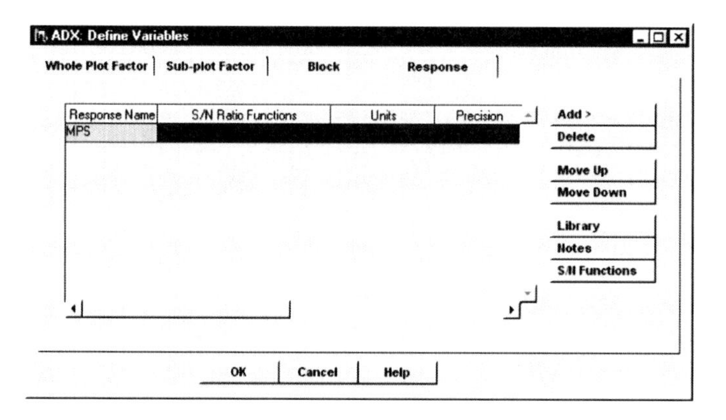

Fig. 1.15. Entering the response information

IV. Exploring Response Data

Before fitting a mixed model, click **Explore** in the main design window. Both box plots and scatter plots are available. Box plots help you visualize the differences in response distribution across levels of different factors, both random and fixed. Scatter plots show the individual responses broken down by factor level or run and can also be used to investigate time dependence.

The box plot is the first graph that is displayed (Figure 1.16). One can explore the distribution of the response broken down by each of the factors or batches that make up the whole plot.

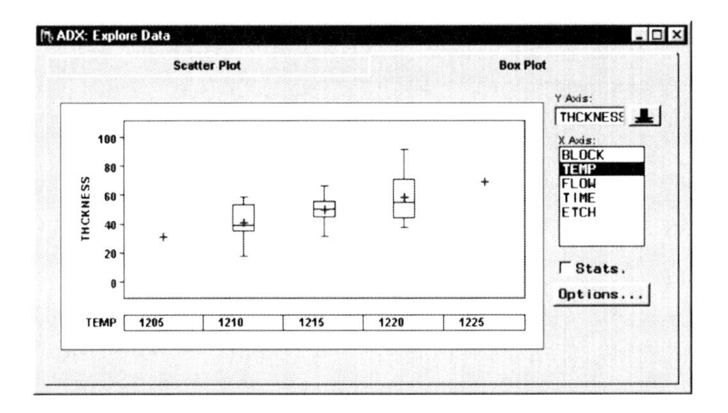

Fig. 1.16. Explore data

Close the window to return to the main design window.

V. Specifying Fixed and Random Effects

ADX does not generate a default master model for a split-plot design, so you must do so before fitting. Click **Fit** in the main design window. The Fit Details for MPS window will open (Figure 1.17). It has sections to define fixed and random effects and classification variables. The fixed effects in a split-plot design are, as usual, the main effects and interactions of interest between the factors. The split-plot structure of the experiment determines the choice of random effects, which in turn determines the proper error term for each fixed effect.

The modeling objectives of a split-plot design are in principle the same as those of a standard screening design. You want to estimate as many effects as possible involving the factors and determine which ones are significant. When this design was analyzed as a standard full factorial, the 64 runs provided enough degrees of freedom to estimate effects and interactions of all orders. However, FLUID and MIX are whole-plot effects and apply only to batches. Therefore, with respect to these two factors, the experiment has only four runs (the four batches). The interaction between the whole-plot effects will estimate the whole-plot error, so this interaction is not included as a fixed effect.

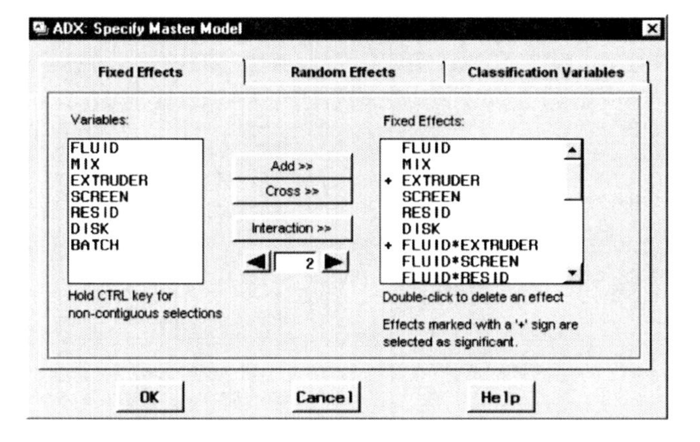

Fig. 1.17. Specifying the fixed effects

1. Click **Fit** in the main design window. Select **Model** ⟶ **Change master model** to open the Specify Master Model window. On the **Fixed Effects** tab (Figure 1.17), specify the following as fixed effects:
   - All main effects. Click and drag or hold down the CTRL key and click to select the FLUID, MIX, EXTRUDER, SCREEN, RESID, and DISK variables. Then click **Add**.
   - All two-factor interactions except FLUID*MIX. Select all six factors and click **Interaction**. Double-click FLUID*MIX in the list on the right to remove it from the master model. (You might have to scroll to find it.).
2. Next, click the **Random Effects** tab (Figure 1.18). Here you specify the whole-plot error, which is BATCH. Select BATCH and click **Add.**

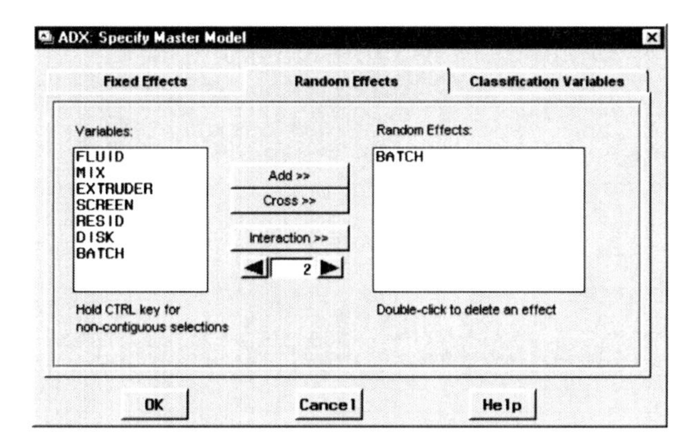

Fig. 1.18. Specifying the random effects

3. Examine the **Classification Variables** tab Figure 1.19. BATCH is automatically entered because it is a random effect. If there were categorical fixed effects, you would enter them on this tab.

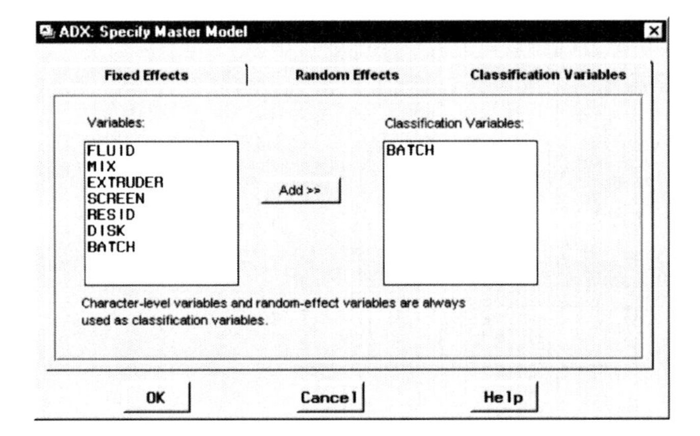

Fig. 1.19. Examining the Classification Variables tab

Click **OK** to close this window and fit the master model.

VI. Fitting a Model

ADX uses the MIXED procedure to estimate the fixed and random effects and variance components of the model. After specifying the model, you will see the Fit Details for MPS window. There are three tabs:

- The **Fixed Effects** tab (Figure 1.20) is similar to the Effects Selector for other designs. The table lists Type 3 tests of the fixed effects. Click an effect to toggle the selection of significant effects.

ADX: Fit details for MPS

Select effects to be included in predictive model.

Fixed Effects | Variance Components | Random Effects

Effect	Num DF	Den DF	F Value	Pr > F
FLUID	1	6.12	2.35	0.1752
MIX	1	6.12	0.59	0.4693
EXTRUDER	1	42	89.78	<.0001
SCREEN	1	42	0.43	0.5149
RESID	1	42	1.79	0.1884
DISK	1	42	3.12	0.0848
FLUID*EXTRUDER	1	42	204.67	<.0001
FLUID*SCREEN	1	42	0.88	0.3548
FLUID*RESID	1	42	2.07	0.1580
FLUID*DISK	1	42	5.86	0.0199

Fig. 1.20. Fixed effects tab

- The **Variance Components** tab provides estimates of the variance of each random effect, including the residual error. The approximate confidence limits for variance components can be calculated easily from this information.

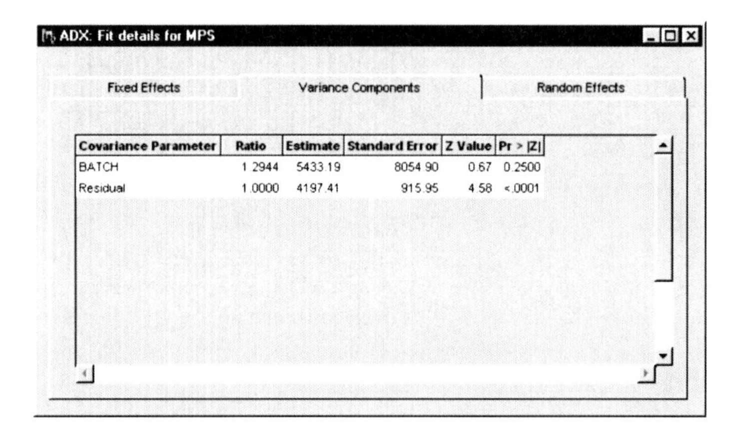

Covariance Parameter	Ratio	Estimate	Standard Error	Z Value	Pr > |Z|
BATCH	1.2944	5433.19	8054.90	0.67	0.2500
Residual	1.0000	4197.41	915.95	4.58	<.0001

Fig. 1.21. Variance components tab

- The **Random Effects** tab lists the random effects broken down by level (for interactions, these are broken down by each combination of levels). These are empirical best linear unbiased predictors (EBLUPs) of the observed random effects. You can use them to screen for unusual whole plots and to assess the normality assumption of the model.

ADX: Fit details for MPS

Fixed Effects | Variance Components | Random Effects

Effect	Batch	Estimate	Std Err Pred	DF	t Value	Pr > |t|	Alpha	Lower	Upper
BATCH	1	-35.9963	64.4192	1	-0.56	0.6756	0.05	-854.52	782.53
BATCH	2	35.9963	64.4192	1	0.56	0.6756	0.05	-782.53	854.52
BATCH	3	35.9963	64.4192	1	0.56	0.6756	0.05	-782.53	854.52
BATCH	4	-35.9963	64.4192	1	-0.56	0.6756	0.05	-854.52	782.53

Fig. 1.22. Random effects tab

Choosing a "best" split-plot design for a given design scenario can be a daunting task, even for a professional statistician [42]. Facilities for construction of split-plot designs are not as yet generally available in software packages (with

SAS/ADX being one exception). Moreover, introductory textbooks frequently consider only very simple, restrictive situations. Unfortunately, in many cases, obtaining an appropriate design requires reading (and understanding) a relevant recent research paper or it requires access to software for generating the design. To make matters even more difficult, construction of split-plot experiments is an active area of research and differences in opinion about the value of the various recent advances exist. Jones and Nachtsheim [42] showed how to make the proper choice of split-plot designs in five areas: (1) two-level full factorial designs; (2) two-level fractional factorial designs; (3) mixture and response surface designs; (4) split-plot designs for robust product experiments; and (5) optimal designs.

1.7.4 Group Method of Data Handling (GMDH)

The important invention of Heisenberg's uncertainty principle [44] from the field of quantum theory has a direct or indirect influence on later scientific developments. Heisenberg's works became popular between 1925 and 1935. According to his principle, a simultaneous direct measurement between the coordinate and momentum of a particle with an exactitude surpassing the limits is impossible; furthermore, a similar relationship exists between time and energy. Since his results were published, various scientists have independently worked on Heisenberg's uncertainty principle.

In 1931, Godel published his works on mathematical logic showing that the axiomatic method itself had inherent limitations and that the principal shortcoming was the so-called inappropriate choice of "external complement." According to his well-known incompleteness theorem [45], it is in principle impossible to find a unique model of an object on the basis of empirical data without using an "external complement" [46]. The regularization method used in solving ill-conditioned problems is also based on this theorem. Hence "external complement" and "regularization" are synonyms expressing the same concept.

In regression analysis, the root mean square (RMS) or least square error determined on the basis of all experimental points monotonically decreases when the model complexity gradually increases. This drops to zero when the number of coefficients n of the model becomes equal to the number of empirical points N. Every equation that possesses n coefficients can be regarded as an absolutely accurate model. It is not possible, in principle, to find a unique model in such a situation. Usually experienced modelers use trial and error techniques to find a unique model without stating that they consciously or unconsciously use an "external complement," necessary in principle for obtaining a unique model. Hence, none of the investigators appropriately selects the "external complement"—the risk involved in using the trial and error methods.

In complex systems modeling statistical probability distributions, like normal distribution, cannot be used if only a few empirical points are available. The important way is to use the inductive approach for sifting various sets of models whose complexity is gradually increased and to test them for their accuracy. The principle of self-organization can be formulated as follows: When the model complexity gradually increases, certain criteria, which are called selection criteria or objective functions and which have the property of "external complement," pass through a minimum. Achievement of a global minimum indicates the existence of a model of optimum complexity (Figure 1.23).

The notion that there exists a unique model of optimum complexity, determinable by the self-organization principle, forms the basis of the inductive approach. The optimum complexity of the mathematical model of a complex object is found by the minimum of a chosen objective function which possesses properties of external supplementation (by the terminology of Godel's incompleteness theorem from mathematical logic). The theory of self-organization modeling [47] is based on the methods of complete, incomplete, and mathematical induction [48]. This has widened the capabilities of system identification, forecasting, pattern recognition, and multicriterial control problems.

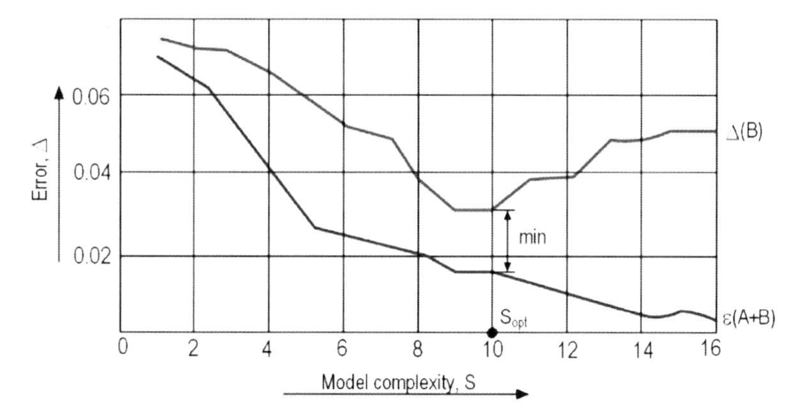

Fig. 1.23. Variation in least square error ε(A + B) and error measure of an "external complement" Δ(B) for a regression equation of increasing complexity S; S_{opt} is the model of optimal complexity

Group Method of Data Handling (hereafter, GMDH) developed as a practical modeling tool based on the theory of self-organization modeling was applied in a great variety of areas for data mining and knowledge discovery, forecasting and systems modeling, optimization, and pattern recognition [47, 49-51]. Inductive GMDH algorithms provide a possibility to find automatically interrelations in data, to select the optimal structure of model or network and to increase the accuracy of existing algorithms. This original self-organizing approach is substantially different from deductive methods used commonly for modeling. It has inductive nature - it finds the best solution by sorting-out possible alternatives and variants. By sorting of different solutions, the inductive modeling approach aims to minimize the influence of the experimentalist on the results of modeling. An algorithm itself can find the structure of the model and the laws, which act in the system. It can be used as an advisor to find new solutions of artificial intelligence (AI) problems.

GMDH is a set of several algorithms for different problems solution. It consists of parametric, clusterization, analogs complexing, rebinarization, and probability algorithms. This self-organizing approach is based on the sorting-out of models of different levels of complicity and selection of the best solution by minimum of external criterion characteristic. Not only polynomials but also nonlinear, probabilistic functions, or clusterizations are used as basic models.

In the author's opinion, the GMDH approach is the most suitable DOE method for experimental studies in manufacturing because:

1. The optimal complexity of model structure is found, adequate to level of noise in data sample. For real problems solution with noisy or short data, simplified forecasting models are more accurate than with any other known methods of DOE.
2. The number of layers and neurons in hidden layers, model structure, and other optimal neutral networks (NN) parameters are determined automatically.
3. It guarantees that the most accurate or unbiased models will be found - method does not miss the best solution during sorting of all variants (in given class of functions).
4. Any non-linear functions or features can be used as input variables are used, which can influence the output variable.
5. It automatically finds interpretable relationships in data and selects effective input variables.
6. GMDH sorting algorithms are rather simple for programming.
7. The method uses information directly from the data samples and minimizes influence of priory researcher assumptions about results of modeling.
8. GMDH neuronets are used to increase the accuracy of other modeling algorithms.
9. The method allows finding an unbiased physical model of object (law or clusterization) - one and the same for all future samples.

There are many published articles and books devoted to GMDH theory and its applications. The GMDH can be considered as a further propagation or extension of inductive self-organizing methods to the solution of more complex practical problems. It solves the problem of how to handle the data samples of observations. The goal is to obtain a mathematical model of the object under study (the problem of identification and pattern recognition) or to describe the processes, which will take place at the object in the future (the problem of process forecasting). GMDH solves, by means of a sorting-out procedure, the multidimensional problem of model optimization

$$g = \arg\min_{g \subset G} CR(g), \quad CR(g) = f\left(P, S, z^2, T_1, V\right) \tag{1.21}$$

where G is a set of considered models, CR is the external criterion of model g quality from this set, P is the number of variables set, S is model complexity, z^2 is the noise dispersion, T_1 is the number of data sample transformation, V is the type of reference function.

For the definite reference function, each set of variables corresponds to definite model structure $P = S$. Problem transforms to much simpler one-dimensional

$$CR(g) = f(S) \tag{1.22}$$

when z^2 = const, T = const, and V = const.

The method is based on the sorting-out procedure, i.e. consequent testing of models, chosen from set of models-candidates in accordance with the given criterion. Most of GMDH algorithms use the polynomial reference functions. General correlation between input and output variables can be expressed by Volterra functional series, discrete analogue of which is Kolmogorov-Gabor polynomial

$$y = b_0 + \sum_{i=1}^{M} b_i x_i + \sum_{i=1}^{M}\sum_{j=1}^{M} b_{ij} x_i x_j + \sum_{i=1}^{M}\sum_{j=1}^{M}\sum_{k=1}^{M} b_{ijk} x_i x_j x_k \tag{1.23}$$

where $X(x_1,x_2,...,x_M)$ is the input variables vector, M is the number of input variables, $A(b_1,b_2,...,b_M)$ is the vector of coefficients.

Components of the input vector X can be independent variables, functional forms or finite difference terms. Other non-linear reference functions, such as difference, probabilistic, harmonic, logistic can also be used. The method allows the finding of *simultaneously* the structure of model and the dependence of modeled system output on the values of most significant inputs of the system.

GMDH, based on the self-organizing principle, requires minimum information about the object under study. As such, all the available information about this object should be used. The algorithm allows the finding of the additional needed information through the sequential analysis of different models using the so-called external criteria. Therefore, GMDH is a combined method: it is used the test data and sequential analysis and estimation of the candidate models. The estimates are found using relatively small part of the test results. The other part of these results is used to estimate the model coefficients and to find the optimal model structure.

Although GMDH and regression analysis use the table of test data, the regression analysis requires the prior formulation of the regression model and its complexity. This is because the row variances used in the calculations are internal criteria. A criterion is called an internal criterion if its determination is based on the same data that is used to develop the model. The use of any internal criterion leads to a false rule: the more complex model is more accurate. This is because the complexity of the model is determined by the number and highest power of its terms. As such, the greater the number of terms, the smaller the variance. GMDH uses external criteria. A criterion is called external if its determination is based on new information obtained using "fresh" points of the experimental table not used in the model development. This allows the selection of the model of optimum complexity corresponding to the minimum of the selected external criterion.

Another significant difference between the regression analysis and GMDH is that the former allows to construct the model only in the domain where the number of the model coefficients is less than then number of point of the design matrix because the examination of the model adequacy is possible only when $f_{ad} > 0$, i.e. when the number of the estimated coefficients of the model, n is less than the number of the points in the design matrix, m. GMDH allows much wider domain where, for example, the number of the model coefficients can be millions and all these are estimated using the design matrix containing only 20 rows. In this new domain, accurate and unbiased models are obtained. GMDH algorithms utilize the minimum input experimental information. This input consists of a table having 10-20 points and the criterion of model selection. The algorithms determine the unique model of optimal complexity by the sorting out of different models using the selected criterion.

The essence of the self-organizing principle in GMDH is that the external criteria pass their minimum when the complexity of the model is gradually increased. When a particular criterion is selected, the computer executing GMDH finds this minimum and the corresponding model of optimal complexity. As such, the value of the selected criterion referred to as the depth of minimum can be considered as the estimate of the accuracy and reliability of this model. If the sufficiently deep minimum is not reached then the model is not found. This might take place when the input data (the experimental data from the design matrix) are: (1) noisy; (2) do not contain essential variables; (3) the basic function (for example, polynomial) is not suitable for the process under consideration, etc.

The following should be clearly understood if one tries to use GMDH:

1. GMDH is not for casual used as, for example the Tagichi method, i.e. it cannot be used readily by everyone with no statistical and programming background. Therefore, it should be deployed in complex manufacturing research programs of high cost and high importance of the results obtained.
2. GMDH is not a part of modern statistical software packages although its algorithms are available, and thus can be programmed.
3. One a research team is gained some experience with GMDH and corresponding algorithms are developed for the application in a certain field of manufacturing studies, GMDH becomes a very powerful method of DOE that can be used at different stages of DOE and, moreover, can be combined with other important DOE methods, for example with split-plot DOE, to adjust GMDH to particular needs.

1.8 Strategy and Principal Steps in Using DOE

The overall strategy of any DOE is presented graphically in Figure 1.24. The basic approach consists of five phases: Conceptual, Discovery, Breakthrough, Optimization, and Validation. The appropriate phase is dependent on the researcher's goals, resources, timing, and subject matter knowledge. Note however, a user could sequence through all five phases within a single project.

Each DOE phase offers different results. For instance, in the discovery phase a researcher's primary goal would include 'screening' for vital input variables. However, the investigator must be aware of the inability to generate a prediction equation using low resolution screening arrays, if any variable interactions are at work in the system. The discovery phase typically focuses on two level, fractional-factorial arrays to identify the "vital few" variables. Fractional-factorial arrays range from resolution III to VI. The lower resolution arrays (III and IV), for instance 2^{3-1}, and 2^{4-1} are limited in application. These arrays are used primarily to screen input variables because of their limited capability to quantify two-factor interactions.

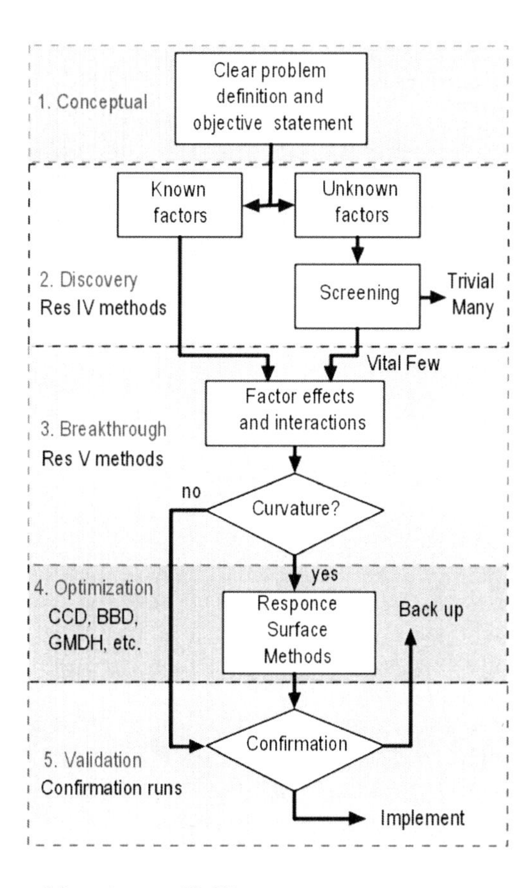

Fig. 1.24. Flow chart of five phases of DOE

Full-factorial and resolution V and VI fractional arrays are more commonly used in the breakthrough phase. All fractional-factorial arrays are subsets of full-factorial arrays. The higher resolution fractional arrays (V and VI), such as 2^{6-1}, offer the opportunity to screen larger numbers of variables with a minimal number of experiments. These designs allow for the identification of some two-factor interactions with a minimal volume of additional experiments. The breakthrough phase is capable of generating first order prediction equations which includes main effects and interaction coefficients. From such data, higher order models from response surface designs can be pursued.

The objective of DOE is to find the correlation between *the response* (for example, tool life, surface finish, etc.) and *the factors* included (in metal cutting, for example, the parameters taken into consideration: the cutting speed, feed, depth of cut, etc.). To achieve this goal, the following steps should be followed [1, 8, 19]:

1. *Select clearly defined problems* that will yield tangible benefits. An unplanned experiment often ends up in total loss of time and money.
2. *Specify the required objective.* To design an effective experiment/process, one must clearly define the experimental objective(s). Objectives may include: screening to identify the critical variables; identification of critical system interactions; or optimization of one or more quality characteristics at several levels. Failure to set good objectives may lead to excessive experiments; failure to identify meaningful quality characteristics; loss of valuable time and resources; unclear results; and poorly defined prediction equations.

   To set good objectives consider: (1) overall project funding and timing, (2) the number of quality characteristics to be monitored, (3) the array design to clarify the need for basic 'main effects' and 'interaction effects' or the need for more detailed quadratic models requiring factor analysis at multiple levels.
3. *Define how the solution delivered* is going to be used.
4. *Understand as much as possible about the problem* and the data set (the domain). Do not rely solely on statistical experts/commercial DOE software manuals: Interaction with subject matter specialists makes professional insight invaluable.
5. *Determine the response variable(s)* of interest that can be measured.
6. *Determine the controllable factors* of interest that might affect the response variables and the levels of each factor to be used in the experiment. It is better to include more factors in the design than to exclude factors, that is, prejudging them to be nonsignificant.
7. *Determine the uncontrollable variables* that might affect the response variables, blocking the known nuisance variables and randomizing the runs to protect against unknown nuisance variables. Replicating an experiment refers to the number of times the combination of input variables are set-up and performed independently. In other words, the defining operations of the experiment must be set-up from scratch for each replicate and not just repeated measurements of the same set-up.

The more times you replicate a given set of conditions, the more precisely you can estimate the response. Replication improves the chance of detecting a statistically significant effect (the signal) in the mist of natural process variation (the noise). The noise of unstable processes can drown out the process signal. Before doing a DOE, it helps to assess the signal-to-noise ratio. The signal-to-noise ratio defines the power of the experiment, allowing the researcher to determine how many replicates will be required for the DOE. Designs reflecting low power require more replicates.

8. *Determine if the data can be measured* (collected) with required accuracy. Do not start statistical analysis without first challenging the validity of the data: What can you expect out of garbage input?
9. *Design the experiments*:
   9.1. *Do a sequential series of experiments.* Designed experiments should be executed in an iterative manner so that information learned in one experiment can be applied to the next. For example, rather than running a very large experiment with many factors and using up the majority of your resources, consider starting with a smaller experiment and then building upon the results. A typical series of experiments consists of a screening design (fractional factorial) to identify the significant factors, a full factorial or response surface design to fully characterize or model the effects, followed up with confirmation runs to verify your results. If you make a mistake in the selection of your factor ranges or responses in a very large experiment, it can be very costly. Plan for a series of sequential experiments so you can remain flexible. A good guideline is not to invest more than 25 percent of your budget in the first DOE.
   9.2. *Remember to randomize* the runs.
   9.3. *Make the model as simple as possible*—but no simpler.
   9.4. *Determine the total number of runs* in the experiment, ideally using estimates of variability, precision required, size of effects expected, etc., but more likely based on available time and resources. Reserve some resources for unforeseen contingencies and follow-up runs.
10. *Perform the experiment strictly according to the experimental design*, including the initial setup for each run in a physical experiment. Do not swap the run order to make the job easier.
11. *Analyze the data from the experiment* using the analysis of variance method. Do not throw away outliers without solid reasoning: Every piece of data stores a hidden story waiting to be opened. Let the problem drive the modeling (i.e., tool selection, data preparation). Stipulate assumptions.
12. *Obtain and statistically verify the model.* Refine the model iteratively.
13. *Define instability in the model* (critical areas where change in output is drastically different for small changes in inputs).
14. *Define uncertainty in the model* (critical areas and ranges in the data set where the model produces low confidence predictions/insights). Do not

underestimate the importance of randomization: The adverse effect of systematic errors can be of vital importance.

15. *Always confirm critical findings.* After all the effort that goes into planning, running and analyzing a designed experiment, it is very exciting to get the results of your work. There is a tendency to eagerly grab the results, rush out to production and say, "We have the answer!" Before doing that, you need to take the time to do a confirmation run and verify the outcome. Good software packages will provide you with a prediction interval to compare the results within some degree of confidence. Remember, in statistics you never deal with absolutes, i.e. there is always uncertainty in your recommendations. Be sure to double-check your results. In other words, do not blindly follow statistical conclusions without taking into account their physical significance, practical implementation, and economic considerations.
16. *Pay particular attention to the presenting of the results.* Do make full use of graphical presentations: A picture can be worth more than hundreds of words. Do avoid statistical jargon in conclusion and report writing: Problem language is the only thing that is universal in a corporation.
17. *Do not think/hope that one iteration of a time experiment can solve the problem* once and for all: The outcome of one experiment often provides a direction for further iterations of exploration.

References

1. Montgomery, D.S., Kowalski, S.M.: Design and Analysis of Experiments, 7th edn. John Wiley & Sons, New York (2009)
2. Antony, Y.: Design of Experiments for Engineers and Scientists. Butterworth-Heinemann, Oxford (2009)
3. Fisher, R.A.: The Design of Experiments, 9th edn. Macmillan, New York (1971)
4. Fisher, R.A.: The Design of Experiments. Oliver and Boyd, Edinburgh (1935)
5. Halt, A.: A History of Mathematical Statistics. Wiley, New York (1998)
6. Student, Tables for estimating the probability that the mean of a unique sample of observations lie between any given distance of the mean of the population from which the sample is drawn. Biometrika 11, p p. 414–417 (1917)
7. Fisher, R.A.: Statistical Methods for Research Workers, 14th edn.(1st edn. 1925). Hafner Press, New York (1973)
8. Telford, J.K.: A brief introduction to design of experiments. Johns Hopkins Apl. Technical Digest 27(3), 224–232 (2007)
9. Roy, R.K.: Design of Experiments Using the Taguchi Approach. Wiley-IEEE, New York (2001)
10. Shina, S.G.: Six Sigma for Electronics Design and Manufacturing. McGraw-Hil, New York (2002)
11. Ross, P.J.: Taguchi Techniques for Quality Engineering. McGraw-Hill, New York (1996)
12. Phadke, M.S.: Quality Engineering Using Robust Design. Pearson Education, Upper Saddle River (2008)

13. Mukerjee, R., Wu, C.-F.: A Modern Theory of Factorial Designs. Springer, London (2006)
14. Kotsireas, I.S., Mujahid, S.N., Pardalos, P.N.: D-Optimal Matrices. Springer, London (2011)
15. Paquete, L.: Experimental Methods for the Analysis of Optimization Algorithms. Springer, London (2010)
16. Farlow, S.J.: Self-organising Methods in Modeling. Marcel Dekker, New York (1984)
17. Madala, H.R.I., Ivakhnenko, A.G.: Inductive Learning Algorithms for Complex Systems Modeling. CRC Press, Boca Raton (1994)
18. Wang, H.S., Chang, C.-P.: Desing of experiments. In: Salvendy, G. (ed.) Handbook of Industrial Engineering. Technology and Operations Management, pp. 2225–2240. John Willey & Sons, New York (2001)
19. Astakhov, V.P.: Tribology of Metal Cutting. Elsevier, London (2006)
20. Mason, R.L., Gunst, R.F., Hess, J.L.: Statistical Design and Analysis of Experiments with Application to Engineering and Science. John Wiley and Sons, New York (1989)
21. Montgomery, D.C.: Design and Analysis of Experiments, 5th edn. John Wiley & Sons, New York (2000)
22. Box, G.E.P., Draper, N.R.: Empirical Model Building and Response Surfaces. Wiley, Hoboken (1987)
23. Astakhov, V.P., Osman, M.O.M., Al-Ata, M.: Statistical design of experiments in metal cutting - Part 1: Methodology. Journal of Testing and Evaluation 25(3), 322–327 (1997)
24. Astakhov, V.P., Al-Ata, M., Osman, M.O.M.: Statistical design of experiments in metal cutting. Part 2: Application. Journal of Testing and Evaluation, JTEVA 25(3), 328–336 (1997)
25. Gopalsamy, B.M., Mondal, B., Ghosh, S.: Taguchi method and ANOVA: An approach for process parameters optimization of hard machining while machining hardened steel. Journal of Scietific & Indistrial Research 68, 659–686 (2009)
26. Nalbant, M., Gökkaya, H., Sur, G.: Application of Taguchi method in the optimization of cutting parameters for surface roughness in turning. Materials & Design 28(4), 1379–1385 (2007)
27. Lin, T.R.: The use of reliability in the Taguchi method for the optimization of the polishing ceramic gauge block. The International Journal of Advanced Manufacturing Technology 22(3-4), 237–242 (2003)
28. Yang, W.H., Tarng, Y.S.: Design optimization of cutting parameters for turning operations based on the Taguchi method. Journal of Material Processing Technology 84, 122–129 (1998)
29. Ghani, J.A., Choudhury, I.A., Hassan, H.H.: Application of Taguchi method in the optimization of end milling operations. Journal of Material Processing Technology 145, 84–92 (2004)
30. Plackett, R.L., Burman, J.P.: The design of optimum multifactorial experiments. Biometrica 33, 305–328 (1946)
31. Astakhov, V.P.: An application of the random balance method in conjunction with the Plackett-Burman screening desing in metal cutting tests. Journal of Testing and Evaluation 32(1), 32–39 (2004)
32. Bashkov, V.M., Katsev, P.G.: Statistical Fundamental of Cutting Tool Tests. Machinostroenie, Moscow (1985) (in Russian)
33. Astakhov, V.P.: Geometry of Single-Point Turning Tools and Drills: Fundamentals and Practical Applications. Springer, London (2010)

34. Holman, J.P.: Experimental Methods for Engineers, 6th edn. McGraw-Hill (1994)
35. Box, G., Hunter, W., Hunter, S.: Statistics for Experimenters: Design, Innovation, and Discovery, 2nd edn. Wiley, New York (2005)
36. Fisher, R.A.: Statistical Methods for Research Workers. Oliver and Boyd, Edinburgh (1925)
37. Yates, F.: Complex experiments, with discussion. Journal of the Royal Statistical Society, Series B2, 181–223 (1935)
38. Anbari, F.T., Lucas, J.M.: Designing and running super-efficient experiments: Optimum blocking with one hard-to-change factor. Journal of Quality Technology 40, 31–45 (2008)
39. Ganju, J., Lucas, J.M.: Randomized and random run order experiments. Journal of Statistical Planning and Inference 133, 199–210 (2005)
40. Ju, H.L., Lucas, J.M.: Lk factorial experiments with hard-to-change and easy-to-change factors. Journal of Quality Technology 34, 411–421 (2002)
41. Webb, D.F., Lucas, J.M., Borkowski, J.J.: Factorial experiments when factor levels are not necessarily reset. Journal of Quality Technology 36, 1–11 (2004)
42. Jones, B., Nachtsheim, C.J.: Split-plot designs: What, why, and how. Journal of Quality Technology 41(4) (2009)
43. Getting Started with the SAS® 9.2 ADX Interface for Design of Experiments. SAS Institute Inc., Cary, NC (2008)
44. Heisenberg, W.: The Physical Principles of Quantum Theory. University of Chicago Press, Chicago (1930)
45. von Neumann, J.: Theory of Self Reproducing Automata. University of Illinois Press, Urbana (1966)
46. Beer, S.: Cybernetics and Management, 2nd edn. Wiley, New York (1959)
47. Madala, H.R., Ivakhnenko, A.G.: Inductive Learning Algorithms for Complex System Modeling. CRC Press, Boca Raton (1994)
48. Arbib, M.A.: Brains, Machines, and Mathematics, 2nd edn. Springer, New York (1987)
49. Ivachnenko, A.G.: Polynomial theory of complex systems. IEEE Transactions on Systems, Man, and Cybernetics SMC-1(4), 284–378 (1971)
50. Ivakhnenko, A.G., Ivakhnenko, G.A.: Problems of further development of the group method of data handling algorithms. Pattern Recognition and Image Analysis 110(2), 187–194 (2000)
51. Ivakhnenko, A.G., Ivaknenko, G.A.: The review of problems solvable by algorithms of the group method of data handling. Pattern Recognition and Image Analysis 5(4), 527–535 (1995)

2

Stream-of-Variation Based Quality Assurance for Multi-station Machining Processes – Modeling and Planning

J.V. Abellan-Nebot[1], J. Liu[2], and F. Romero Subiron[1]

[1] Department of Industrial Systems Engineering and Design,
Universitat Jaume I, Av. Vicent Sos Baynat s/n. Castelló, C.P. 12071. Spain
abellan@esid.uji.es
[2] Department of Systems and Industrial Engineering,
The University of Arizona, Tucson, AZ 85704, USA
jianliu@sie.arizona.edu, fromero@esid.uji.es

In the effort of quality assured product design and implementation, a reliable 3D manufacturing variation propagation model for multi-station machining processes (MMPs) is a key enabler to evaluate the output of geometric and dimensional product quality. Recently, the extension of the stream-of-variation (SoV) methodology provides quality engineers with a tool to model the propagation of machining-induced variations, together with fixture- and datum-induced variations, along multiple stations in MMPs. In this chapter, we present a generic framework of building the extended SoV model for MMPs. Its application in manufacturing process planning is introduced and demonstrated in detail through a 3D case study.

2.1 Introduction

Conventionally, product design has been separated from manufacturing process design in the product development cycle. This product-oriented approach is often referred to as the *over-the-wall design* due to its sequential nature of the design activities. It prevents the integration of design and manufacturing activities, and causes the increase of production ramp-up time, product change cost and the degradation of product quality, which is inversely related to the geometrical and dimensional variations of the key product characteristics (KPCs). In order to overcome the limitation of this approach, manufacturers have begun to investigate means of simultaneously evaluating product designs and manufacturing processes in an attempt to proactively address the potential quality problems in manufacturing phase, reduce ramp-up times and ensure geometrical product

quality. For this purpose, research efforts have been conducted to develop reliable three-dimensional (3D) variation propagation models that integrate both product and manufacturing process information. Such models can be applied in a variety of fields, such as manufacturing variation sources diagnosis, process planning, process oriented tolerancing.

To illustrate the importance of 3D manufacturing variation propagation models, consider the part design shown in Fig. 2.1 and the given manufacturing process plan (with its corresponding fixture layouts, locators specifications, cutting-tool used, thermal conditions of machine-tools, etc.) shown in Fig. 2.2. We consider two KPCs, i.e., the distance between surfaces S_0 and S_3, named KPC_1, and the distance between surfaces S_6 and S_8, named KPC_2. In order to ensure the quality of these two KPCs, the engineers should formulate the following questions:

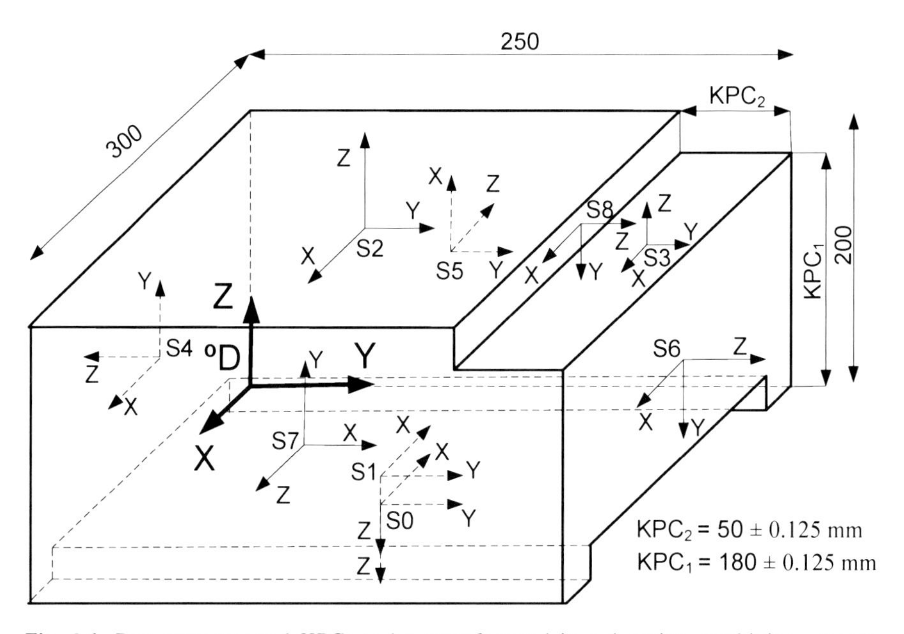

Fig. 2.1. Part geometry and KPCs to be manufactured in a 4-station machining process. Dimensions in mm

- Is this multi-station machining process (MMP) able to produce KPCs whose dimensions are within specifications?
- Which are the critical components of the process, named key control characteristics (KCCs), that have significant impacts on the quality of the KPCs?
- How can we improve the manufacturing process to generate acceptable KPC variations? Or in other words, how can we modify the process to be more robust?
- How can we assign the tolerance of KCCs to ensure the deliver of acceptable KPC variations?

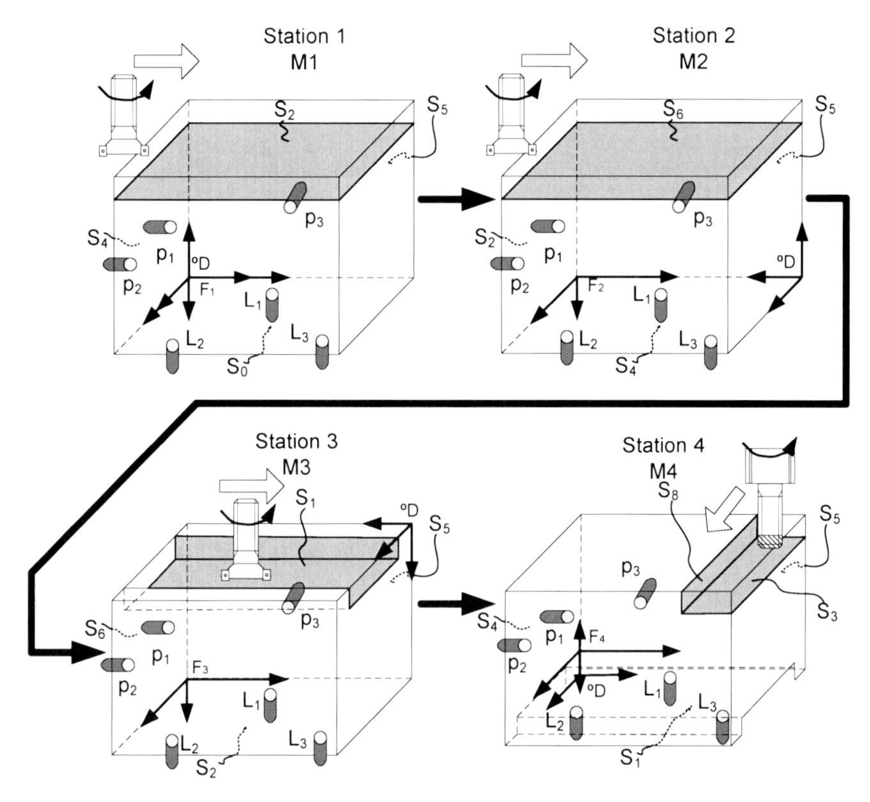

Fig. 2.2. 4-machining station process to manufacture the part

In order to answer these questions, engineers should be able to define analytically how each potential source of variations affects the KPCs during machining. In MMPs, the mathematical models that describe the relationships between dimensional KPCs and KCCs are commonly named 3D manufacturing variation propagation models. These models provide the engineers with a great tool for a large number of applications such as manufacturing and product tolerance allocation, geometric product validation, fault diagnosis, process planning evaluation and selection, etc. In this chapter, a 3D variation propagation model for MMPs, named the stream-of-variation (SoV) model, will be presented, together with its application in process planning. The organization of this chapter is as follows. Section 2.2 introduces the SoV model in its extended version, where datum-, fixture- and machining-induced variations are included into the variation and propagation model. Section 2.3 shows the application of the extended SoV model in process planning. We present the SoV-based methodology that improves the robustness of a given process plan. In Section 2.4 previous questions related to the example shown in Figs. 2.1 and 2.2 are discussed through the application of the extended SoV model. Finally, Section 2.5 concludes the chapter.

2.2 3D Variation Propagation Modeling

2.2.1 Fundamentals

Manufacturing variability in MMPs and its impacts on part quality can be modeled by capturing the mathematical relationships between the KCCs and the KPCs. These relationships can be modeled with non-linear functions. For instance, a function f_1 can be defined such that $y_1=f_1(\mathbf{u})$, where y_1 is the value of a KPC and $\mathbf{u}=[u_1, u_2, \ldots, u_n]^T$ are the KCCs in a MMP. By the assumption of small variations, the non-linear function can be linearized through a Taylor series expansion and the value of a KPC can be defined as

$$y_1 = f_1(\bar{\mathbf{u}}) + \left.\frac{\delta f_1(\mathbf{u})}{\delta u_1}\right|_{\mathbf{u}=\bar{\mathbf{u}}} \cdot (u_1 - \bar{u}_1) + \ldots + \left.\frac{\delta f_1(\mathbf{u})}{\delta u_n}\right|_{\mathbf{u}=\bar{\mathbf{u}}} \cdot (u_n - \bar{u}_n) + \varepsilon_1, \qquad (2.1)$$

where ε_1 contains the high-order non-linear residuals of the linearization, and the linearization point is defined by $\bar{\mathbf{u}} = [\bar{u}_1, \bar{u}_2, \ldots, \bar{u}_n]^T$. This linear approximation can be considered good enough for many MMPs [1]. From Eq. (2.1), the dimensional variation of a KPC from its nominal value is defined as

$$\Delta y_1 = \left.\frac{\delta f_1(\mathbf{u})}{\delta u_1}\right|_{\mathbf{u}=\bar{\mathbf{u}}} \cdot \Delta u_1 + \ldots + \left.\frac{\delta f_1(\mathbf{u})}{\delta u_n}\right|_{\mathbf{u}=\bar{\mathbf{u}}} \cdot \Delta u_n + \varepsilon_1, \qquad (2.2)$$

where $\Delta y_1 = y_1 - f_1(\bar{\mathbf{u}})$, defines the variations of the KPC, whereas $\Delta u_j = u_j - \bar{u}_j$, for $j = 1, \ldots, n$, defines the small variations of the KCCs in a MMP. Considering that there are M KPCs in the part whose variations are stacked in the vector $\mathbf{Y} = [\Delta y_1, \Delta y_2, \ldots, \Delta y_M]^T$, Eq. (2.2) can be rewritten in a matrix form as

$$\mathbf{Y} = \mathbf{\Gamma} \cdot \mathbf{U} + \boldsymbol{\varepsilon}, \qquad (2.3)$$

where $\mathbf{U} = [\Delta u_1, \Delta u_2, \ldots, \Delta u_n]^T$, $\boldsymbol{\varepsilon}$ is the stacked vector of the high-order non-linear residuals defined as $\boldsymbol{\varepsilon} = [\varepsilon_1, \varepsilon_2, \ldots, \varepsilon_n]^T$ and $\mathbf{\Gamma}$ is the matrix $\left[\left[\left.\frac{\delta f_1(\mathbf{u})}{\delta u_1}\right|_{\mathbf{u}=\bar{\mathbf{u}}}, \ldots, \left.\frac{\delta f_1(\mathbf{u})}{\delta u_n}\right|_{\mathbf{u}=\bar{\mathbf{u}}}\right]^T; \ldots; \left[\left.\frac{\delta f_M(\mathbf{u})}{\delta u_n}\right|_{\mathbf{u}=\bar{\mathbf{u}}}, \ldots, \left.\frac{\delta f_M(\mathbf{u})}{\delta u_n}\right|_{\mathbf{u}=\bar{\mathbf{u}}}\right]^T\right]^T$.

For MMPs, the derivation of Eq. (2.3) is a challenging task. At the end of the nineties, researchers from the University of Michigan proposed the adoption of the well-known state space model from control theory [2, Chapter 11] to mathematically represent the relationship between the variation sources and the variations of the machined surfaces at each station, including how the variation of the surfaces generated at upstream stations influence the surfaces generated at downstream stations when the upstream surfaces are used as locating datums. In this representation, dimensional variations of the machined surfaces from nominal values at station k are defined by a series of 6-by-1 vectors named differential motion vectors (DMVs), in the form of $\mathbf{x}_{k,i} = [(\mathbf{d}_i^R)^T, (\boldsymbol{\theta}_i^R)^T]^T$, where $\mathbf{d}_i^R = [d_{ix}^R, d_{iy}^R, d_{iz}^R]^T$ is a vector of translational variation and $\boldsymbol{\theta}_i^R = [\theta_{ix}^R, \theta_{iy}^R, \theta_{iz}^R]^T$ is a vector of orientation variation of the coordinate system (CS) of the ith part surface with respect to (w.r.t.) a reference CS (R), as it is shown in Fig. 2.3.

According to this representation, ^{o}i denotes the nominal CS and i to the current CS. The variation of all part surfaces at station k are stacked in the vector $\mathbf{x}_k = [\mathbf{x}_{k,1}^T, \ldots, \mathbf{x}_{k,i}^T, \ldots]^T$. By the adoption of the state space model, the variation of all part surfaces is related to the potential sources of variation at each machining station. In a machining station, three main sources of variation are distinguished: datum-induced variations, fixture-induced variations and machining-induced variations. These sources of variation produce random deviations between the cutting-tool paths and the workpiece location on the machine-tool table and thus, random deviations of the machined surfaces occur.

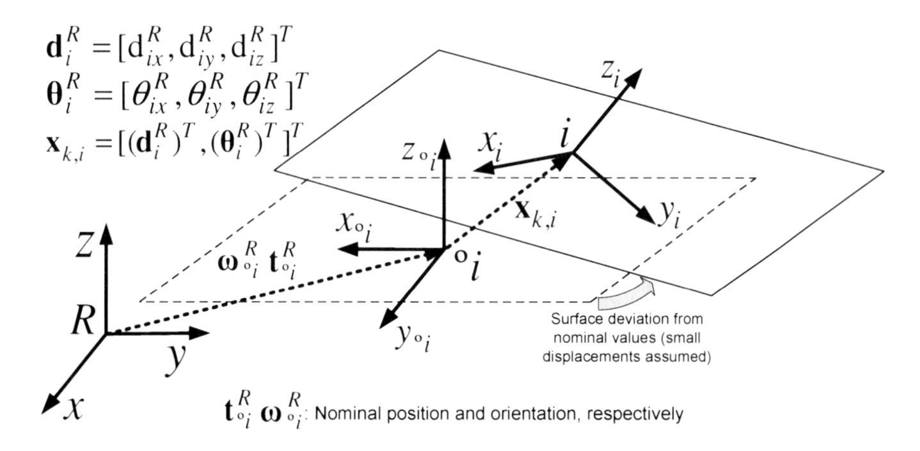

Fig. 2.3. Dimensional variation of a plane surface at station k using a DMV

To illustrate how these three main sources of variation influence on part quality, we consider an N-station machining process shown in Fig. 2.4 and the kth machining station with the workpiece and the fixture device shown in Fig. 2.5. At this kth station, the following variation sources exist.

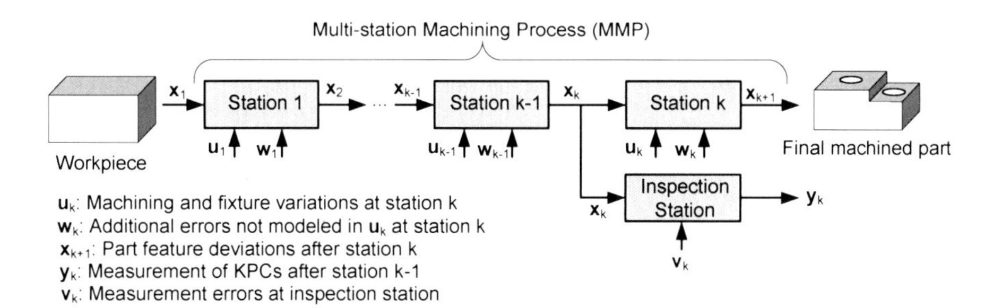

Fig. 2.4. Manufacturing variation propagation in a MMP

First, the variations of datum surfaces used for locating the workpiece deviate the workpiece location on the machine-tool table. This term can be estimated as $\mathbf{x}_{k+1}^{d} = \mathbf{A}_k \cdot \mathbf{x}_k$, where $\mathbf{x}_k$ is the vector of part surfaces variations from upstream machining stations and $\mathbf{A}_k$ linearly relates the datum variations with the machined surface variation due to the locating deviation of the workpiece.

Secondly, the fixture-induced variations deviate the workpiece location on the machine-tool table and thus, a machined surface variation is produced after machining. This term can be estimated as $\mathbf{x}_{k+1}^{f} = \mathbf{B}_k^{f} \cdot \mathbf{u}_k^{f}$, where $\mathbf{u}_k^{f}$ is the vector that defines the KCCs related to fixture-induced variations and $\mathbf{B}_k^{f}$ is a matrix that linearly relates locator variations with variations of machined surface.

Thirdly, the machining-induced variations such as those due to geometrical and kinematic errors, tool-wear errors, etc., deviates the cutting-tool tip and thus, the machined surface is deviated from its nominal values. This term is modeled as $\mathbf{x}_{k+1}^{m} = \mathbf{B}_k^{m} \cdot \mathbf{u}_k^{m}$, where $\mathbf{u}_k^{m}$ is the vector that defines the KCCs related to machining-induced variations and $\mathbf{B}_k^{m}$ is a matrix that linearly relates these KCCs with the machined surface variations.

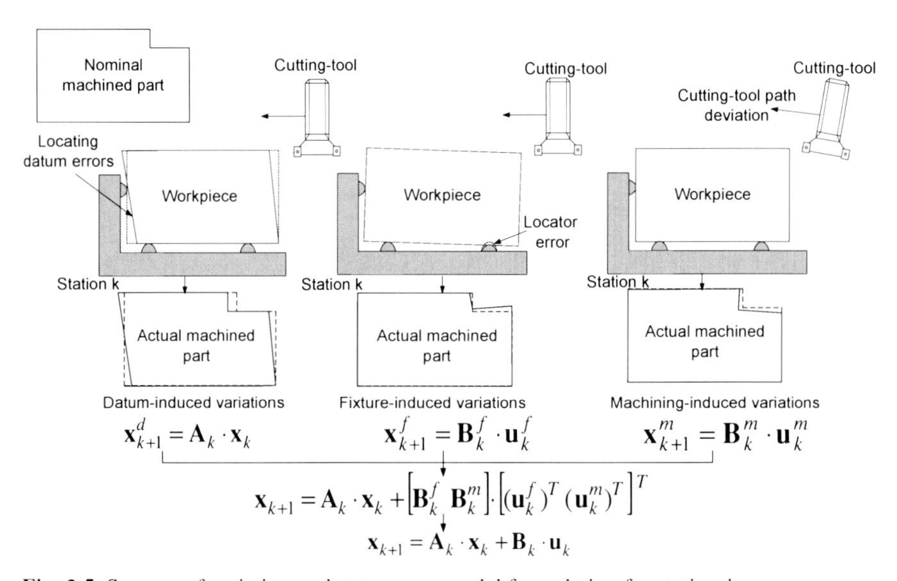

Fig. 2.5. Sources of variation and state space model formulation for station k

Therefore, for an N-station machining process the derivation of the state space model can be defined in a generic form as

$$\mathbf{x}_{k+1} = \mathbf{A}_k \cdot \mathbf{x}_k + \mathbf{B}_k \cdot \mathbf{u}_k + \mathbf{w}_k, \quad k = 1, \ldots, N, \tag{2.4}$$

where $\mathbf{B}_k \cdot \mathbf{u}_k$ represents the variations introduced within station k due to the KCCs and it is defined as $[\mathbf{B}_k^{f} \quad \mathbf{B}_k^{m}] \cdot [(\mathbf{u}_k^{f})^T \quad (\mathbf{u}_k^{m})^T]^T$; and $\mathbf{w}_k$ is the unmodeled system noise and linearization errors.

Eq. (2.4) shows the relationship between KCCs and KPCs along a MMP. Considering that an inspection station is placed after station $k-1$ in order to verify if the workpiece/part is within specifications, then, following the state space model formulation from control theory, the KPC measurements can be expressed as

$$\mathbf{y}_k = \mathbf{C}_k \cdot \mathbf{x}_k + \mathbf{v}_k, \tag{2.5}$$

where $\mathbf{y}_k$ represents the variations of the inspected KPCs; $\mathbf{C}_k \cdot \mathbf{x}_k$ are the linear combinations of the variations of workpiece features after the kth station that define the KPCs; and $\mathbf{v}_k$ is the measurement noise of the inspection process. Similar to $\mathbf{x}_k$, the vector $\mathbf{y}_k$ is defined as $[y_{k,1},\ldots,y_{k,q},\ldots,y_{k,M}]^T$, where $y_{k,q}$ is the inspected variation of the qth KPC and M is the number of KPCs inspected.

For a MMP, Eqs. (2.4) and (2.5) form the generic math-based state space representation, named in the literature as the SoV model, which allows for integration of KPCs and KCCs through product and process information such as fixture layout, part geometry, sequence of machining and inspection operations, etc. Based on this model, the use of advanced control theory, multivariate statistics and Monte Carlo simulations enables a large number of applications along product-life cycle (Fig. 2.6). In the literature, interesting research works can be found about manufacturing fault identification [3-6], part quality estimation [1, 7], active control for quality variation reduction [8-11] and process planning and process-oriented tolerancing [12-16].

The SoV model can be presented in its conventional version [1, 7, 17] and in its extended version [18]. The conventional SoV model includes datum-, fixture- and machining-induced variations but defines the machining-induced variations as a generic cutting-tool path variation defined by three translational and three orientation deviations. The extended SoV model expands the conventional model by including specific machining-induced variations due to geometric, kinematic errors, thermal distortions, cutting-tool wear and cutting-tool deflections, etc. In the next subsections, the derivation of the extended SoV model will be introduced.

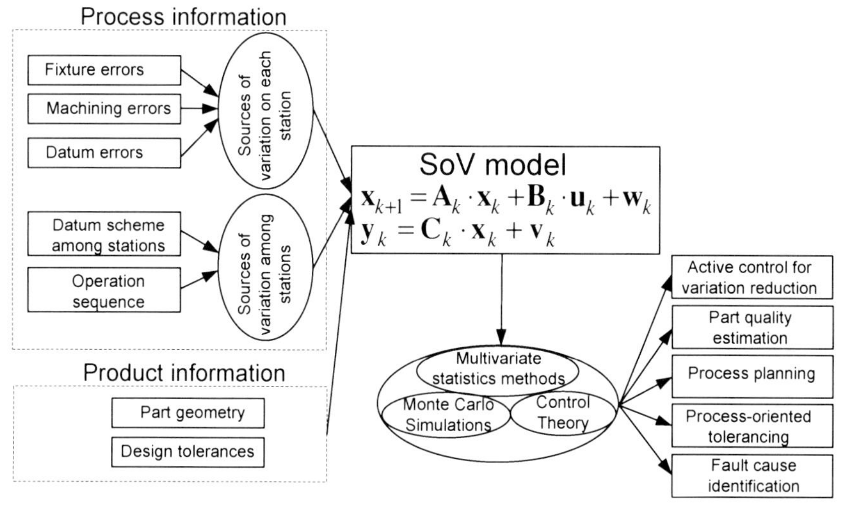

Fig. 2.6. Diagram of the SoV model derivation and its applications

2.2.2 Definition of Coordinate Systems

As explained previously, Eqs. (2.4) and (2.5) define the SoV model in MMPs. The derivation of the SoV model is achieved by analyzing the KCCs that may be present in the MMP and relating them with the resulting KPC variations. In order to define the relationships between KCCs and KPCs, it is necessary to identify the elements that compose each machining station component (machine-tool, fixture device and workpiece) and their corresponding CSs. For this purpose, we consider the *k*th machining station of a MMP, composed of a 3-axis vertical machine-tool, a fixture device and a workpiece, as it is shown in Fig. 2.7. For the machining station *k*, the following CSs can be defined:

Design Coordinate System (DCS). The nominal DCS, denoted as ${}^{\circ}D$, define the reference for the workpiece features during design. The definition of ${}^{\circ}D$ usually depends on the nominal geometry of the part and it is usually defined at an accessible corner. As this CS is only used in design, this CS cannot be deviated.

Reference Coordinate System (RCS). The nominal and true RCS, denoted as ${}^{\circ}R_k$ and R_k, respectively, define the reference for the workpiece features (Fig. 2.7d) at station *k*. To facilitate the model derivation, the R_k is defined as the local coordinate system of the primary datum feature at station *k*. In a 3-2-1 fixture layout, the primary datum is the main workpiece surface used to locate the part at the machine-tool table [19, Chapter 3]. The R_k is defined similarly according to the actual part geometry.

Fixture Coordinate System (FCS). The nominal and true FCS at station *k*, denoted as ${}^{\circ}F_k$ and F_k, respectively, define the physical position and orientation of the fixture device according to the fixture layout. Fig. 2.7d shows the FCS for a fixture layout based on the 3-2-1 principle.

Local Coordinate System (LCS^j). The nominal and true LCS^j at station *k*, denoted as ${}^{\circ}L_k^j$ and L_k^j, define the physical position and orientation of the *j*th nominal and actual machined surface of the part respectively (Fig. 2.7c). For planar surfaces the Z-axis of ${}^{\circ}L_k^j$ is commonly defined normal to the surface.

Machine-Tool Coordinate System (MCS). The MCS at station *k*, denoted as ${}^{\circ}M_k$, define the physical position and orientation of the reference CS for machine-tool movements. The origin of the ${}^{\circ}M_k$ is located at the locating origin of the nominal machine-tool table, with its Z-axis normal to the table and pointing upward, its X-axis parallel to the long axis of the table and pointing to its positive direction, and its Y-axis defined according to the right hand rule, as shown in Fig. 2.7a. In this chapter, it is assumed that ${}^{\circ}M_k$ serves as the reference at station k and thus will not deviate.

Axis Coordinate System (ACS^i). The nominal and true ACS^i of the i-axis used at station k, denoted as ${}^{\circ}A_k^i$ and A_k^i, respectively, define the physical position and orientation of the i-axis of the machine-tool. The origin of the ${}^{\circ}A_k^i$ is located at the geometrical center of the joint of the i-axis. For prismatic joints, the axes of ${}^{\circ}A_k^i$ have the same orientation as that of the ${}^{\circ}M_k$. An example of ${}^{\circ}A_k^i$'s of a 3-axis vertical machine-tool is shown in Fig. 2.7a. The A_k^i is similarly defined for an actual axis.

Spindle Coordinate System (SCS). The nominal and true SCS at station k, denoted as ${}^{\circ}S_k$ and S_k, respectively, define the physical position and orientation of the spindle during machining. The origin of the ${}^{\circ}S_k$ is located at the geometrical center of the spindle and the orientations of axes are identical to that of the Z-axis of the machine-tool, as shown in Fig. 2.7b. The S_k is defined similarly for the actual spindle.

Cutting-Tool Coordinate System (CCS). The nominal and true CCS at station k, denoted as ${}^{\circ}C_k$ and C_k, respectively, define the physical position and orientation of the cutter tip center during machining. The origin of the ${}^{\circ}C_k$ is located at the cutter tip center and the orientations of its axes are identical to that of a S_k, as shown in Fig. 2.7c. The C_k is defined similarly for the actual cutting-tool.

Cutting-Tool Tip Coordinate System (TPCS). The nominal and true TPCS at station k, denoted as ${}^{\circ}P_k$ and P_k, respectively, defines the physical position and orientation of the cutting-tool tip. The origin of the ${}^{\circ}P_k$ is located at the center of the cutting edge that is used to generate feature j, and the orientations of its axes are identical to that of ${}^{\circ}L_k^j$. Please note that, when machining feature j at station k, the cutting-tool tip removes material generating the machined feature which is defined by the L_k^j. Thus, the position and orientation of the P_k defines the position and orientation of L_k^j, as shown in Fig. 2.7c.

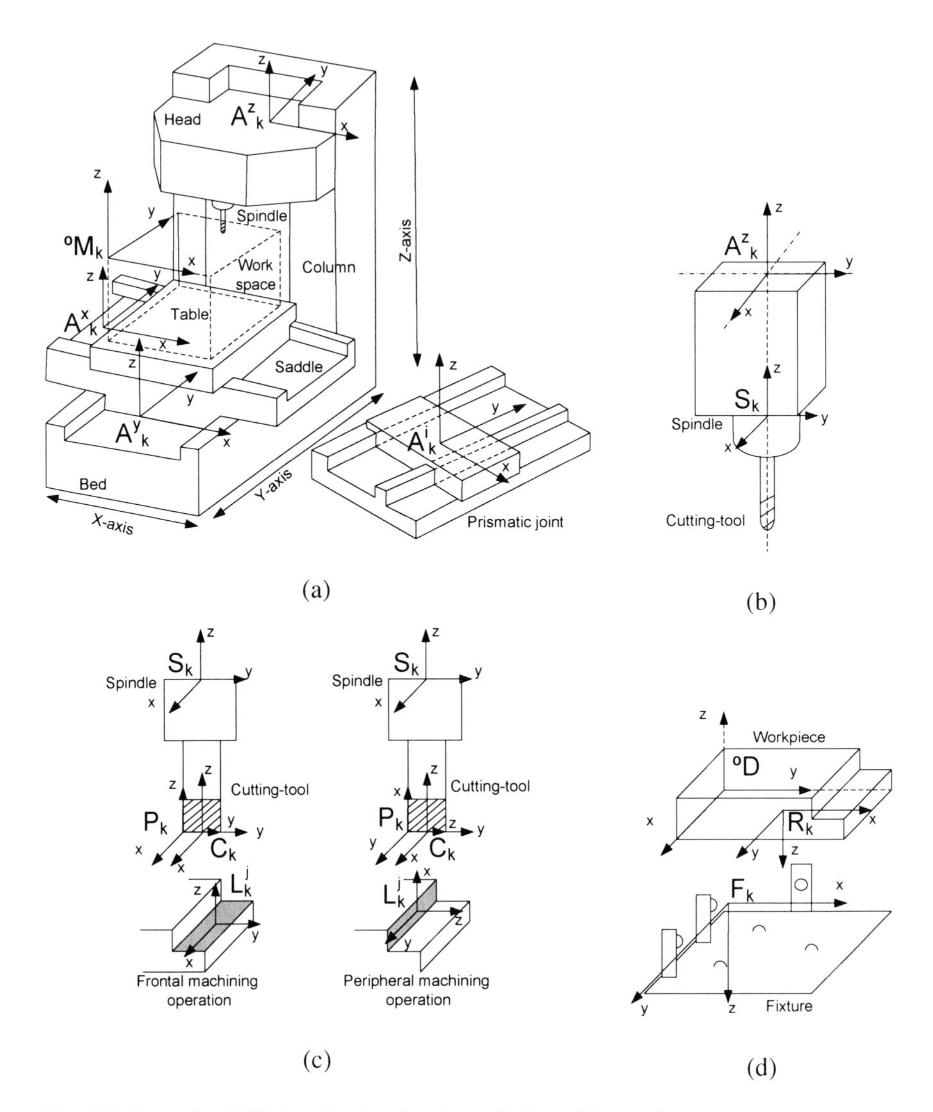

Fig. 2.7. Example of CSs involve in a 3-axis vertical machine-tool

2.2.3 Derivation of the DMVs

In previous subsection, it has been shown the different elements with their corresponding CSs that are involved during machining feature *j*. In this subsection, the extended SoV model is derived by defining the relationships among these CSs with their deviations from nominal values. For a 3-axis vertical machine-tool, a generic illustration of these relationships is shown in Fig. 2.8. From this figure, two main subchains of CSs can be identified. The first chain, defined from ${}^{\circ}M_k$ to P_k, represents how the machining-induced variations

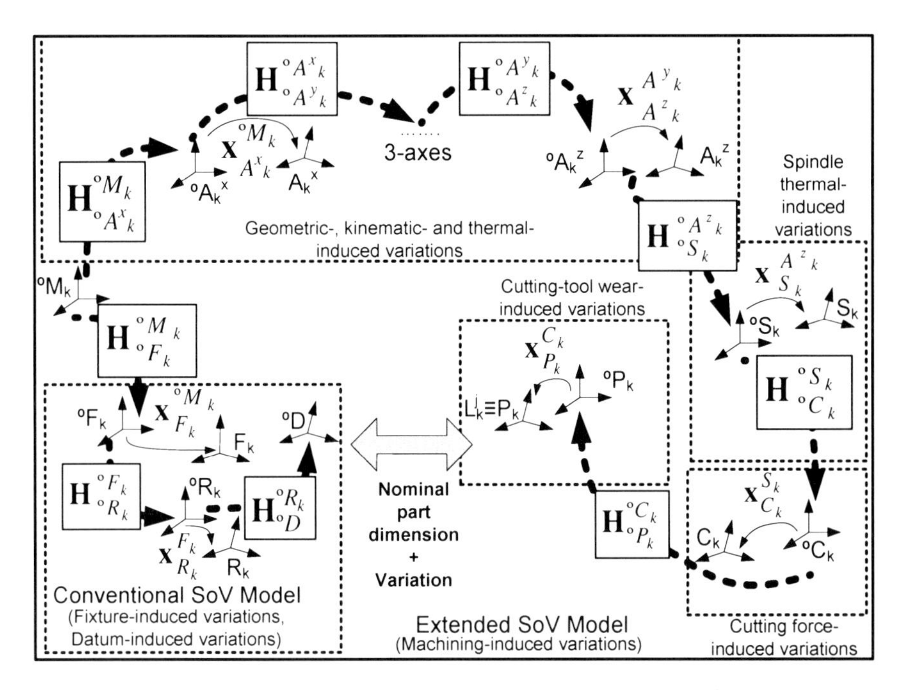

Fig. 2.8. Relationships between the different CSs in a *3*-axis vertical machine-tool

deviate the cutting-tool tip w.r.t. the machine-tool CS. The most common machining-induced variations are due to geometric and kinematic errors of machine-tool axes, thermal distortions, cutting-tool deflections and cutting-tool wear, which induce deviations of the CSs A_k^i, S_k, C_k and P_k from their respective nominal values. The second chain, defined from ${}^{\circ}M_k$ to R_k, represents how fixture- and datum-induced variations deviate the workpiece location w.r.t. the MCS. Note that in order to represent one CS w.r.t. another CS, a homogeneous transformation matrix (HTM) is used. How to derive the HTMs is explained in Appendix 2.1.

In order to derive the machined surface variation as a function of the variations of all CSs involved in these chains, we consider the following Corollary from [1].

Corollary 1: Consider the CS R, 1 and 2 as shown in Fig. 2.9. Consider now that CS 1 and CS 2 are deviated from nominal values. Noting the variation of CS 1 and CS 2 w.r.t. R as $\mathbf{x}_1^R$ and $\mathbf{x}_2^R$, respectively, then, the variation of CS 2 w.r.t. 1 in vector form can be formulated as

$$\mathbf{x}_2^1 = \begin{pmatrix} -\left(\mathbf{R}_{\circ 2}^{\circ 1}\right)^T & \left(\mathbf{R}_{\circ 2}^{\circ 1}\right)^T \cdot \left(\hat{\mathbf{t}}_{\circ 2}^{\circ 1}\right) & \mathbf{I}_{3\times 3} & 0 \\ 0 & -\left(\mathbf{R}_{\circ 2}^{\circ 1}\right)^T & 0 & \mathbf{I}_{3\times 3} \end{pmatrix} \cdot \begin{pmatrix} \mathbf{x}_1^R \\ \mathbf{x}_2^R \end{pmatrix}, \tag{2.6}$$

$$= \mathbf{D}_2^1 \cdot \mathbf{x}_1^R + \mathbf{x}_2^R,$$

where $\mathbf{R}_{^\circ 2}^{^\circ 1}$ is the rotation matrix of $^\circ 2$ w.r.t. $^\circ 1$, $\mathbf{I}_{3\times 3}$ is a 3×3 identity matrix and $\hat{\mathbf{t}}_{^\circ 2}^{^\circ 1}$ is the skew matrix of vector $\mathbf{t}_{^\circ 2}^{^\circ 1}$ (see Appendix 2.2). The proof of this Corollary can be found in [1].

Applying Corollary 1, the machined surface variation at station k defined as the variation of CS L_k^j (denoted as L_k hereafter) w.r.t. R_k, denoted as $\mathbf{x}_{L_k}^{R_k}$, can be obtained as

$$\mathbf{x}_{L_k}^{R_k} = \mathbf{D}_{L_k}^{R_k} \cdot \mathbf{x}_{R_k}^{^\circ M_k} + \mathbf{x}_{L_k}^{^\circ M_k}, \tag{2.7}$$

where $\mathbf{x}_{R_k}^{^\circ M_k}$ represents the variation of the position and orientation of the workpiece w.r.t. the MCS due to fixture- and datum-induced variations, and $\mathbf{x}_{L_k}^{^\circ M_k}$ represents the overall cutting-tool path variation due to machining-induced variations when manufacturing feature j. Note that, according to the previous CS definition, $\mathbf{x}_{L_k}^{^\circ M_k} \equiv \mathbf{x}_{P_k}^{^\circ M_k}$.

As shown in Fig. 2.8, $\mathbf{x}_{R_k}^{^\circ M_k}$ will depend on the CSs variations that define the chain from $^\circ M_k$ to R_k. And similarly, $\mathbf{x}_{P_k}^{^\circ M_k}$ will depend on the variations of the CSs from $^\circ M_k$ to P_k. In order to express these DMVs as functions of the DMVs of all the CSs for each chain, the following Corollary from [1] is applied.

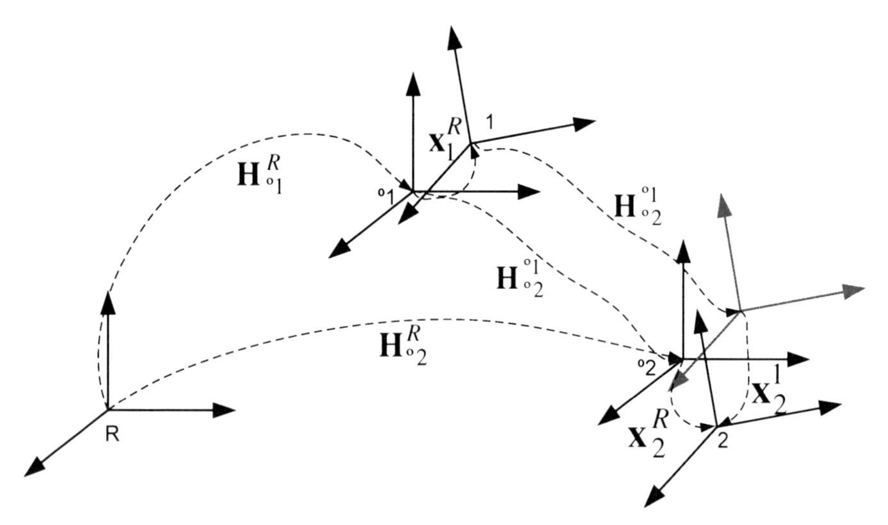

Fig. 2.9. Differential motion vector from CS 2 to CS 1 if both CSs deviate from nominal values

Corollary 2: Consider the CSs R , 1 and 2 as shown in Fig. 2.9, with CSs 1 and 2 deviating from their nominal positions and orientations. Noting the variation of CS 1 w.r.t. R as $\mathbf{x}_1^R$ and the variation of CS 2 w.r.t. CS 1 as $\mathbf{x}_2^1$, then, the variation of CS 2 w.r.t. R can be formulated as

$$\mathbf{x}_2^R = \begin{pmatrix} \left(\mathbf{R}_{\circ 2}^{\circ 1}\right)^T & -\left(\mathbf{R}_{\circ 2}^{\circ 1}\right)^T \cdot \left(\hat{\mathbf{t}}_{\circ 2}^{\circ 1}\right) & \mathbf{I}_{3\times 3} & 0 \\ 0 & \left(\mathbf{R}_{\circ 2}^{\circ 1}\right)^T & 0 & \mathbf{I}_{3\times 3} \end{pmatrix} \cdot \begin{pmatrix} \mathbf{x}_1^R \\ \mathbf{x}_2^1 \end{pmatrix}, \tag{2.8}$$
$$= \mathbf{T}_2^1 \cdot \mathbf{x}_1^R + \mathbf{x}_2^1.$$

The proof of this Corollary can be found in [1]. Applying repeatedly the Corollary 2, the DMVs $\mathbf{x}_{R_k}^{\circ M_k}$ and $\mathbf{x}_{P_k}^{\circ M_k}$ can be respectively rewritten as

$$\mathbf{x}_{R_k}^{\circ M_k} = \mathbf{T}_{R_k}^{F_k} \cdot \mathbf{x}_{F_k}^{\circ M_k} + \mathbf{x}_{R_k}^{F_k}, \tag{2.9}$$

$$\mathbf{x}_{P_k}^{\circ M_k} = \mathbf{T}_{P_k}^{C_k} \cdot (\mathbf{T}_{C_k}^{S_k} \cdot (\mathbf{T}_{S_k}^{A_k^z} \cdot \mathbf{x}_{A_k^z}^{\circ M_k} + \mathbf{x}_{S_k}^{A_k^z}) + \mathbf{x}_{C_k}^{S_k}) + \mathbf{x}_{P_k}^{C_k}. \tag{2.10}$$

Substituting Eqs. (2.10) and (2.9) in Eq. (2.7), the variation of the feature L_k w.r.t. R_k can be expressed as a function of the DMVs of all the CSs that are involved between L_k and R_k CS, such as $\mathbf{x}_{A_k^z}^{\circ M_k}$, $\mathbf{x}_{S_k}^{A_k^z}$, $\mathbf{x}_{C_k}^{S_k}$ and $\mathbf{x}_{P_k}^{C_k}$. These DMVs are functions of different sources of variation such as fixture locator variations, machine-tool inaccuracy, thermal state of the machine-tool, cutting-tool wear, cutting-tool deflections, etc. In the next subsection, the derivation of the DMVs for each CS will be presented in detail.

2.2.3.1 Fixture-Induced Variations

The DMV $\mathbf{x}_{F_k}^{\circ M_k}$ refers to the variations of the FCS due to fixture inaccuracies, and it can also be described as the DMV $\mathbf{x}_{F_k}^{\circ F_k}$ since $^{\circ}M_k$ is not deviated. In order to derive $\mathbf{x}_{F_k}^{\circ F_k}$ as a function of fixture layout and fixture variations, the following assumptions are considered: i) locating surfaces are assumed to be perfect in form (without form errors) and there is no deformations of locators; ii) locators are assumed to be punctual and distributed according to the 3-2-1 workholding principle [19, Chapter 3]; iii) only variations based on small displacements from nominal values are considered.

Let us consider a 3-2-1 fixture device where $\mathbf{r}_i$ defines the contact point between the ith locator and the workpiece, and $\mathbf{n}_i$ defines the normal vector of the workpiece surface at the ith contact point, all expressed w.r.t. ${}^{\circ}F_k$. Following the research work in [20], a small perturbation in the location of F_k ($\mathbf{x}_{F_k}^{\circ F_k}$) due to a small variation in a fixture locator in the direction of movement constraint (direction normal to the locating surface defined by $\mathbf{n}_i$), denoted as Δl_i, can be mathematically expressed as

$$\Delta l_i = \overline{\mathbf{w}}_i^T \cdot \mathbf{x}_{F_k}^{\circ F_k}, \tag{2.11}$$

where

$$\overline{\mathbf{w}}_i^T = [\mathbf{n}_i^T, (\mathbf{r}_i \times \mathbf{n}_i)^T], \tag{2.12}$$

and $\times$ is the cross product operator. Δl_i is defined as

$$\Delta l_i = \mathbf{n}_i^T \cdot \Delta \mathbf{r}_i, \tag{2.13}$$

where $\Delta \mathbf{r}_i$ is the position variation of the locator i. The deterministic localization condition (a unique solution of $\mathbf{x}_{F_k}^{\circ F_k}$) requires that Eq. (2.11) is satisfied for all locators. Thus, considering all locators, Eq. (2.11) becomes

$$\Delta \mathbf{l} = \mathbf{G}^T \cdot \mathbf{x}_{F_k}^{\circ F_k}, \tag{2.14}$$

where $\Delta \mathbf{l} = [\Delta l_1, \Delta l_2, \ldots \Delta l_6]^T$ and $\mathbf{G} = [\overline{\mathbf{w}}_1, \overline{\mathbf{w}}_2, \ldots \overline{\mathbf{w}}_6]^T$. Matrix $\mathbf{G}$ is called *locator matrix* in [20]. In order to ensure a deterministic localization, the locator matrix $\mathbf{G}$ must not be singular, indicating that all $\{\overline{\mathbf{w}}_1, \ldots, \overline{\mathbf{w}}_6\}$ are linearly independent [20]. Thus, for a deterministic localization, matrix $\mathbf{G}$ is invertible, and the variation of ${}^{\circ}F_k$ is related to the variation of fixture locators by the following expression

$$\mathbf{x}_{F_k}^{\circ F_k} = (\mathbf{G}^T)^{-1} \cdot \Delta \mathbf{l}. \tag{2.15}$$

Eq. (2.15) can be rewritten in the form

$$\mathbf{x}_{F_k}^{\circ F_k} = \mathbf{x}_{F_k}^{\circ M_k} = \mathbf{B}_k^{f_1} \cdot [\Delta l_1, \Delta l_2, \ldots, \Delta l_6]^T. \tag{2.16}$$

For a generic 3-2-1 locating scheme shown in Fig. 2.10a, matrix $\mathbf{B}_k^{f_1}$ can be straight forward derived following the methodology explained above, resulting in the matrix:

$$\mathbf{B}_k^{f_1}=\begin{pmatrix} \frac{(l_{2y}-l_{3y})\cdot l_{5z}}{C} & \frac{-(l_{1y}-l_{3y})\cdot l_{5z}}{C} & \frac{(-l_{2y}+l_{1y})\cdot l_{5z}}{C} & \frac{-l_{5y}}{(-l_{5y}+l_{4y})} & \frac{l_{4y}}{(-l_{5y}+l_{4y})} & 0 \\ \frac{-(l_{2x}-l_{3x})\cdot l_{6z}}{C} & \frac{(l_{1x}-l_{3x})\cdot l_{6z}}{C} & \frac{-(-l_{2x}+l_{1x})\cdot l_{6z}}{C} & \frac{l_{6x}}{(-l_{5y}+l_{4y})} & \frac{-l_{6x}}{(-l_{5y}+l_{4y})} & 1 \\ \frac{(l_{3y}l_{2x}-l_{2y}l_{3x})}{C} & \frac{-(-l_{3x}l_{1y}+l_{3y}l_{1x})}{C} & \frac{(-l_{1y}l_{2x}+l_{2y}l_{1x})}{C} & 0 & 0 & 0 \\ \frac{-(l_{2x}-l_{3x})}{C} & \frac{(l_{1x}-l_{3x})}{C} & \frac{-(-l_{2x}+l_{1x})}{C} & 0 & 0 & 0 \\ \frac{-(l_{2y}-l_{3y})}{C} & \frac{(l_{1y}-l_{3y})}{C} & \frac{-(-l_{2y}+l_{1y})}{C} & 0 & 0 & 0 \\ 0 & 0 & 0 & \frac{-1}{(-l_{5y}+l_{4y})} & \frac{1}{(-l_{5y}+l_{4y})} & 0 \end{pmatrix}, \quad (2.17)$$

where $C=l_{3x}l_{1y}-l_{1y}l_{2x}+l_{3y}l_{2x}+l_{2y}l_{1x}-l_{2y}l_{3x}-l_{3y}l_{1x}$. Note that the position of locator i is described by the vector $\mathbf{l}_i=[l_{ix},l_{iy},l_{iz}]$ expressed in the FCS.

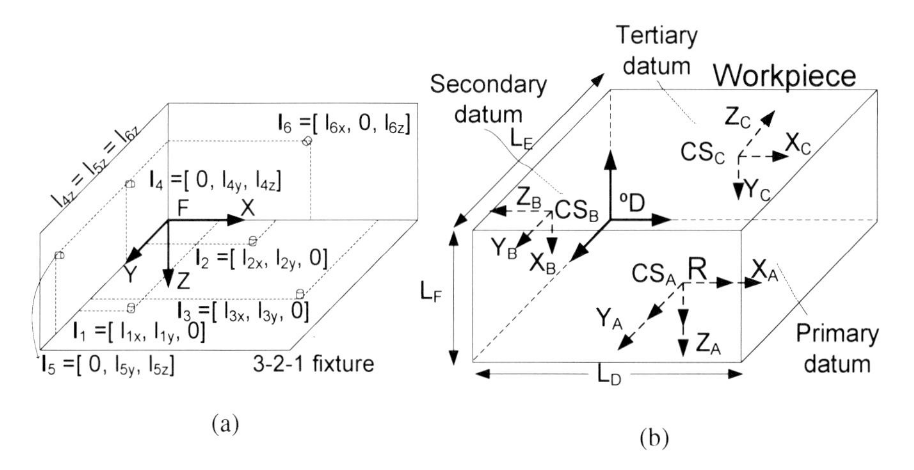

Fig. 2.10. a) 3-2-1 fixture layout based on locators; b) Workpiece datums (primary, secondary and tertiary). CSs centered on each face.

2.2.3.2 Datum-Induced Variations

For 3-2-1 fixture layouts based on locators (Fig. 2.10a), the deterministic location of the part is ensured when workpiece touches the six locators. Due to datum inaccuracies, workpiece location can be deviated from its nominal values. Assuming prismatic surfaces, the influence of the datum variations on workpiece location can be obtained as follows [1].

Considering a 3-2-1 fixture device and a prismatic workpiece as shown in Fig. 2.10a and 2.10b, respectively, the following constrains apply:

- Locators l_4 and l_5 touch workpiece surface B, and thus the third component of the coordinate of l_4 and l_5 w.r.t. the CS B are zero. Mathematically, this is expressed as $\left\lfloor \tilde{\mathbf{p}}_{l_4}^{B} \right\rfloor_{(3)} = 0$ and $\left\lfloor \tilde{\mathbf{p}}_{l_5}^{B} \right\rfloor_{(3)} = 0$, where $\tilde{\mathbf{p}}_{l_i}^{B} = [l_{ix}^{B}, l_{iy}^{B}, l_{iz}^{B}, 1]^T$ and $\lfloor \cdot \rfloor_{(3)}$ denotes the third component of the vector.
- Similarly, locator l_6 touches workpiece surface C, so $\left\lfloor \tilde{\mathbf{p}}_{l_6}^{C} \right\rfloor_{(3)} = 0$.

According to these constrains, the contact points can be mathematically expressed as

$$\tilde{\mathbf{p}}_{l_4}^{B} = \mathbf{H}_{A}^{B} \cdot \mathbf{H}_{F}^{A} \cdot \tilde{\mathbf{p}}_{l_4}^{F}, \tag{2.18}$$

$$\tilde{\mathbf{p}}_{l_5}^{B} = \mathbf{H}_{A}^{B} \cdot \mathbf{H}_{F}^{A} \cdot \tilde{\mathbf{p}}_{l_5}^{F}, \tag{2.19}$$

$$\tilde{\mathbf{p}}_{l_6}^{C} = \mathbf{H}_{A}^{C} \cdot \mathbf{H}_{F}^{A} \cdot \tilde{\mathbf{p}}_{l_6}^{F}. \tag{2.20}$$

According to [1], $\mathbf{H}_{A}^{B}$ is defined as:

$$\begin{aligned} \mathbf{H}_{A}^{B} &= \left(\delta\mathbf{H}_{B}^{A}\right)^{-1} \cdot \mathbf{H}_{\circ A}^{\circ B}, \\ &= (\mathbf{I}_{4\times 4} - \mathbf{\Delta}_{B}^{A}) \cdot \mathbf{H}_{\circ A}^{\circ B}, \end{aligned} \tag{2.21}$$

where $\delta\mathbf{H}_{B}^{A}$ is the HTM that defines the small translational and orientational deviations of CS B w.r.t. A due to the variation from nominal values of B and A; and $\mathbf{\Delta}_{B}^{A}$ is the differential transformation matrix (DTM) of B w.r.t. A (see Appendix 2).

$\mathbf{H}_{F}^{A}$ it is defined as

$$\begin{aligned} \mathbf{H}_{F}^{A} &= \mathbf{H}_{\circ F}^{\circ A} \cdot \delta\mathbf{H}_{F}^{A}, \\ &= \mathbf{H}_{\circ F}^{\circ A} \cdot (\mathbf{I}_{4\times 4} + \mathbf{\Delta}_{F}^{A}). \end{aligned} \tag{2.22}$$

Substituting Eqs. (2.21) and (2.22) in Eq. (2.18),

$$\tilde{\mathbf{p}}_{l_4}^{B} = (\mathbf{I}_{4\times 4} - \mathbf{\Delta}_{B}^{A}) \cdot \mathbf{H}_{\circ A}^{\circ B} \cdot \mathbf{H}_{\circ F}^{\circ A} \cdot (\mathbf{I}_{4\times 4} + \mathbf{\Delta}_{F}^{A}) \cdot \tilde{\mathbf{p}}_{l_4}^{F}. \tag{2.23}$$

By neglecting the second-order small values and considering the contact between surfaces ($\left[\tilde{\mathbf{p}}_{l_4}^{B}\right]_{(3)} = 0$), the following equation applies

$$\left[\tilde{\mathbf{p}}_{l_4}^{B}\right]_{(3)} = \left[(-\boldsymbol{\Delta}_B^A \cdot \mathbf{H}_{\circ F}^{\circ B} + \mathbf{H}_{\circ F}^{\circ B} \cdot \boldsymbol{\Delta}_F^A + \mathbf{H}_{\circ F}^{\circ B}) \cdot \tilde{\mathbf{p}}_{l_4}^{F}\right]_{(3)} = 0. \tag{2.24}$$

As the X coordinate of the location of locator 4 w.r.t. the CS F is zero, through the HTM $\mathbf{H}_{\circ F}^{\circ B}$, the term $\left[\mathbf{H}_{\circ F}^{\circ B} \cdot \tilde{\mathbf{p}}_{l_4}^{F}\right]_{(3)}$ becomes zero and thus, Eq. (2.24) is rewritten as

$$\left[(-\boldsymbol{\Delta}_B^A \cdot \mathbf{H}_{\circ F}^{\circ B} + \mathbf{H}_{\circ F}^{\circ B} \cdot \boldsymbol{\Delta}_F^A) \cdot \tilde{\mathbf{p}}_{l_4}^{F}\right]_{(3)} = 0, \tag{2.25}$$

and thus,

$$\left[(\mathbf{H}_{\circ F}^{\circ B} \cdot \boldsymbol{\Delta}_F^A) \cdot \tilde{\mathbf{p}}_{l_4}^{F}\right]_{(3)} = \left[\boldsymbol{\Delta}_B^A \cdot \mathbf{H}_{\circ F}^{\circ B} \cdot \tilde{\mathbf{p}}_{l_4}^{F}\right]_{(3)}. \tag{2.26}$$

Following the same procedure, Eqs. (2.27) and (2.28) can be derived for locator l_5 and l_6, respectively.

$$\left[(\mathbf{H}_{\circ F}^{\circ B} \cdot \boldsymbol{\Delta}_F^A) \cdot \tilde{\mathbf{p}}_{l_5}^{F}\right]_{(3)} = \left[\boldsymbol{\Delta}_B^A \cdot \mathbf{H}_{\circ F}^{\circ B} \cdot \tilde{\mathbf{p}}_{l_5}^{F}\right]_{(3)}, \tag{2.27}$$

$$\left[(\mathbf{H}_{\circ F}^{\circ C} \cdot \boldsymbol{\Delta}_F^A) \cdot \tilde{\mathbf{p}}_{l_6}^{F}\right]_{(3)} = \left[\boldsymbol{\Delta}_C^A \cdot \mathbf{H}_{\circ F}^{\circ C} \cdot \tilde{\mathbf{p}}_{l_6}^{F}\right]_{(3)}. \tag{2.28}$$

Note that for Eqs. (2.26-2.28), we are interested in evaluating the DTM $\boldsymbol{\Delta}_F^A$, which shows the effect of datum variations on workpiece location. Furthermore, note that from the six parameters of $\boldsymbol{\Delta}_F^A$ (three translational and three orientation deviations), only three are unknown since datum variations of surface B and C only influence on the X and Y positioning of the workpiece and on the rotation about the Z-axis, all expressed from CS A. Thus, the three unknown parameters that depend on datum fixture variations of surfaces B and C can be obtained solving Eqs. (2.26-2.28). After solving, the variation of FCS w.r.t. CS A can be rewritten by the effects of each datum feature in a vector form as

$$\mathbf{x}_F^A = \mathbf{A}^1 \cdot \mathbf{x}_B^A + \mathbf{A}^2 \cdot \mathbf{x}_C^A. \tag{2.29}$$

As the primary datum (CS A) also defines CS R_k, Eq. (2.29) is rewritten, considering the station subscript, as

$$\mathbf{x}_{F_k}^{R_k} = \mathbf{A}_k^1 \cdot \mathbf{x}_{B_k}^{R_k} + \mathbf{A}_k^2 \cdot \mathbf{x}_{C_k}^{R_k}. \tag{2.30}$$

For the 3-2-1 locating scheme based on locators shown in Fig. 2.10a, Eqs. (2.26-2.28) were solved obtaining the following matrices $\mathbf{A}_k^1$ and $\mathbf{A}_k^2$. These results can also be found in [1].

$$A_k^1 = \begin{pmatrix} 0 & 0 & -1 & L_E/2 & l_{5z} + L_F/2 + \dfrac{l_{5y} \cdot (l_{5z} - l_{4z})}{(l_{4y} - l_{5y})} & 0 \\ 0 & 0 & 0 & -l_{6x} & \dfrac{l_{6x} \cdot (l_{4z} - l_{5z})}{l_{4y} - l_{5y}} & 0 \\ 0 & 0 & 0 & 0 & 0 & 0 \\ 0 & 0 & 0 & 0 & 0 & 0 \\ 0 & 0 & 0 & 0 & 0 & 0 \\ 0 & 0 & 0 & 1 & \dfrac{-(l_{4z} - l_{5z})}{l_{4y} - l_{5y}} & 0 \end{pmatrix}, \tag{2.31}$$

$$A_k^2 = \begin{pmatrix} 0 & 0 & 0 & 0 & 0 & 0 \\ 0 & 0 & -1 & -l_{6z} - L_F/2 & l_{6x} - L_D/2 & 0 \\ 0 & 0 & 0 & 0 & 0 & 0 \\ 0 & 0 & 0 & 0 & 0 & 0 \\ 0 & 0 & 0 & 0 & 0 & 0 \\ 0 & 0 & 0 & 0 & 0 & 0 \end{pmatrix}. \tag{2.32}$$

2.2.3.3 Geometric, Kinematic and Thermal-Induced Variations

Machine-tools are typically composed of a kinematic chain of multiple translational and rotational axes that are designed to have only one degree of freedom of moving. However, due to geometric inaccuracy, misalignments and thermal effects, all these axes have actually three translational and three rotational deviations. Intensive research has been conducted to estimate the positioning errors of a cutting-tool in multi-axis machine-tools using rigid-body kinematics and HTMs [21-23]. With these methods, each axis of a machine-tool, relative to each other or to a reference CS, is modeled using HTM. After determining the HTMs of all the movable links, they are multiplied successively from the last axis to the reference CS of the machine-tool, the MCS, resulting in the actual position of the CS of the last axis w.r.t. the MCS. Considering a 3-axis vertical machine-tool, the final position of A_k^z w.r.t. ${}^{\circ}M_k$ can be defined as

$$\begin{aligned} \mathbf{H}_{A_k^z}^{{}^{\circ}M_k} &= \mathbf{H}_{{}^{\circ}A_k^z}^{{}^{\circ}M_k} \cdot \delta\mathbf{H}_{A_k^z}^{{}^{\circ}M_k}, \\ &= \mathbf{H}_{{}^{\circ}A_k^x}^{{}^{\circ}M_k} \cdot \delta\mathbf{H}_{A_k^x}^{{}^{\circ}M_k} \cdot \mathbf{H}_{{}^{\circ}A_k^y}^{{}^{\circ}A_k^x} \cdot \delta\mathbf{H}_{A_k^y}^{A_k^x} \cdot \mathbf{H}_{{}^{\circ}A_k^z}^{{}^{\circ}A_k^y} \cdot \delta\mathbf{H}_{A_k^z}^{A_k^y}. \end{aligned} \tag{2.33}$$

With the assumption of rigid-body kinematics and small-angle approximation, any $\delta\mathbf{H}_{A_k^i}^{A_k^j}$ can be generally defined as a product of three HTMs [21-23], i.e.,

$$\delta\mathbf{H}_{A_k^i}^{A_k^j} = \begin{pmatrix} 1 & -\varepsilon_{zi} & \varepsilon_{yi} & \delta_{xi} \\ \varepsilon_{zi} & 1 & -\varepsilon_{xi} & \delta_{yi} \\ -\varepsilon_{yi} & \varepsilon_{xi} & 1 & \delta_{zi} \\ 0 & 0 & 0 & 1 \end{pmatrix} \cdot \begin{pmatrix} 1 & -\varepsilon_z(i) & \varepsilon_y(i) & \delta_x(i) \\ \varepsilon_z(i) & 1 & -\varepsilon_x(i) & \delta_y(i) \\ -\varepsilon_y(i) & \varepsilon_x(i) & 1 & \delta_z(i) \\ 0 & 0 & 0 & 1 \end{pmatrix} \cdot$$
$$\cdot \begin{pmatrix} 1 & -\varepsilon_z^t(t,T_1,...,T_m,i) & \varepsilon_y^t(t,T_1,...,T_m,i) & \delta_x^t(t,T_1,...,T_m,i) \\ \varepsilon_z^t(t,T_1,...,T_m,i) & 1 & -\varepsilon_z^t(t,T_1,...,T_m,i) & \delta_y^t(t,T_1,...,T_m,i) \\ -\varepsilon_y^t(t,T_1,...,T_m,i) & \varepsilon_x^t(t,T_1,...,T_m,i) & 1 & \delta_z^t(t,T_1,...,T_m,i) \\ 0 & 0 & 0 & 1 \end{pmatrix}. \quad (2.34)$$

The first HTM describes the mounting errors of the i-axis w.r.t. the previous j-axis. The mounting errors are position and orientation errors due to assembly errors and they are not dependent on the carriage position. Mounting errors can be represented by three possible angular variations, ε_{xi} (rotation around the X-axis), ε_{yi} (rotation around the Y-axis) and ε_{zi} (rotation around the Z-axis), and three offsets (δ_{xi}, δ_{yi}, δ_{zi}), as it is shown in Fig. 2.11 for the X-axis carriage. The second HTM represents the motional variations, which include the terms $\delta_p(q)$ and $\varepsilon_p(q)$. $\delta_p(q)$ refers to the positional variation in the p-axis direction when the prismatic joint moves along the q-axis and is a function of the position of the q-axis. $\varepsilon_p(q)$ refers to the angular variation around the p-axis when the q-axis moves and it is also a function of the position of the q-axis. The third HTM describes the geometrical variations due to thermal effects, whose components are defined as $\delta_p^t(t,T_1,...,T_m,q)$ and $\varepsilon_p^t(t,T_1,...,T_m,q)$ for position and angular variations around the p-axis when the q-axis moves, respectively, and include scalar thermal components and position-dependent thermal components [23]. Mathematically, $\delta_p^t(t,T_1,...,T_m,q)$ and $\varepsilon_p^t(t,T_1,...,T_m,q)$ are generally defined as

$$\begin{aligned} \delta_p^t(t,T_1,\ldots,T_m,q) &= f_0^{pq}(T_1,\ldots,T_m,t) + f_1^{pq}(T_1,\ldots,T_m,t)\cdot q + f_2^{pq}(T_1,\ldots,T_m,t)\cdot q^2 + \ldots, \\ \varepsilon_p^t(t,T_1,\ldots,T_m,q) &= g_0^{pq}(T_1,\ldots,T_m,t) + g_1^{pq}(T_1,\ldots,T_m,t)\cdot q + g_2^{pq}(T_1,\ldots,T_m,t)\cdot q^2 + \ldots \end{aligned} \quad (2.35)$$

The term $f_0^{pq}(T_1,\ldots,T_m,t)$ and $g_0^{pq}(T_1,\ldots,T_m,t)$ are scalar thermal components that model the position variation on the p-axis when the q-axis moves and are function of operation time, t, and the temperatures $T_1,\ldots,T_m$ at different locations on the machine-tool structure. The position-dependent thermal components are defined by the terms $f_1^{pq}(T_1,\ldots,T_m,t)\cdot q + f_2^{pq}(T_1,\ldots,T_m,t)\cdot q^2 + \ldots$ and $g_1^{pq}(T_1,\ldots,T_m,t)\cdot q + g_2^{pq}(T_1,\ldots,T_m,t)\cdot q^2 + \ldots$ and they model the position variation on the p-axis when the q-axis moves.

From Eq. (2.34), it is shown that geometrical variations due to kinematic and thermal effects may present non-linear relationships. In order to include these sources of variation into the SoV model, a linearization should be conducted based on three important assumptions. Firstly, it is assumed that the geometric-thermal variations are modeled when the machine-tool is warmed-up adequately and thus, the effect of the time on the thermal variations can be neglected. Secondly, it is assumed that the workpiece is repeatedly placed in the same region inside the allowable work space of the machine-tool table, so it is only expected small variations in the placement of the workpiece. Thirdly, it is assumed that geometric, kinematic and thermal variations do not change drastically along the travels at any i-axis for the region where the workpiece is repeatedly placed on the machine-tool table (the experimentation in [23] holds this assumption), so the geometric-thermal variations in the machine-tool axis can be linearized without significant loss of precision. Under these assumptions, the motional variations $\delta_p(q)$ and $\varepsilon_p(q)$ are linearized as $\delta_p(q_0)+\left.\frac{\delta(\delta_p(q))}{\delta q}\right|_{q=q_0}\cdot\Delta q$ and $\varepsilon_p(q_0)+\left.\frac{\delta(\varepsilon_p(q))}{\delta q}\right|_{q=q_0}\cdot\Delta q$, respectively, where q_0 is the nominal placement of the workpiece on the q-axis, and Δq is the admissible variation range of the workpiece placement along the q-axis. The thermal-induced variations $\delta_p^t(t,T_1,...,T_m,q)$ and $\varepsilon_p^t(t,T_1,...,T_m,q)$ from Eq. (2.35) can be linearized as

$$\begin{aligned}\delta_p^t(\Delta T_1,...,\Delta T_m,\Delta q) &= C_0^{pq}+C_1^{pq}\cdot\Delta T_1+\ldots+C_m^{pq}\cdot\Delta T_m+C_{m+1}^{pq}\cdot\Delta q,\\ \varepsilon_p^t(\Delta T_1,...,\Delta T_m,\Delta q) &= D_0^{pq}+D_1^{pq}\cdot\Delta T_1+\ldots+D_m^{pq}\cdot\Delta T_m+D_{m+1}^{pq}\cdot\Delta q,\end{aligned} \tag{2.36}$$

where $C_{(\cdot)}^{pq}$ and $D_{(\cdot)}^{pq}$ are constants, ΔT_c is the variation of the cth temperature at the machine-tool structure from its nominal values T_{c_0} where $c=1,\ldots,m$.

As a conclusion, the resulting position and orientation deviation defined by the matrix $\delta\mathbf{H}_{A_k^z}^{^\circ M_k}$, considering the linearization, can be obtained from Eq. (2.33). This matrix will be defined in the form of

$$\delta\mathbf{H}_{A_k^z}^{^\circ M_k}=\begin{pmatrix}1 & -\theta_{A_k^z z}^{^\circ M_k} & \theta_{A_k^z y}^{^\circ M_k} & d_{A_k^z x}^{^\circ M_k}\\ \theta_{A_k^z z}^{^\circ M_k} & 1 & -\theta_{A_k^z x}^{^\circ M_k} & d_{A_k^z y}^{^\circ M_k}\\ -\theta_{A_k^z y}^{^\circ M_k} & \theta_{A_k^z x}^{^\circ M_k} & 1 & d_{A_k^z z}^{^\circ M_k}\\ 0 & 0 & 0 & 1\end{pmatrix}. \tag{2.37}$$

For a generic 3-axis vertical machine-tool, Eq. (2.37) can be rewritten in a DMV as

$$\mathbf{x}_{A_k^z}^{^\circ M_k} = \mathbf{B}_k^{m_1} \cdot [\Delta T_1^k, \ldots, \Delta T_m^k, \Delta \mathbf{r}^k]^T. \tag{2.38}$$

In Eq. (2.38), $\Delta \mathbf{r}^k$ refers to the admissible variation range of workpiece placement at the three axes and thus, $\Delta \mathbf{r}^k = [\Delta x^k, \Delta y^k, \Delta z^k]$. Please note that we discard the systematic variations due to linealization such as the terms C_0^{pq}, D_0^{pq}, $\delta_p(q_0)$, $\varepsilon_p(q_0)$, since they can be eliminated by an initial calibration.

For practical purposes, geometric and thermal parameters related to geometric-thermal induced variations can be estimated from experimental data. Research works [23-26] show in detail the derivation of these parameters.

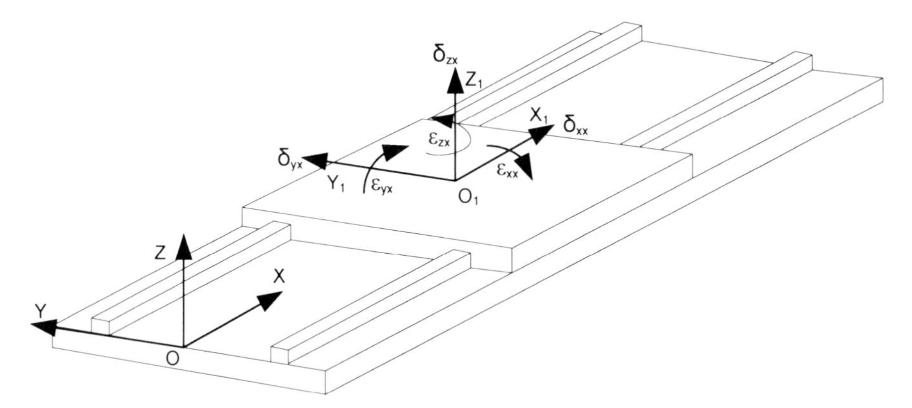

Fig. 2.11. Position and orientation deviations of a machine-tool carriage system due to mounting errors

2.2.3.4 Spindle Thermal-Induced Variations

The spindle thermal variations are an important contributor to the total thermal-induced variations during machining due to the large amounts of heat generated at high-speed revolutions [27]. Spindle thermal expansion produces three translational and two rotational drifts to the spindle CS [23]. This variation is represented as the deviation of $^\circ S_k$ w.r.t. A_k^z and is proportional to the increase of the spindle temperature, denoted as ΔT_s, from nominal conditions. At station *k*, this variation is defined by the DMV

$$\begin{aligned} \mathbf{x}_{S_k}^{A_k^z} &= \left[f_1^k(\Delta T_s^k) \quad f_2^k(\Delta T_s^k) \quad f_3^k(\Delta T_s^k) \quad f_4^k(\Delta T_s^k) \quad f_5^k(\Delta T_s^k) \quad f_6^k(\Delta T_s^k) \right]^T \\ &\approx \left[Cf_x^k \quad Cf_y^k \quad Cf_z^k \quad Cf_\alpha^k \quad Cf_\beta^k \quad 0 \right]^T \cdot \Delta T_s^k = \mathbf{B}_k^{m_2} \cdot \Delta T_s^k, \end{aligned} \tag{2.39}$$

where $f_i^k(\cdot)$ is a function that relates position and orientation deviations of S_k with the variation of the spindle temperature from nominal conditions. Cf_i^k are proportional coefficients linearizing the $f_i^k(\cdot)$ functions and the $f_6^k(\Delta T_s^k)$ is considered as 0 since the Z-axis is the cutting-tool rotation axis. These coefficients can be obtained through experimentation [23].

2.2.3.5 Cutting Force-Induced Variations

The geometric variations of the machined workpieces due to cutting force-induced variations can be modeled considering the cutting-tool deflection occur during the machining process (Fig. 2.12). Various methods for cutting-tools with different complexity have been applied to model the deflection variation [28-30]. To represent these variations in the SoV model, it is necessary to describe position and orientation deviations of the cutting-tool due to deflection. Assuming finishing operations, where the depth of cut is small and it can be considered insignificant in comparison to the length of the cutting-tool overhang, and simplifying that the cutting force acts at the tool tip, the cutting-tool can be modeled as a unique cantilever beam defined by the equivalent tool diameter [29]. Then, the displacement of the cutting-tool perpendicular to the cutting-tool axis is proportional to the cutting force according to the following equation [31, Chapter 7]

$$\delta_r = \frac{F \cdot L^3}{3 \cdot E \cdot I} = \frac{64 \cdot F \cdot L^3}{3 \cdot \pi \cdot E \cdot D^4}, \tag{2.40}$$

where E is the Young's Modulus for the material tool; L^3/D^4 is the tool slenderness parameter, where D is the equivalent tool diameter [29] and L is the overhang length; and F is the cutting force perpendicular to the tool axis. Furthermore, the rotation of the tool tip around the θ-axis perpendicular to the cutting-tool axis is defined as [31, Chapter 7],

$$\delta_\theta = \frac{F \cdot L^2}{2 \cdot E \cdot I} = \frac{64 \cdot F \cdot L^2}{2 \cdot \pi \cdot E \cdot D^4}, \tag{2.41}$$

where F is the force applied at the tool tip perpendicular to the plane defined by the θ-axis and the cutting-tool axis. As a conclusion, C_k is deviated due to the cutting force-induced deflection. The variation of C_k w.r.t. S_k can be expressed by the DMV

$$\mathbf{x}_{C_k}^{S_k} = \mathbf{B}_k^{m_3} \cdot [\Delta F_x^k \quad \Delta F_y^k]^T, \tag{2.42}$$

where $\mathbf{B}_k^{m_3} = [C_1 \quad 0; 0 \quad C_1; 0 \quad 0; 0 \quad C_2; C_2 \quad 0; 0 \quad 0]$, and C_1 and C_2 are defined as $C_1 = \dfrac{64L^3}{3\pi ED^4}$ and $C_2 = \dfrac{3C_1}{2L}$. ΔF_x^k and ΔF_y^k are the variation of the cutting force in X and Y direction from nominal conditions, respectively.

The parameters required to model the force-induced variations can be obtained directly by the geometry of the cutting-tool. In order to obtain the Young's modulus, an experimentation may be required since the manufacturers of cutting-tools do not usually provide this value. In [29], two experimental methods are proposed to obtain the value of Young's modulus for each cutting-tool.

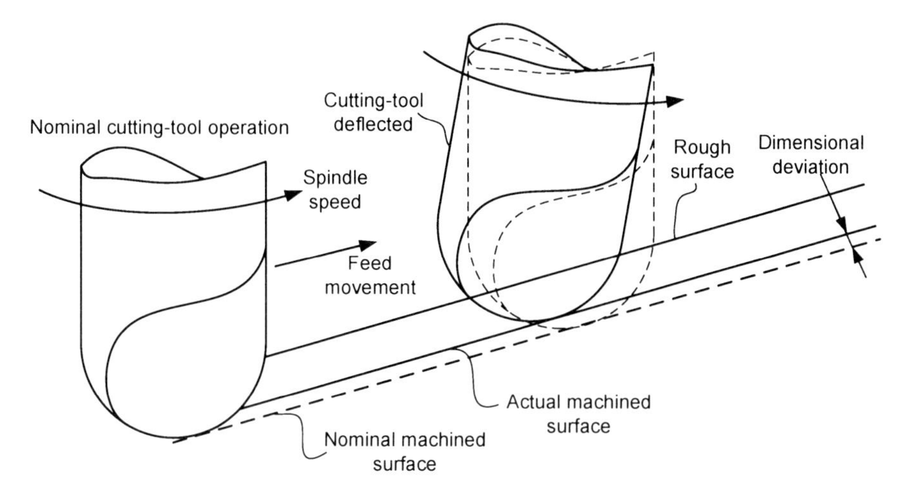

Fig. 2.12. Example of cutting force-induced deviation in a ball end-milling operation

2.2.3.6 Cutting-Tool Wear-Induced Variations

Cutting-tool wear is another important factor affecting part dimensional quality. Different types of wear can be defined in cutting-tools, such as crater wear, flank wear, etc. Among them, the flank wear (V_B) is the most common tool wear measurement in industry to keep parts within dimensional specifications [32]. However, the impact of V_B on part dimensional quality highly depends on the type of machining operation and the geometry of the cutting-tool.

According to the CSs defined previously, the V_B modifies the geometry of the tool tip and causes the variation of P_k from ${}^{\circ}P_k$. The wear of the cutting-tool tip will result in a loss of effective radial and axial depth of cut that generates dimensional variations (Fig. 2.13). In order to model the effect of V_B on part quality, the following assumptions are considered: i) flank wear is homogeneous; ii) there is no other factors on the cutting-tool edge such as the generation of a built-up edge or similar; and iii) the cutting-tool edge is a sharp edge. Under these assumptions, the variation of the machined surface in its normal direction (and thus, in Z direction of CS L_k) from its nominal value is formulated as follows

$$\delta_z = \frac{\tan(\alpha)}{(1-\tan(\gamma)\cdot\tan(\alpha))}\cdot V_B, \tag{2.43}$$

where α is the clearance angle and γ is the rake angle of the cutting inserts. According to Eq. (2.43), dimensional variations are proportional to the flank wear magnitude and thus, the dimensional quality variation can be described by a proportional coefficient that relates the influence of tool flank wear with the dimensional variation of a manufacturing feature for a specific cutting operation and cutting-tool geometry.

Assuming that tool flank wear remains constant during the same cutting operation of one workpiece, the cutting-tool tip presents a constant variation which is modeled as the DMV of P_k w.r.t. C_k by the expression

$$\mathbf{x}_{P_k}^{C_k} = \mathbf{B}_k^{m_4} \cdot V_{B_{ij}}^k, \tag{2.44}$$

where $\mathbf{B}_k^{m_4} = [0 \;\; 0 \;\; Cf_{V_{B_{ij}}}^k \;\; 0 \;\; 0 \;\; 0]^T$, $V_{B_{ij}}^k$ refers to the flank wear of the ith cutting edge of the jth cutting-tool at the kth machining station and $Cf_{V_{B_{ij}}}^k$ is the proportional coefficient.

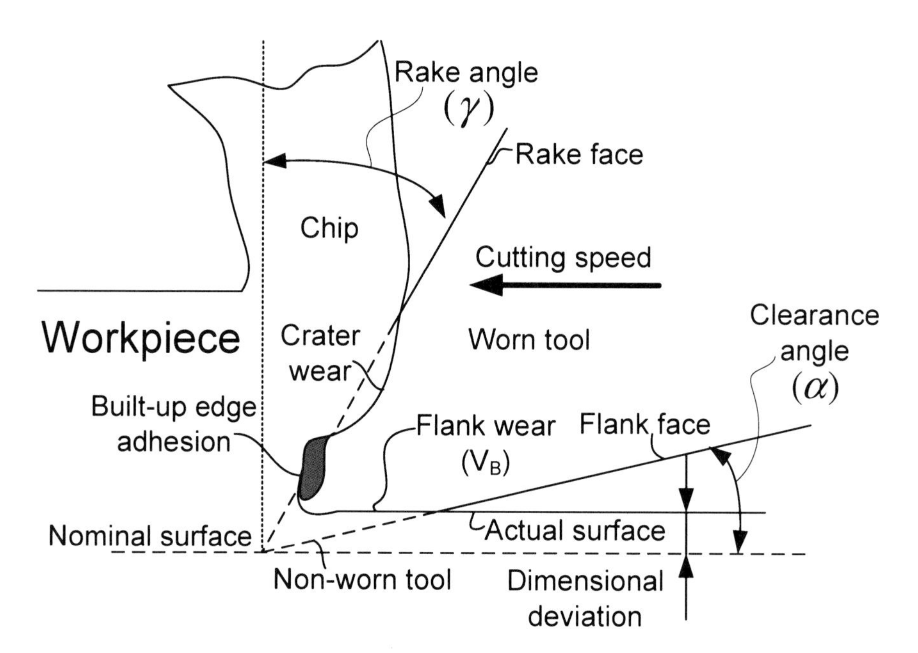

Fig. 2.13. Example of cutting-tool wear-induced deviation in machining

2.2.4 Derivation of the SoV Model

Considering the derivation of the DMVs presented above, the resulting DMV that defines the dimensional variation of part features after station k can be obtained.

The resulting DMV depends on two components: the feature variation from previous stations, and the variation added in the station k itself, which can be due to variations in datum surfaces, fixture locators and machining operations.

The first component is named the relocation term. If there is no variation added at station k, the resulting DMV after station k, denoted as $\mathbf{x}_k^{R_k}$, is the same as the resulting DMV from station $k\text{-}1$ but expressed w.r.t. R_k, denoted as $\mathbf{x}_k^{R_{k-1}}$. For a single feature S_i, the relationship between feature variation i w.r.t. R_{k-1} and the same variation but w.r.t. R_k is defined by applying Corollary 1 as

$$\mathbf{x}_{S_i}^{R_k} = \mathbf{D}_{S_i}^{R_k} \cdot \mathbf{x}_{R_k}^{R_{k-1}} + \mathbf{x}_{S_i}^{R_{k-1}}. \tag{2.45}$$

If one considers all features of the workpiece, the relationship is defined as

$$\mathbf{x}_k^{R_k} = [(\mathbf{x}_{S_1}^{R_{k-1}})^T, \ldots, (\mathbf{x}_{S_M}^{R_{k-1}})^T]^T + [\mathbf{D}_{S_1}^{R_k}, \ldots, \mathbf{D}_{S_M}^{R_k}]^T \cdot \mathbf{x}_{R_k}^{R_{k-1}}, \tag{2.46}$$

and rewritten Eq. (2.46),

$$\begin{aligned}\mathbf{x}_k^{R_k} &= \begin{pmatrix} \mathbf{I}_{6\times6} & \cdots & \mathbf{0}_{6\times6} & \cdots & \mathbf{D}_{S_1}^{R_k} & \cdots & \mathbf{0}_{6\times6} \\ \cdots & & \cdots & & \cdots & & \cdots \\ \mathbf{0}_{6\times6} & \cdots & \mathbf{I}_{6\times6} & \cdots & \mathbf{D}_{S_i}^{R_k} & \cdots & \mathbf{0}_{6\times6} \\ \cdots & & \cdots & & \cdots & & \cdots \\ \mathbf{0}_{6\times6} & \cdots & \mathbf{0}_{6\times6} & \cdots & \mathbf{D}_{S_M}^{R_k} & \cdots & \mathbf{I}_{6\times6} \end{pmatrix} \cdot [(\mathbf{x}_{S_1}^{R_{k-1}})^T, \ldots, (\mathbf{x}_{S_M}^{R_{k-1}})^T]^T, \\ &= \mathbf{A}_k^3 \cdot \mathbf{x}_k^{R_{k-1}}.\end{aligned} \tag{2.47}$$

On the other hand, the second term for deriving the resulting DMV for dimensional variation of part features is related to the variation added at station k due to the machining operation itself and thus, they only affect on the features machined at station k. As it was described above, the variation of a machined feature is defined as $\mathbf{x}_{L_k}^{R_k}$, which depends on the DMVs $\mathbf{x}_{P_k}^{{}^\circ M_k}$ and $\mathbf{x}_{R_k}^{{}^\circ M_k}$, as it is shown in Fig. 2.8. Applying Corollary 1, the total variation added to the workpiece at the station k due to datum-, fixture- and machining-induced variations is defined as

$$\mathbf{x}_{L_k}^{R_k} = \mathbf{D}_{L_k}^{R_k} \cdot \mathbf{x}_{R_k}^{{}^\circ M_k} + \mathbf{x}_{P_k}^{{}^\circ M_k}. \tag{2.48}$$

The DMV $\mathbf{x}_{R_k}^{{}^\circ M_k}$ defined in Eq. (2.9) can be rewritten as

$$\begin{aligned}\mathbf{x}_{R_k}^{{}^\circ M_k} &= \mathbf{T}_{R_k}^{F_k} \cdot \mathbf{B}_k^{f_1} \cdot [\Delta l_1, \Delta l_2, \ldots, \Delta l_6]^T - \mathbf{A}_k^1 \cdot \mathbf{x}_{B_k}^{R_k} - \mathbf{A}_k^2 \cdot \mathbf{x}_{C_k}^{R_k}, \\ &= \mathbf{B}_k^{f_2} \cdot \mathbf{u}_k^f - \mathbf{A}_k^4 \cdot \mathbf{x}_k^{R_k},\end{aligned} \tag{2.49}$$

where $\mathbf{A}_k^4 = [\mathbf{0}_{1\times 6},\ldots,\mathbf{A}_k^1,\mathbf{0}_{1\times 6},\ldots,\mathbf{A}_k^2,\mathbf{0}_{1\times 6},\ldots]$; $\mathbf{x}_k^{R_k} = [\mathbf{0}_{1\times 6},\ldots,(\mathbf{x}_{B_k}^{R_k})^T,\mathbf{0}_{1\times 6},\ldots,(\mathbf{x}_{C_k}^{R_k})^T,$ $\mathbf{0}_{1\times 6},\ldots]^T$, and $\mathbf{u}_k^f = [\Delta l_1, \Delta l_2, \ldots, \Delta l_6]^T$. The DMV $\mathbf{x}_k^{R_k}$ defines the variations of the workpiece w.r.t. the reference CS R_k at the station k. Since features may be deviated after machining at station $k-1$, the input of feature variations at station k is the resulting feature variations after station $k-1$ but expressed in R_k CS. As the relationship between $\mathbf{x}_k^{R_{k-1}}$ and the current $\mathbf{x}_k^{R_k}$ is defined in Eq. (2.47) by a relocation matrix, Eq. (2.49) can be rewritten adding the feature variations from station $k-1$ as

$$\mathbf{x}_{R_k}^{{}^\circ M_k} = \mathbf{B}_k^{f_2} \cdot \mathbf{u}_k^f - \mathbf{A}_k^4 \cdot \mathbf{A}_k^3 \cdot \mathbf{x}_k^{R_{k-1}}. \tag{2.50}$$

On the other hand, the DMV $\mathbf{x}_{P_k}^{{}^\circ M_k}$ defined in Eq. (2.10) can be rewritten to include the machining sources of variation as

$$\begin{aligned}\mathbf{x}_{P_k}^{{}^\circ M_k} &= [\mathbf{T}_{P_k}^{C_k} \cdot \mathbf{T}_{C_k}^{S_k} \cdot \mathbf{T}_{S_k}^{A_k^z} \cdot \mathbf{B}_k^{m_1} \;\vdots\; \mathbf{T}_{P_k}^{C_k} \cdot \mathbf{T}_{C_k}^{S_k} \cdot \mathbf{B}_k^{m_2} \;\vdots\; \mathbf{T}_{P_k}^{C_k} \cdot \mathbf{B}_k^{m_3} \;\vdots \mathbf{B}_k^{m_4}] \cdot \\ &\quad \cdot [\Delta T_1^k \;\; \ldots \;\; \Delta T_m^k \;\; \Delta x^k \;\; \Delta y^k \;\; \Delta z^k \;\vdots\; \Delta T_s^k \;\vdots\; \Delta F_x^k \;\; \Delta F_y^k \;\vdots\; V_{B_{ij}}^k]^T \\ &= \mathbf{B}_k^{m_5} \cdot \mathbf{u}_k^m.\end{aligned} \tag{2.51}$$

Substituting Eqs. (2.50) and (2.51) into Eq. (2.48), $\mathbf{x}_{L_k}^{R_k}$ is rewritten as

$$\begin{aligned}\mathbf{x}_{L_k}^{R_k} &= \mathbf{D}_{L_k}^{R_k} \cdot [\mathbf{B}_k^{f_2} \cdot \mathbf{u}_k^f - \mathbf{A}_k^4 \cdot \mathbf{A}_k^3 \cdot \mathbf{x}_k^{R_{k-1}}] + \mathbf{B}_k^{m_5} \cdot \mathbf{u}_k^m \\ &= \mathbf{A}_k^5 \cdot \mathbf{x}_k^{R_{k-1}} + \mathbf{B}_k^{f_3} \cdot \mathbf{u}_k^f + \mathbf{B}_k^{m_5} \cdot \mathbf{u}_k^m,\end{aligned} \tag{2.52}$$

where $\mathbf{A}_k^5 = [-\mathbf{D}_{L_k}^{R_k} \cdot \mathbf{A}_k^4 \cdot \mathbf{A}_k^3]$ and $\mathbf{B}_k^{f_3} = [\mathbf{D}_{L_k}^{R_k} \cdot \mathbf{B}_k^{f_2}]$. As $\mathbf{x}_{L_k}^{R_k}$ refers only to the feature machined, Eq. (2.52) can be rewritten in a more general form to include all the features of the part as

$$\mathbf{x}_k^{R_k} = \mathbf{A}_k^6 \cdot [\mathbf{A}_k^5 \cdot \mathbf{x}_k^{R_{k-1}} + \mathbf{B}_k^{f_3} \cdot \mathbf{u}_k^f + \mathbf{B}_k^{m_5} \cdot \mathbf{u}_k^m], \tag{2.53}$$

where $\mathbf{A}_k^6 = [\mathbf{0}_{6\times 6},\ldots,\mathbf{I}_{6\times 6},\ldots,\mathbf{0}_{6\times 6}]^T$ is a selector matrix that indicates the feature machined at station k.

Finally, the resulting DMV after station k that defines the dimensional variation of part features is obtained summing up both components: the feature variation from previous stations, and the variation added at the station k itself. Thus, the resulting DMV after station k is defined as

$$\mathbf{x}_{k+1}^{R_k} = \mathbf{A}_k^3 \cdot \mathbf{x}_k^{R_{k-1}} + \mathbf{A}_k^6 \cdot [\mathbf{A}_k^5 \cdot \mathbf{x}_k^{R_{k-1}} + \mathbf{B}_k^{f_3} \cdot \mathbf{u}_k^f + \mathbf{B}_k^{m_5} \cdot \mathbf{u}_k^m], \tag{2.54}$$

and reorganizing terms

$$\mathbf{x}_{k+1}^{R_k} = [\mathbf{A}_k^3 + \mathbf{A}_k^6 \cdot \mathbf{A}_k^5] \cdot \mathbf{x}_k^{R_{k-1}} + \mathbf{A}_k^6 \cdot \mathbf{B}_k^{f_3} \cdot \mathbf{u}_k^f + \mathbf{A}_k^6 \cdot \mathbf{B}_k^{m_5} \cdot \mathbf{u}_k^m, \tag{2.55}$$

which becomes the so-called SoV model in its generic version

$$\mathbf{x}_{k+1} = \mathbf{A}_k \cdot \mathbf{x}_k + \mathbf{B}_k^f \cdot \mathbf{u}_k^f + \mathbf{B}_k^m \cdot \mathbf{u}_k^m. \tag{2.56}$$

Besides Eq. (2.56), the SoV model is also composed of the observation equation, which represents the inspection of KPCs. This equation is formulated as follows

$$\begin{aligned} \mathbf{y}_k &= [\mathbf{C}_k^1 \cdot \mathbf{A}_k^3] \cdot \mathbf{x}_k + \mathbf{v}_k, \\ &= \mathbf{C}_k \cdot \mathbf{x}_k + \mathbf{v}_k, \end{aligned} \tag{2.57}$$

where matrix $\mathbf{C}_k$ defines the features that are inspected after the station k which define the KPCs and $\mathbf{v}_k$ is a vector that represents the measurement errors during inspection. The inspection station can be viewed as a special machining station where only a relocation is conducted since there is no machining operation. Therefore, $\mathbf{C}_k^1$ is the selector matrix similar to $\mathbf{A}_k^6$ to indicate the features inspected, and $\mathbf{A}_k^3$ is the relocating matrix that relates the primary datum surface at station k with the primary datum surface at the inspection station, placed after station k.

2.3 Process Planning

An important application of the extended SoV model can be found in the field of process planning. The integration of the KCCs and the KPCs let the engineers evaluate the manufacturing capability of a process plan and detect the critical KCCs that could be modified in order to reduce manufacturing variability. To describe this application, the SoV model defined by Eqs. (2.56) and (2.57) can be rewritten in an input-output form at the Nth station as

$$\begin{aligned} \mathbf{y}_N = &\sum_{i=1}^{N} \mathbf{C}_N \cdot \mathbf{\Phi}_{N,i}^{(\bullet)} \cdot \mathbf{B}_i^f \cdot \mathbf{u}_i^f + \sum_{i=1}^{N} \mathbf{C}_N \cdot \mathbf{\Phi}_{N,i}^{(\bullet)} \cdot \mathbf{B}_i^m \cdot \mathbf{u}_i^m + \mathbf{C}_N \cdot \mathbf{\Phi}_{N,0}^{(\bullet)} \cdot \mathbf{x}_0 \\ &+ \sum_{i=1}^{N} \mathbf{C}_N \cdot \mathbf{\Phi}_{N,i}^{(\bullet)} \cdot \mathbf{w}_i + \mathbf{v}_N, \end{aligned} \tag{2.58}$$

where

$$\mathbf{\Phi}_{N,i}^{(\bullet)} = \begin{cases} \mathbf{A}_{N-1} \cdot \mathbf{A}_{N-2} \cdot \ldots \cdot \mathbf{A}_i, & i < N \\ \mathbf{I} & i = N \end{cases} \quad (2.59)$$

and $\mathbf{I}$ is the identity matrix. The vector $\mathbf{x}_0$ represents the original variations of workpiece features in its raw state. These original variations are generated by previous manufacturing processes (i.e. bulk forming processes). Without loss of generality, it is assumed that the impact of initial workpiece variations on part quality is negligible in comparison with fixture- and machining-induced variations and other unmodeled errors. Thus, Eq. (2.58) can be rewritten as

$$\mathbf{Y}_N = \mathbf{\Gamma}_N^f \cdot \mathbf{U}_N^f + \mathbf{\Gamma}_N^m \cdot \mathbf{U}_N^m + \mathbf{\Gamma}_N^w \cdot \mathbf{W}_N + \mathbf{v}_N, \quad (2.60)$$

where

$$\mathbf{\Gamma}_N^f = [[\mathbf{C}_N \cdot \mathbf{\Phi}_{N,1}^{(\bullet)} \cdot \mathbf{B}_1^f] \;\; [\mathbf{C}_N \cdot \mathbf{\Phi}_{N,2}^{(\bullet)} \cdot \mathbf{B}_2^f] \;\; \cdots \;\; [\mathbf{C}_N \cdot \mathbf{\Phi}_{N,N}^{(\bullet)} \cdot \mathbf{B}_N^f]], \quad (2.61)$$

$$\mathbf{\Gamma}_N^m = [[\mathbf{C}_N \cdot \mathbf{\Phi}_{N,1}^{(\bullet)} \cdot \mathbf{B}_1^m] \;\; [\mathbf{C}_N \cdot \mathbf{\Phi}_{N,2}^{(\bullet)} \cdot \mathbf{B}_2^m] \;\; \cdots \;\; [\mathbf{C}_N \cdot \mathbf{\Phi}_{N,N}^{(\bullet)} \cdot \mathbf{B}_N^m]], \quad (2.62)$$

$$\mathbf{\Gamma}_N^w = [[\mathbf{C}_N \cdot \mathbf{\Phi}_{N,1}^{(\bullet)}] \;\; [\mathbf{C}_N \cdot \mathbf{\Phi}_{N,2}^{(\bullet)}] \;\; \cdots \;\; [\mathbf{C}_N \cdot \mathbf{\Phi}_{N,N}^{(\bullet)}]], \quad (2.63)$$

and $\mathbf{U}_N^f = [(\mathbf{u}_1^f)^T, \ldots, (\mathbf{u}_N^f)^T]^T$, $\mathbf{U}_N^m = [(\mathbf{u}_1^m)^T, \ldots, (\mathbf{u}_N^m)^T]^T$, $\mathbf{W}_N = [(\mathbf{w}_1)^T, \ldots, (\mathbf{w}_N)^T]^T$, $\mathbf{Y}_N = \mathbf{y}_N$.

Eq. (2.60) can be used by the process planner to predict the part quality variation of a process plan according to the fixtures and machine-tools capabilities and to detect critical KCCs whose elimination could notably improve the final part quality. In the next subsections, the evaluation of a process plan and how to improve its robustness will be described in detail.

2.3.1 Process Plan Evaluation

Assuming that $\mathbf{U}_N^f$, $\mathbf{U}_N^m$, $\mathbf{W}_N$ and $\mathbf{v}_N$ are independent of each other, Eq. (2.60) can be analyzed in terms of covariances as

$$\mathbf{\Sigma}_{\mathbf{Y}_N} = \mathbf{\Gamma}_N^f \cdot \mathbf{\Sigma}_{\mathbf{U}_N^f} \cdot (\mathbf{\Gamma}_N^f)^T + \mathbf{\Gamma}_N^m \cdot \mathbf{\Sigma}_{\mathbf{U}_N^m} \cdot (\mathbf{\Gamma}_N^m)^T + \mathbf{\Gamma}_N^w \cdot \mathbf{\Sigma}_{\mathbf{W}_N} \cdot (\mathbf{\Gamma}_N^w)^T + \mathbf{\Sigma}_{\mathbf{v}_N}, \quad (2.64)$$

where $\mathbf{\Sigma}_{\mathbf{Y}_N}$ is the covariance matrix of the KPCs, $\mathbf{\Sigma}_{\mathbf{U}_N^f}$ is the covariance matrix of fixture variations; $\mathbf{\Sigma}_{\mathbf{U}_N^m}$ is the covariance matrix of machining variations; $\mathbf{\Sigma}_{\mathbf{W}_N}$ is the covariance matrix of unmodeled source of variations; and $\mathbf{\Sigma}_{\mathbf{v}_N}$ is the covariance matrix of measurement noise. Note that, since process planners are more focused on the variation propagation during process planning, it is assumed that there is no mean-shift in any source of variation.

By applying Eq. (2.64), the process planner can estimate the expected part quality variation given a manufacturing process plan. In practice, the evaluation and selection among different process plans is conducted by analyzing a capability ratio that indicates the capability of the manufacturing process to manufacture the parts within specifications. Since products possess multiple, rather than a single, KPCs, process capability analysis is based on a multivariate capability index. A variety of multivariate capability indices, such as those presented in [33-37], have been proposed for assessing capability.

In this chapter, the multivariate process capability ratio proposed by Chen [34] is adopted for evaluating the manufacturing capability of a MMP. This capability ratio was successfully implemented by previous authors [17, Chapter 13] in similar MMPs with the use of the SoV methodology. Focusing only on the variations in the KPC measurements, $\mathbf{Y}_N$, the multivariate process capability index for the process plan χ can be expressed as

$$MC_p^{\chi} = \frac{1}{r_0}, \tag{2.65}$$

where r_0 is a value that has the probability, $Pr(max(|Y_{i,N}^c|/r_{i,N}, i = 1,\ldots,M) \leq r_0)$, equal to $1-\alpha$, and α is the allowable expected proportion of non-conforming products from the process. $Y_{i,N}^c$ is the c th component vector of the ith KPC measured at station N, and $r_{i,N}$ is the tolerance value for this component. The r_0 value can be obtained by Monte Carlo simulations of Eq. (2.60), where the fixture-induced variations ($\mathbf{U}_N^f$) are assumed normally distributed with the standard deviation defined from the fixture tolerances and the machining-induced variations ($\mathbf{U}_N^m$) are distributed according to the nature of the machining error. The unmodeled errors ($\mathbf{W}_N$) and the measurement errors ($\mathbf{v}_N$) are set according to process planner's knowledge about the process and measurement devices, respectively.

Through this formulation, an index value of $MC_p^{\chi} = 1.0$ corresponds to an expected proportion of conforming products of exactly $1-\alpha$. A larger MC_p^{χ} indicates a lower expected proportion of non-conforming products or a more capable process plan. Therefore, these interpretations enable the MC_p^{χ} ratio to have a similar interpretation as the univariate ratio C_p [17, Chapter 13].

2.3.2 Process Plan Improvement

In order to improve a MMP and reduce the final part quality variability, the critical KCCs that influence the most on part quality should be identified. For this purpose, a series of sensitivity indices can be evaluated from Eq. (2.60), detecting the fixture component or the machining source of variation that contribute significant amount of variations to KPCs. Then, these critical KCCs' effects can be minimized or eliminated to improve the robustness of the process.

The process plan improvement can be accomplished through evaluating different sensitivity indices at different levels following the flowchart shown in Fig. 2.14. According to this flowchart, at phase I, the process sensitivity index considering fixture-induced variations, $S^Y_{f,process}$, and the process sensitivity index considering machining-induced variations, $S^Y_{m,process}$, are evaluated to determine the main source of variation (fixturing or machining) in the process plan. At phase II, additional sensitivity indices of individual KPCs are evaluated and compared with their corresponding process sensitivity index scaled by a factor C_b to identify whether there is a critical KPC whose variability is notably higher than that from other KPCs. Finally, at phase III, it is further investigated which KCCs are the causes of the high variation of KPCs, detecting critical locators or critical machining sources of variation. Such information can lead the process planner to redefine the process plan and increase its robustness.

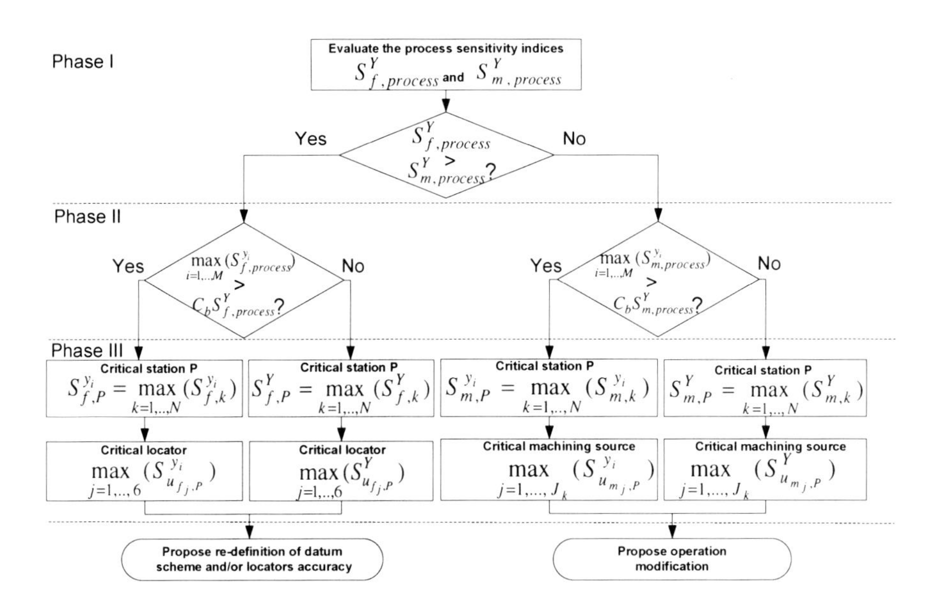

Fig. 2.14. Flowchart for process plan improvement

The different sensitivity indices applied in the proposed flowchart are extensions of widely accepted indices [38]. Basically, a sensitivity index of a generic variable H w.r.t. a generic parameter φ can be described by the differential equation

$$S_{\varphi}^{H} = \frac{\delta H}{\delta \varphi} \frac{\Delta \varphi}{\Delta H}, \tag{2.66}$$

where ΔH and $\Delta \varphi$ denote the admissible variation of variable H and the expected variation of parameter φ, respectively. For a MMP design problem, the admissible variation of a variable H is related to a KPC and it can be defined by its tolerance from the design drawing, denoted by H^{tol}. The expected variation of a generic parameter φ is related to fixture-induced or machining-induced variations, and it can be defined by its expected range of variation denoted as φ^{tol} and defined as $6\sigma_{\varphi}$ (99.73% of confidence level), where σ_{φ} is the standard deviation of this source of variation. Thus, the generic sensitivity index is defined as

$$S_{\varphi}^{H} = \frac{\delta H}{\delta \varphi} \frac{\varphi^{tol}}{H^{tol}}. \tag{2.67}$$

We consider a MMP composed of N stations, each of which is equipped with a 3-2-1 fixture device. For this MMP, the locator variations at station k are defined as $\mathbf{u}_k^f = [u_{f_1}, \ldots, u_{f_6}]^T$, and the machining variations are defined as $\mathbf{u}_k^m = [u_{m_1}, \ldots, u_{m_{J_k}}]^T$, where J_k is the number of machining sources of variation. Consider that this MMP will be used to manufacture a product composed of M KPCs, and each KPC is denoted by the index i ($i = 1, \ldots, M$). According to Eq. (2.67), a series of sensitivity indices can be defined as shown in the following subsections. In addition to this definition, a summary of these indices and their respective interpretations is shown in Table 2.1.

2.3.2.1 Sensitivity Indices Related to Fixtures

<u>Locator sensitivity index</u>: This type of indices evaluates the impact of the variability of each fixture locator at station k. Two indices can be distinguished: that w.r.t. an individual KPC, as defined in Eq. (2.68), and that w.r.t. a product, as defined in Eq. (2.69).

$$S_{u_{f_j,k}}^{y_i} = abs\left(\frac{\delta y_i}{\delta u_{f_j,k}} \frac{u_{f_j,k}^{tol}}{y_i^{tol}} \right) = abs\left(\Gamma_{i,j}^{f} \frac{u_{f_j,k}^{tol}}{y_i^{tol}} \right), \tag{2.68}$$

$$S_{u_{f_j,k}}^{Y} = \frac{1}{M} \sum_{i=1}^{M} S_{u_{f_j,k}}^{y_i}, \tag{2.69}$$

where $abs(A)$ denotes the absolute value of A; $\Gamma_{i,j}^{f}$ denotes the element at the ith row and the jth column of the matrix $\mathbf{\Gamma}_N^f$; y_i^{tol} is the product dimensional tolerance of the KPC y_i; and $u_{f_j,k}^{tol}$ is the tolerance of the jth fixture locator component at station k. This sensitivity index is defined according to Eq. (2.67) by substituting H, φ, H^{tol} and φ^{tol} with y_i, $u_{f_j,k}$, y_i^{tol}, and $u_{f_j,k}^{tol}$, respectively. $S_{u_{f_j,k}}^{y_i}$ refers to the impact of the variability of the j th locator at station k on the i th KPC, whereas $S_{u_{f_j,k}}^{Y}$ refers to the impact on the average of all KPCs.

Fixture sensitivity index: This type of indices evaluates the average impact of locator variations on part quality at station k. The fixture sensitivity index w.r.t. a KPC ($S_{f,k}^{y_i}$) and that w.r.t. the product ($S_{f,k}^{Y}$) are defined respectively as

$$S_{f,k}^{y_i} = \frac{1}{6}\sum_{j=1}^{6} S_{u_{f_j,k}}^{y_i}, \tag{2.70}$$

$$S_{f,k}^{Y} = \frac{1}{M}\sum_{i=1}^{M} S_{f,k}^{y_i}. \tag{2.71}$$

2.3.2.2 Sensitivity Indices Related to Machining Operations

Machining source of variation sensitivity index: This type of indices evaluate the impact of the variability of a specific machining source of variation when conducting a specific machining operation at station k. Two indices are defined w.r.t. a KPC ($S_{u_{m_j,k}}^{y_i}$) and w.r.t. a product ($S_{u_{m_j,k}}^{Y}$), respectively, as

$$S_{u_{m_j,k}}^{y_i} = abs\left(\frac{\delta y_i}{\delta u_{m_j,k}} \frac{u_{m_j,k}^{tol}}{y_i^{tol}}\right) = abs\left(\Gamma_{i,j}^{m} \frac{u_{m_j,k}^{tol}}{y_i^{tol}}\right), \tag{2.72}$$

$$S_{u_{m_j,k}}^{Y} = \frac{1}{M}\sum_{i=1}^{M} S_{u_{m_j,k}}^{y_i}, \tag{2.73}$$

where $\Gamma_{i,j}^{m}$ denote the element in the i th row and j th column of the matrix $\mathbf{\Gamma}_N^m$; and $u_{m_j,k}^{tol}$ defines the variability of the j th machining source of variation at the k th station and $j = 1,\ldots,J_k$.

Operation sensitivity index: This type of indices evaluates the average impact of machining sources of variations at station k. Two indices are defined w.r.t. a KPC ($S_{m,k}^{y_i}$) and w.r.t. a product ($S_{m,k}^{Y}$), respectively, as

$$S_{m,k}^{y_i} = \frac{1}{J_k} \sum_{j=1}^{J_k} S_{u_{m_j},k}^{y_i}, \tag{2.74}$$

$$S_{m,k}^{Y} = \frac{1}{M} \sum_{i=1}^{M} S_{m,k}^{y_i}. \tag{2.75}$$

2.3.2.3 Sensitivity Index Related to Stations

Station sensitivity index: This type of indices evaluates the average impact of the variability of both fixture and machining-induced variations at station k. Two indices are defined w.r.t. a KPC ($S_{u_k}^{y_i}$) and w.r.t. a product ($S_{u_k}^{Y}$), respectively, as

$$S_{u_k}^{y_i} = \frac{S_{f,k}^{y_i} + S_{m,k}^{y_i}}{2}, \tag{2.76}$$

$$S_{u_k}^{Y} = \frac{S_{f,k}^{Y} + S_{m,k}^{Y}}{2}. \tag{2.77}$$

2.3.2.4 Sensitivity Indices Related to Manufacturing Process

Process sensitivity index considering fixture-induced variations: This type of indices evaluates the average impact of fixture-induced variations along a MMP. Two indices are defined w.r.t. a KPC ($S_{f,process}^{y_i}$) and w.r.t. a product ($S_{f,process}^{Y}$), respectively, as

$$S_{f,process}^{y_i} = \frac{1}{N} \sum_{k=1}^{N} S_{f,k}^{y_i}, \tag{2.78}$$

$$S_{f,process}^{Y} = \frac{1}{M} \sum_{i=1}^{M} S_{f,process}^{y_i}. \tag{2.79}$$

Process sensitivity index considering machining-induced variations: This type of indices evaluates the impact of machining-induced variations along a MMP. Two indices are defined w.r.t. a KPC ($S^{y_i}_{m,process}$) and w.r.t. a product ($S^{Y}_{m,process}$), respectively, as:

$$S^{y_i}_{m,process} = \frac{1}{N}\sum_{k=1}^{N} S^{y_i}_{m,k}, \tag{2.80}$$

$$S^{Y}_{m,process} = \frac{1}{M}\sum_{i=1}^{M} S^{y_i}_{m,process}. \tag{2.81}$$

Note that in the flowchart for process planning improvement shown in Fig. 2.14, the indices $S^{y_i}_{u_k}$ and $S^{Y}_{u_k}$ are not evaluated since the purpose is to analyze separately the variation propagation in the MMP due to fixture- and machining-induced variations in order to propose process plan adjustments.

Table 2.1. Summary of the sensitivity indices and physical meaning

Sensitivity index	KPC	Product	Rationale for definition
Locator index	$S^{y_i}_{u_{f_j},k}$	$S^{Y}_{u_{f_j},k}$	Identifies the most influent locators on KPC/product variability
Fixture index	$S^{y_i}_{u_f,k}$	$S^{Y}_{u_f,k}$	Identifies the most influent fixtures on KPC/product variability
Machining source of variation index	$S^{y_i}_{u_{m_j},k}$	$S^{Y}_{u_{m_j},k}$	Identifies the most influent machining source of variation on KPC/product variability
Machining operation index	$S^{y_i}_{m,k}$	$S^{Y}_{m,k}$	Identifies the most influent machining operation on KPC/product variability
Station index	$S^{y_i}_{u_k}$	$S^{Y}_{u_k}$	Identifies the most influent station on KPC/product variability
Process index due to fixture-induced variations	$S^{y_i}_{f,process}$	$S^{Y}_{f,process}$	Evaluates the influence on KPC/product variability due to all fixture-induced variations
Process index due to machining-induced variations	$S^{y_i}_{m,process}$	$S^{Y}_{m,process}$	Evaluates the influence on KPC/product due to all machining-induced variations

2.4 Case Study

In this section, the SoV model is derived in order to analyze the MMP shown in Fig. 2.2 and its capability to manufacture the part geometry shown in Fig. 2.1. For this case study, the final part geometry is shown in Table 2.2. The positions of the locators that compose the fixture at different stations are shown in Table 2.3, and the characteristics of the MMP in terms of locator tolerances, expected thermal variations of the machine-tool spindle at each station, and admissible cutting-tool wear at each cutting-tool are shown in Table 2.4.

For the sake of simplicity and without loss of generality, the extended SoV model applied in this case study only deals with cutting-tool wear-induced variations, thermal-induced variations from the machine-tool spindle, fixture-induced variations and datum-induced variations. The empirical coefficients that relate the spindle thermal variations with the dimensional spindle expansion and the cutting-tool wear effects with the cutting-tool tip loss of cut are obtained from [18]. According to this research work and the results of a set of experiments, the cutting-tool wear coefficients were adjusted to $Cf_{V_B}^k = 0.125$ and $Cf_{V_B}^k = 0.135$ for frontal and peripheral machining operations, respectively, and the thermal coefficients were adjusted to $Cf_x^k = Cf_y^k = Cf_\alpha^k = Cf_\beta^k = Cf_\gamma^k \approx 0$ and $Cf_z^k = -0.0052 \;\; mm/^\circ C$.

Table 2.2. Product design information. Nominal position ($\mathbf{t}_{S_i}^{\circ D}$) and orientation ($\boldsymbol{\omega}_{S_i}^{\circ D}$) of each feature w.r.t. DCS.

Feature	$\boldsymbol{\omega}_{S_i}^{\circ D}$ (rad)	$\mathbf{t}_{S_i}^{\circ D}$ (mm)
S_0	$[0,\pi,0]$	$[150,125,0]$
S_1	$[0,\pi,0]$	$[150,125,20]$
S_2	$[0,0,0]$	$[150,75,200]$
S_3	$[0,0,0]$	$[150,225,180]$
S_4	$[\pi/2,-\pi/2,-\pi/2]$	$[150,0,100]$
S_5	$[0,-\pi/2,0]$	$[0,75,100]$
S_6	$[\pi/2,\pi/2,-\pi/2]$	$[150,250,75]$
S_7	$[0,\pi/2,\pi/2]$	$[300,125,100]$
S_8	$[\pi/2,\pi/2,-\pi/2]$	$[150,200,190]$

Table 2.3. Nominal position ($\mathbf{t}_{F_k}^{\circ D}$) and orientation ($\boldsymbol{\omega}_{F_k}^{\circ D}$) of FCS w.r.t. DCS at each station and fixture layout

St.	$\boldsymbol{\omega}_{F_k}^{\circ D}$ (rad)	$\mathbf{t}_{F_k}^{\circ D}$ (mm)	Locator position w.r.t. FCS (mm)
1	$[-\pi/2,\pi,0]$	[0,0,0]	$L_{1x}=125, L_{1y}=50, L_{2x}=50, L_{2y}=250, L_{3x}=200, L_{3y}=250$
			$p_{1y}=50, p_{1z}=-100, p_{2y}=250, p_{2z}=-100, p_{3x}=125, p_{3z}=-100$
2	$[\pi/2,-\pi/2,\pi]$	[0,0,200]	$L_{1x}=100, L_{1y}=50, L_{2x}=50, L_{2y}=150, L_{3x}=150, L_{3y}=150$
			$p_{1y}=50, p_{1z}=-125, p_{2y}=250, p_{2z}=-125, p_{3x}=100, p_{3z}=-125$
3	$[-\pi/2,0,0]$	[0,250,200]	$L_{1x}=125, L_{1y}=50, L_{2x}=50, L_{2y}=250, L_{3x}=200, L_{3y}=250$
			$p_{1y}=50, p_{1z}=-100, p_{2y}=250, p_{2z}=-100, p_{3x}=125, p_{3z}=-100$
4	$[-\pi/2,\pi,0]$	[0,0,20]	$L_{1x}=125, L_{1y}=50, L_{2x}=50, L_{2y}=250, L_{3x}=200, L_{3y}=250$
			$p_{1y}=50, p_{1z}=-100, p_{2y}=250, p_{2z}=-100, p_{3x}=125, p_{3z}=-100$
5	$[-\pi/2,\pi,0]$	[0,0,0]	Inspection station

Table 2.4. Process description of the case study

Station	Datum features	Features machined	Δl_j^{tol} (mm)	V_B (mm)	ΔT_s (ºC)
1	$S_0-S_4-S_5$	S_2	0.035	[0,0.5]	[−5,+5]
2	$S_4-S_2-S_5$	S_6	0.035	[0,0.5]	[−5,+5]
3	$S_2-S_6-S_5$	S_1	0.035	[0,0.5]	[−5,+5]
4	$S_1-S_4-S_5$	S_3, S_8	0.035	[0,0.5]	[−5,+5]
5	$S_0-S_4-S_5$	-CMM-	-	-	-
Δl_j^{tol} for $j=4,5,6$ refer to tolerances at locators p_1, p_2 and p_3, respectively					

2.4.1 Process Plan Evaluation

In order to estimate the capability of the MMP to manufacture the part geometry defined in Fig. 2.1, 1,000 Monte Carlo simulations were conducted. Each component of the fixture variations $\mathbf{U}_k^f$ was assumed independent and normally distributed with a standard deviation of $\Delta l_j^{tol}/6$; each component related to the thermal variation of the machine-tool spindle in $\mathbf{U}_k^m$ was assumed independent of other components and uniformly distributed within the expected range of thermal variations; each component related to cutting-tool wear in $\mathbf{U}_k^m$ was assumed independent of other components and uniformly distributed within the expected range of the admissible cutting-tool wear; the measurement error and the term related to non-linear errors were assumed to be negligible in comparison with fixture- and machining-induced variations. Eq. (2.60) was evaluated for 1,000 simulations. The resulting distribution of the KPC variations was used to evaluate the multivariate process capability ratio defined in Eq. (2.65).

After Monte Carlo simulations and considering the expected proportion of conforming products as $1-\alpha=0.99$, the capability index obtained was $MC_p=0.82$. Indeed, Monte Carlo simulations showed that 4.4% of the simulated parts have nonconforming KPC_1's, whereas 0.5% of them have nonconforming KPC_2.

2.4.2 Process Plan Improvement

As shown above, the analyzed MMP is not able to produce 99% of conforming parts ($MC_p=0.82<1$). In order to investigate how to improve the process and reduce part quality variability, a group of sensitivity indices are evaluated following the flowchart shown in Fig. 2.14. According to the sensitivity indices evaluated, the manufacturing variability due to fixture-induced variations is higher than that due to machining-induced variations ($S_{m,process}^Y=0.044$ and $S_{f,process}^Y=0.094$). Since the fixture-induced variations have a higher impact on product quality than the machining-induced variations, at the second phase of the flowchart it is analyzed the sources of fixture-induced variations. At this phase, the sensitivity indices show that there is no special KPC with an important variability w.r.t. other KPCs and thus, the sensitivity indices should be analyzed w.r.t. product. In fact, the sensitivity indices for each KPC are similar ($S_{f,process}^{KPC1}=0.101$ and $S_{f,process}^{KPC2}=0.087$), which means that both KPCs present a similar manufacturing variability. By analyzing in more detail the sensitivity indices related to fixture-induced variations, it can be observed that the fourth station is the most critical station in the MMP. Indeed, the fourth station is responsible for 35% of the part quality variability due to fixture-induced variations. Following the third phase of the flowchart, a further investigation is conducted at the fourth station. At this station, the locator sensitivity indices show the particular impact of each fixture locator within the fixture-induced variations. These indices are $S_{u_{f_1},4}^Y=0.14$, $S_{u_{f_2},4}^Y=0.25$, $S_{u_{f_3},4}^Y=0.32$, $S_{u_{f_4},4}^Y=0.03$, $S_{u_{f_5},4}^Y=0.03$, $S_{u_{f_6},4}^Y=0$. Thus, 41% of part fixture-induced quality variability at this station is due to the low precision of locator number 3.

In order to improve the MMP, it is reasonable to reduce the impact of fixture-induced variations at station 4. Additionally, the sensitivity analysis shows that station 2 also contributes to the 23.4% of part variability due to fixture-induced variations. Therefore, a process planning modification is proposed to increase the locator precision at both stations. For this example, the locator tolerances at station 2 and 4 are reduced to 0.015 mm. By this modification, the multivariate capability ratio is increased up to $MC_p = 0.91$. For this MMP with a first improvement, the percentage of part quality variability due to fixture-induced variations is 61% whereas due to machining-induced variations is 39% ($S^Y_{m,process} = 0.047$ and $S^Y_{f,process} = 0.075$).

For illustrative purposes, we propose a second improvement of the MMP but now being focused on reducing the effect of machining-induced variations. Analyzing the sensitivity indices it is observed that the variability of KPC_1 depends more on machining-induced variations than that of the KPC_2 ($S^{KPC_1}_{m,process} > C_b \cdot S^Y_{m,process}$, with $C_b = 1.3$). Furthermore, it is observed that 33.3% of the variability of KPC_1 due to machining-induced variations is generated at station 1, 3 and 4 in the same proportion. In any of these stations, 55% of the variability is due to the cutting-tool wear-induced variations whereas 45% of the variability is due to the thermal expansion of the machine-tool spindle. According to this analysis, it seems reasonable to reduce the admissible level of cutting-tool wear during machining at these three stations in order to reduce the impact of machining-induced variations. For this example, it is proposed to reduce the admissible flank wear at stations 1, 3 and 4 to a value of $V_B = 0.2$ mm. By this modification, the multivariate capability ratio is increased up to $MC_p = 1.02$ and thus, the MMP is more capable to produce more than the 99% of parts within specifications.

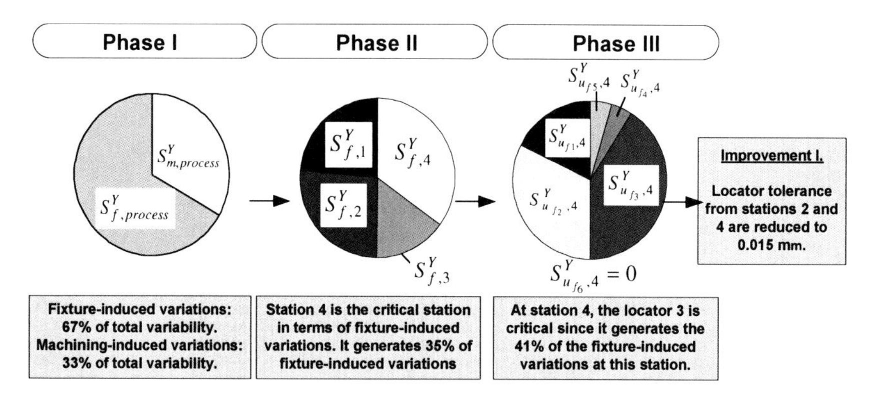

Fig. 2.15. Sensitivity analysis results of fixture-induced variations

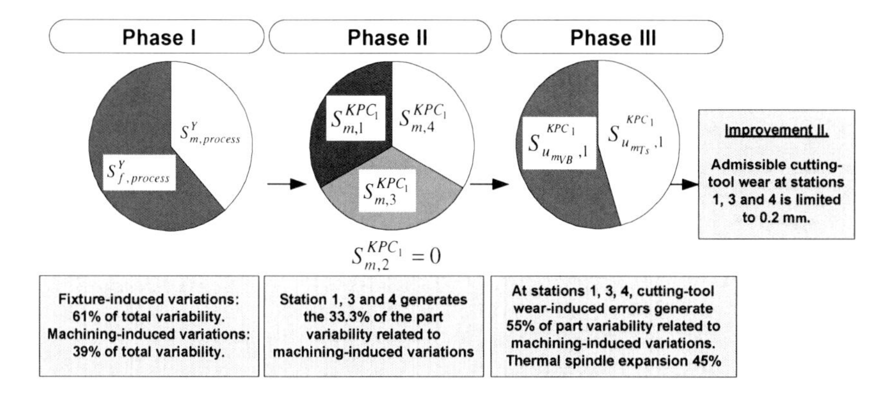

Fig. 2.16. Sensitivity analysis results of machining-induced variations

A summary of the results obtained after the sensitivity analysis is shown in Fig. 2.15 and 2.16. The resulting probability density functions (p.d.f.) of the KPCs for the initial MMP and the MMP with the proposed improvements are also shown in Fig. 2.17.

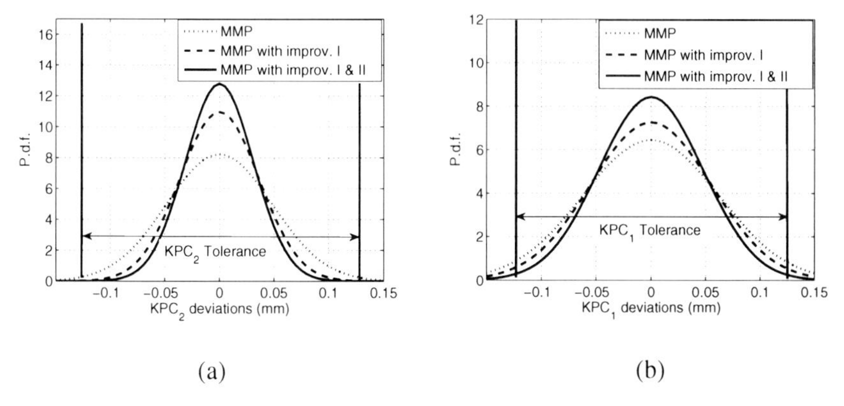

Fig. 2.17. P.d.f. obtained from Monte Carlo simulation results for a) KPC_1 and b) KPC_2 variations according to the initial MMP and the MMP with improvement I and I and II

2.5 Conclusions

In this chapter, the derivation of the 3D stream of variation (SoV) model for MMPs is presented in its extended version. With this model, engineers can consider machining-induced variations such as those related to machine-tool inaccuracy, cutting-tool deflections, thermal distortions or cutting-tool wear, as well as fixture- and datum-induced variations in order to estimate dimensional and geometrical part variations and their propagation in MMPs. The potential applicability of the extended SoV model is extensive, especially in fields such as manufacturing fault diagnosis, process planning, process oriented tolerancing and dimensional quality control.

Besides the derivation of the extended SoV model, this chapter shows the applicability of the model on process planning activities. Firstly, it was described how to evaluate the robustness of a manufacturing process plan when fixture-induced and machining-induced variations are present. Secondly, it was presented a series of sensitivity indices and a flowchart to detect which are the sources of variations that affect part quality and at which stations they can be shown up. Through these indices, process plan improvements can be straight forward conducted. The application of the SoV model on process planning was also demonstrated through a 3D case study with a product design with two KPCs and a MMP composed of 4 machining stations.

Acknowledgement. This work has been partially supported by Fundació Caixa-Castelló Bancaixa (Research Promotion 2007, 2009 and 2011) and the Spanish National Project number DPI2007-66871-C02-01 (PIA12007-83).

Appendix 2.1

A Homogeneous Transformation Matrix (HTM) in the 3D space is a 4×4 matrix. It can be used to represent one CS w.r.t. another CS. Let us consider two CS 1 and 2. Given the position and orientation vector of CS 1 w.r.t. CS 2, the CS 1 can be expressed in 2 by a nominal HTM as [39]

$$\mathbf{H}_1^2 = \overline{\mathbf{T}}_1^2 \cdot \overline{\mathbf{R}}_1^2, \tag{2.82}$$

where $\overline{\mathbf{T}}_1^2$ and $\overline{\mathbf{R}}_1^2$ are the translational and rotational matrix, respectively. For a HTM, the superscript represents the CS we want the results to be represented in, and the subscript represents the CS we are transferring from. In Eq. (2.82) the translational matrix is defined as

$$\overline{\mathbf{T}}_1^2 = \begin{pmatrix} 1 & 0 & 0 & t_{1x}^2 \\ 0 & 1 & 0 & t_{1y}^2 \\ 0 & 0 & 1 & t_{1z}^2 \\ 0 & 0 & 0 & 1 \end{pmatrix} = \begin{pmatrix} \mathbf{I}_{3\times3} & \mathbf{t}_1^2 \\ \mathbf{0}_{1\times3} & 1 \end{pmatrix}, \tag{2.83}$$

where $\mathbf{t}_1^2$ is the position vector of the origin of CS 1 expressed in 2, defined as $\mathbf{t}_1^2 = [t_{1x}^2, t_{1y}^2, t_{1z}^2]^T$. The rotational matrix is defined according to the rotation representation used. According to the Euler angles conversion, three Euler angles ($\overline{\phi}$, $\overline{\theta}$, and $\overline{\psi}$) are used to define the rotational matrix. The orientation of CS 1 w.r.t. CS 2 can be obtained by three successive rotations as follows. First, CS 2 is rotated about the Z-axis by the angle $\overline{\phi}$. Then, the resulting CS is rotated around the new Y-axis by angle $\overline{\theta}$, and finally, the resulting CS is rotated around the new Z-axis by $\overline{\psi}$. According to this representation, the rotational matrix is formulated as [40, Chapter 2]

$$\overline{\mathbf{R}}_1^2 = \begin{pmatrix} \mathbf{R}_1^2 & \mathbf{0}_{3\times1} \\ \mathbf{0}_{1\times3} & 1 \end{pmatrix}, \tag{2.83}$$

where

$$\mathbf{R}_1^2 = \mathbf{R}_{z,\overline{\phi}} \cdot \mathbf{R}_{y,\overline{\theta}} \cdot \mathbf{R}_{z,\overline{\psi}}, \tag{2.84}$$

and

$$\mathbf{R}_{z,\overline{\phi}} = \begin{pmatrix} cos\overline{\phi} & -sin\overline{\phi} & 0 \\ sin\overline{\phi} & cos\overline{\phi} & 0 \\ 0 & 0 & 1 \end{pmatrix}, \tag{2.85}$$

$$\mathbf{R}_{y,\overline{\theta}} = \begin{pmatrix} cos\overline{\theta} & 0 & sin\overline{\theta} \\ 0 & 1 & 0 \\ -sin\overline{\theta} & 0 & cos\overline{\theta} \end{pmatrix}, \tag{2.86}$$

$$\mathbf{R}_{z,\overline{\psi}} = \begin{pmatrix} cos\overline{\psi} & -sin\overline{\psi} & 0 \\ sin\overline{\psi} & cos\overline{\psi} & 0 \\ 0 & 0 & 1 \end{pmatrix}. \tag{2.87}$$

For this rotation representation, the rotational matrix is defined as

$$\mathbf{R}_1^2 = \begin{pmatrix} c\overline{\phi}c\overline{\theta}c\overline{\psi} - s\overline{\phi}s\overline{\psi} & -c\overline{\phi}c\overline{\theta}s\overline{\psi} - s\overline{\phi}c\overline{\psi} & c\overline{\phi}s\overline{\theta} \\ s\overline{\phi}c\overline{\theta}c\overline{\psi} + c\overline{\phi}s\overline{\psi} & -s\overline{\phi}c\overline{\theta}s\overline{\psi} + c\overline{\phi}c\overline{\psi} & s\overline{\phi}s\overline{\theta} \\ -s\overline{\theta}c\overline{\psi} & s\overline{\theta}s\overline{\psi} & c\overline{\theta} \end{pmatrix}, \tag{2.88}$$

where c and s refers to cos and sin respectively. As a result, Eq. (2.82) can be rewritten as

$$\mathbf{H}_1^2 = \begin{pmatrix} \mathbf{R}_1^2 & \mathbf{t}_1^2 \\ \mathbf{0}_{1\times3} & 1 \end{pmatrix}. \tag{2.89}$$

Using the HTM, an ith point in the CS 1, defined as $\mathbf{p}_i^1 = [p_{ix}^1, p_{iy}^1, p_{iz}^1]$, is related to the same point expressed in the CS 2, defined as $\mathbf{p}_i^2$, by the following equation

$$\tilde{\mathbf{p}}_i^2 = \mathbf{H}_1^2 \cdot \tilde{\mathbf{p}}_i^1, \tag{2.90}$$

where $\tilde{\mathbf{p}}$ is equal to $[\mathbf{p},1]^T$.

Appendix 2.2

A Differential Transformation Matrix (DTM) in the 3D space is a 4 × 4 matrix that is used to represent the small position and orientation deviation of one CS w.r.t. another CS. For illustrative purposes, we consider two CSs, 1 and 2. If both CSs are deviated from nominal values, the HTM between the actual CS 1 and CS 2 can be defined as

$$\mathbf{H}_2^1 = \mathbf{H}_{\circ 2}^{\circ 1} \cdot \delta\mathbf{H}_2^1, \tag{2.91}$$

where $\mathbf{H}_{\circ 2}^{\circ 1}$ is the HTM between the nominal CSs 1 and 2, and $\delta\mathbf{H}_2^1$ is the HTM that defines the small position and orientation deviations of CS 2 w.r.t. 1 due to the deviation from their nominal values, and it is defined as

$$\delta\mathbf{H}_2^1 = \begin{pmatrix} 1 & -\theta_{2z}^1 & \theta_{2y}^1 & d_{2x}^1 \\ \theta_{2z}^1 & 1 & -\theta_{2x}^1 & d_{2y}^1 \\ -\theta_{2y}^1 & \theta_{2x}^1 & 1 & d_{2z}^1 \\ 0 & 0 & 0 & 1 \end{pmatrix}. \tag{2.92}$$

Note that the rotational matrix in $\delta\mathbf{H}_2^1$ is defined as Eq. (2.88) using the approximation of $cos(\theta) \approx 1$ and $sin(\theta) \approx \theta$ and neglecting second-order small values since only small position and orientation variations are considered. From Eq. (2.92), the small position and orientation deviation of CS 2 w.r.t. 1 is defined as $\mathbf{d}_2^1 = [d_{2x}^1, d_{2y}^1, d_{2z}^1]^T$ and $\boldsymbol{\theta}_2^1 = [\theta_{2x}^1, \theta_{2y}^1, \theta_{2z}^1]^T$, respectively.

The small position and orientation deviations defined by $\delta\mathbf{H}_2^1$ can be expressed as

$$\delta\mathbf{H}_2^1 = \mathbf{I}_{4\times 4} + \boldsymbol{\Delta}_2^1, \tag{2.93}$$

where $\boldsymbol{\Delta}_2^1$ is named the DTM, and is defined as

$$\boldsymbol{\Delta}_2^1 = \begin{pmatrix} \hat{\boldsymbol{\theta}}_2^1 & \mathbf{d}_2^1 \\ \mathbf{0}_{1\times 3} & 0 \end{pmatrix}, \tag{2.94}$$

where $\hat{\boldsymbol{\theta}}_2^1$ is the skew matrix of $\boldsymbol{\theta}_2^1$ and it is defined as

$$\hat{\boldsymbol{\theta}}_2^1 = \begin{pmatrix} 0 & -\theta_{2z}^1 & \theta_{2y}^1 \\ \theta_{2z}^1 & 0 & -\theta_{2x}^1 \\ -\theta_{2y}^1 & \theta_{2x}^1 & 0 \end{pmatrix}. \tag{2.95}$$

It is important to remark that any DTM can also be expressed as a differential motion vector (DMV) in a vector form. Indeed, given the DTM of CS 2 w.r.t. 1, denoted as $\mathbf{\Delta}_2^1$, a DMV is then defined as

$$\mathbf{x}_2^1 = \begin{pmatrix} \mathbf{d}_2^1 \\ \mathbf{\theta}_2^1 \end{pmatrix}. \tag{2.96}$$

References

[1] Zhou, S., Huang, Q., Shi, J.: State Space Modeling of Dimensional Variation Propagation in Multistage Machining Process Using Differential Motion Vectors. IEEE Transactions on Robotics and Automation 19, 296–309 (2003)
[2] Ogata, K.: Modern Control Engineering. Prentice Hall, New Jersey (2001)
[3] Ding, Y., Shi, J., Ceglarek, D.: Diagnosability analysis of multi-station manufacturing processes. Journal of Dynamic Systems Measurement and Control-Transactions of the Asme 124, 1–13 (2002)
[4] Zhou, S., Chen, Y., Ding, Y., Shi, J.: Diagnosability Study of Multistage Manufacturing Processes Based on Linear Mixed-Effects Models. Technometrics 45, 312–325 (2003)
[5] Zhou, S., Chen, Y., Shi, J.: Statistical estimation and testing for variation root-cause identification of multistage manufacturing processes. IEEE Transactions on Automation Science and Engineering 1, 73–83 (2004)
[6] Li, Z.G., Zhou, S., Ding, Y.: Pattern matching for variation-source identification in manufacturing processes in the presence of unstructured noise. IIE Transactions 39, 251–263 (2007)
[7] Djurdjanovic, D., Ni, J.: Dimensional errors of fixtures, locating and measurement datum features in the stream of variation modeling in machining. Journal of Manufacturing Science and Engineering-Transactions of the Asme 125, 716–730 (2003)
[8] Djurdjanovic, D., Zhu, J.: Stream of Variation Based Error Compensation Strategy in Multi-Stage Manufacturing Processes. In: ASME Conference Proceedings, vol. 42231, pp. 1223–1230 (2005)
[9] Djurdjanovic, D., Ni, J.: Online stochastic control of dimensional quality in multistation manufacturing systems. Proceedings of the Institution of Mechanical Engineers Part B-Journal of Engineering Manufacture 221, 865–880 (2007)
[10] Jiao, Y., Djurdjanovic, D.: Joint allocation of measurement points and controllable tooling machines in multistage manufacturing processes. IIE Transactions 42, 703–720 (2010)
[11] Zhong, J., Liu, J., Shi, J.: Predictive Control Considering Model Uncertainty for Variation Reduction in Multistage Assembly Processes. IEEE Transactions on Automation Science and Engineering 7, 724–735 (2010)
[12] Liu, Q., Ding, Y., Chen, Y.: Optimal coordinate sensor placements for estimating mean and variance components of variation sources. IIE Transactions 37, 877–889 (2005)
[13] Liu, J., Shi, J., Hu, J.S.: Quality-assured setup planning based on the stream-of-variation model for multi-stage machining processes. IIE Transactions 41, 323–334 (2009)

[14] Chen, Y., Ding, Y., Jin, J., Ceglarek, D.: Integration of Process-Oriented Tolerancing and Maintenance Planning in Design of Multistation Manufacturing Processes. IEEE Transactions on Automation Science and Engineering 3, 440–453 (2006)
[15] Ding, Y., Jin, J., Ceglarek, D., Shi, J.: Process-oriented tolerancing for multi-station assembly systems. IIE Transactions 37, 493–508 (2005)
[16] Abellan-Nebot, J.V., Liu, J., Romero, F.: Design of multi-station manufacturing processes by integrating the stream-of-variation model and shop-floor data. Journal of Manufacturing Systems 30, 70–82 (2011)
[17] Shi, J.: Stream of Variation Modeling and Analysis for Multistage. CRC Press Taylor and Francis Group (2007)
[18] Abellan-Nebot, J.V., Liu, J., Romero, F.: State Space Modeling of Variation Propagation in Multi-station Machining Processes Considering Machining-Induced Variations. Journal of Manufacturing Science and Engineering-Transactions of the Asme (in press, 2012)
[19] Joshi, P.H.: Jigs and Fixtures Design Manual. McGraw-Hill, New York (2003)
[20] Wang, M.Y.: Characterizations of positioning accuracy of deterministic localization of fixtures. In: IEEE International Conference on Robotics and Automation, pp. 2894–2899 (2002)
[21] Choi, J.P., Lee, S.J., Kwon, H.D.: Roundness Error Prediction with a Volumetric Error Model Including Spindle Error Motions of a Machine Tool. International Journal of Advanced Manufacturing Technology 21, 923–928 (2003)
[22] Lei, W.T., Hsu, Y.Y.: Accuracy test of five-axis CNC machine tool with 3D probe-ball. Part I: design and modeling, International Journal of Machine Tools and Manufacture 42, 1153–1162 (2002)
[23] Chen, J.S., Yuan, J., Ni, J.: Thermal Error Modelling for Real-Time Error Compensation. International Journal of Advanced Manufacturing Technology 12, 266–275 (1996)
[24] Chen, G., Yuan, J., Ni, J.: A displacement measurement approach for machine geometric error assessment. International Journal of Machine Tools and Manufacture 41, 149–161 (2001)
[25] Yang, S.H., Kim, K.H., Park, Y.K., Lee, S.G.: Error analysis and compensation for the volumetric errors of a vertical machining centre using a hemispherical helix ball bar test. International Journal of Advanced Manufacturing Technology 23, 495–500 (2004)
[26] Tseng, P.C.: A real-time thermal inaccuracy compensation method on a machining centre. International Journal of Advanced Manufacturing Technology 13, 182–190 (1997)
[27] Haitao, Z., Jianguo, Y., Jinhua, S.: Simulation of thermal behavior of a CNC machine tool spindle. International Journal of Machine Tools and Manufacture 47, 1003–1010 (2007)
[28] Kim, G.M., Kim, B.H., Chu, C.N.: Estimation of cutter deflection and form error in ball-end milling processes. International Journal of Machine Tools and Manufacture 43, 917–924 (2003)
[29] López de Lacalle, L.N., Lamikiz, A., Sanchez, J.A., Salgado, M.A.: Effects of tool deflection in the high-speed milling of inclined surfaces. International Journal of Advanced Manufacturing Technology 24, 621–631 (2004)
[30] Dow, T.A., Miller, E.L., Garrard, K.: Tool force and deflection compensation for small milling tools. Precision Engineering 28, 31–45 (2004)

[31] Gere, J.M., Goodno, B.J.: Mechanics of Materials, 7th edn. Nelson Engineering (2008)
[32] ISO 8688-1:1989, Tool-life testing in milling – Part 1: face milling
[33] Taam, W., Subbaiah, P., Liddy, J.W.: A note on multivariate capability indices. Journal of Applied Statistics 20, 339–351 (1993)
[34] Chen, H.F.: A Multivariate Process Capability Index Over A Rectangular Solid Tolerance Zone. Statistica Sinica 4, 749–758 (1994)
[35] Wang, F.K., Chen, J.C.: Capability indices using principle components analysis. Quality Engineering 11, 21–27 (1998)
[36] Wang, F.K., Du, T.C.T.: Using principle component analysis in process performance for multivariate data. Omega 28, 185–194 (2000)
[37] Pearn, W.L., Kotz, S.: Encyclopedia And Handbook of Process Capability Indices: A Comprehensive Exposition of Quality Control Measures. World Scientific Publishing Company (2006)
[38] Zhang, M., Djurdjanovic, D., Ni, J.: Diagnosibility and sensitivity analysis for multistation machining processes. International Journal of Machine Tools and Manufacture 47, 646–657 (2007)
[39] Okafor, A.C., Ertekin, Y.M.: Derivation of machine tool error models and error compensation procedure for three axes vertical machining center using rigid body kinematics. International Journal of Machine Tools and Manufacture 40, 1199–1213 (2000)
[40] Spong, M.W., Hutchinson, S., Vidyasagar, M.: Robot modeling and control. John Wiley & Sons, Inc., New York (2006)

3

Finite Element Modeling of Chip Formation in Orthogonal Machining

Amrita Priyadarshini, Surjya K. Pal, and Arun K. Samantaray

Department of Mechanical Engineering, Indian Institute of Technology,
Kharagpur, Kharagpur- 721 302, West Bengal, India
amrita@mech.iitkgp.ernet.in,
skpal@mech.iitkgp.ernet.in,
samantaray@mech.iitkgp.ernet.in

Finite element method has gained immense popularity in the area of metal cutting for providing detailed insight in to the chip formation process. This chapter presents an overview of the application of finite element method in the study of metal cutting process. The basics of both metal cutting and finite element methods, being the foremost in understanding the applicability of finite element method in metal cutting, have been discussed in brief. Few of the critical issues related to finite element modeling of orthogonal machining have been cited through various case studies. This would prove very helpful for the readers not simply because it provides basic steps for formulating FE model for machining but also focuses on the issues that should be taken care of in order to come up with accurate and reliable FE simulations.

3.1 Introduction

Metal cutting or machining is considered as one of the most important and versatile processes for imparting final shape to the preformed blocks and various manufactured products obtained from either casting or forging. Major portion of the components manufactured worldwide necessarily require machining to convert them into finished product. This is the only process in which the final shape of the product is achieved through removal of excess material in the form of chips from the given work material with the help of a cutting tool. Basic chip formation processes include turning, shaping, milling, drilling, etc., the phenomenon of chip formation in all the cases being similar at the point where the cutting edge meets the work material. During cutting, the chip is formed by deforming the work material on the surface of the job using a cutting tool. The technique by which the metal is cut or removed is complex not simply because it involves high straining and heating but it is found that conditions of operation are most varied in the machining process as compared to other manufacturing processes. Although numerous researches are being carried out in the area of metal cutting both for its obvious technical and economical importance, machining still is sometimes referred to as one of the least understood manufacturing processes because of the complexities associated with the process. Efforts are being made continuously to understand the complex mechanism of cutting in a simple and effective way.

The knowledge of basic mechanism of chip formation helps understanding the physics governing the chip formation process. This enables one to solve the problems associated with metal cutting so that the metal removal process can be carried out more efficiently and economically. A significant improvement in process efficiency may be obtained by process parameter optimization that leads to desired outputs with acceptable variations ensuring a lower cost of manufacturing. Researchers have developed various techniques and the associated studies that have contributed remarkably in the area of metal cutting. The age old technique of trial and error experiment has contributed greatly and is still widely used in metal cutting research. Various mathematical models have also been developed which form the core of the metal cutting theory. These models range from simplified analytical models to complex, computer-based models using Finite Element Method (FEM).

Metal cutting tests have been carried out for at least 150 years using a variety of cutting conditions in tremendously increasingly volume [1]. Frederick W. Taylor is regarded as the great historical figure in the field of metal cutting whose works published at the end of the nineteenth century had a great impact on manufacturing industry. In industries, the conventional way to optimize metal cutting process is to conduct experimental tests with varying cutting conditions and then analyze its effect on various outputs, e.g., cutting forces, temperatures, tool wear and surface finish based on optimization techniques such as Design of Experiment (DOE), Taguchi or Response Surface Methodology (RSM). The experimental study is as important as the theoretical study. Unfortunately, the experimental optimization approach requires large number of experiments to achieve desired accuracy, is very time consuming and expensive.

For more than half a century the analysis of the machining problems has been carried out by researchers using different analytical models such as shear plane solution and slip line field method [2]. Pioneering works to determine cutting temperature using analytical approach began in 1951 by Hahn [3] followed by Chao and Trigger [4], Leone [5], Loewan and Shaw [6], Weiner [7] and Rapier [8]. In analytical modeling, especially in earlier works, uniform plane heat source and velocity discontinuity were often assumed by considering that the chip is formed instantaneously at the shear plane. The secondary deformation zone was usually neglected and the tool chip frictional force was taken as uniform. Such simplified assumptions modify the original problem to some extent. Besides, various other important machining features such as strain hardening, temperature dependence, chip curling, etc., are neglected owing to the complexity associated with the machining process.

In case of metal cutting process, taking all the boundary conditions into consideration renders the mathematical equations that make the process so complicated that solutions to such problem seldom exist. One prospect, as done in case of analytical approach, is to make simplifying assumptions in order to ignore the difficulties and reduce the problem to one that can be solved. This, however, modifies the actual problem to great extent and leads to serious inaccuracies. Now that more and more powerful computers have emerged and are being widely used, a more viable alternative is to obtain approximate numerical solutions rather than exact form solutions. This enables one to retain the complexity of the problem on one hand, and the desired accuracy on the other. The most popular numerical technique that has evolved in recent decades for the analysis of metal cutting is FEM. FEM allows the coupled simulation of plastic and thermal process and is capable of considering numerous dependencies of physical constants on each other.

The advantages of FEM over empirical and analytical methods in machining process can be summed up as follows [9]:

- some of the difficult to measure variables, namely, stress, strain, strain rate, temperature, etc., can be obtained quantitatively, in addition to cutting forces and chip geometry.
- non-linear geometric boundaries such as the free surface of the chip can be represented and used.
- material properties can be handled as functions of strain, strain rate and temperature.
- the interaction between the chip and the tool can be modeled as sticking and sliding conditions.

Thus, FEM-based analysis provides detailed qualitative and quantitative insight in to the chip formation process that is very much required for profound understanding of the influence of machining parameters. While experimental tests and analytical models serve as the foundation of metal cutting, FEM leads to the advancement and further refinement of knowledge in the area of metal cutting.

3.2 Basics of Machining

Metal cutting process involves various independent and dependent variables. Independent variables are the input variables over which the machinist has direct control [10]. These include type of workpiece material and tool material, shape and size of workpiece material, cutting tool geometry, type of machining process, cutting parameters (speed, feed and depth of cut) and cutting fluids. The type of input parameters selected during machining process decides much about the dependent variables. The important dependent or output variables are cutting force and power, surface finish, tool wear and tool failure, size and properties of the finished product. A small change in input variables, say, cutting parameters, tool geometry and workpiece or tool material may alter the forces to great extent. Surface finish is again a function of tool geometry, tool material, workpiece material and cutting parameters. Tool wear is also a crucial aspect of machining pertaining to the economics of machining since longer tool life leads to higher productivity. Moreover, as the tool wear takes place, it changes in both geometry and size. Such a change can result in increased cutting forces which in turn increase deflection in the workpiece and may create a chatter condition. Again due to increased power consumption, there can be increased heat generation, thus accelerating the wear rate. The enormous variety in input variables leads to infinite combinations and understanding the interrelationship between these input variables and output variables again becomes an arduous task.

3.2.1 Orthogonal Cutting Model

During cutting as the tool touches the work material, severe plastic deformation occurs within the workpiece material and the excess of material starts flowing over the rake face of the cutting tool in the form of chip. The zone of plastic deformation known as shear plane lies between the chip and the undeformed work material. Based on the position of cutting edge, the cutting process can be classified as orthogonal cutting and oblique cutting. In orthogonal cutting, also known as two-dimensional cutting, the cutting edge is perpendicular to the cutting velocity vector. On the other hand, in oblique cutting, the edge is inclined with the cutting velocity by a certain angle called the inclination angle. Oblique cutting is a common type of three-dimensional cutting used in machining process. Orthogonal cutting is only a particular case of oblique cutting such that any analysis of orthogonal cutting can be applied to oblique cutting [11]. As far as practical requirements of rake and other angles are concerned, the ideal conditions of orthogonal cutting are rarely achieved. But analysis of oblique cutting is much more difficult, so focus has been mainly given on the orthogonal cutting with very few exceptions dealing with the mechanics of oblique cutting. The important theories based on orthogonal cutting model which have paved the way for the analysis of chip formation process include Merchant's model [12], Lee and Shaffer's model [13], Oxley's model [14], just to name a few.

The significant geometrical parameters involved in chip formation (as shown in Fig. 3.1) are described as follows:

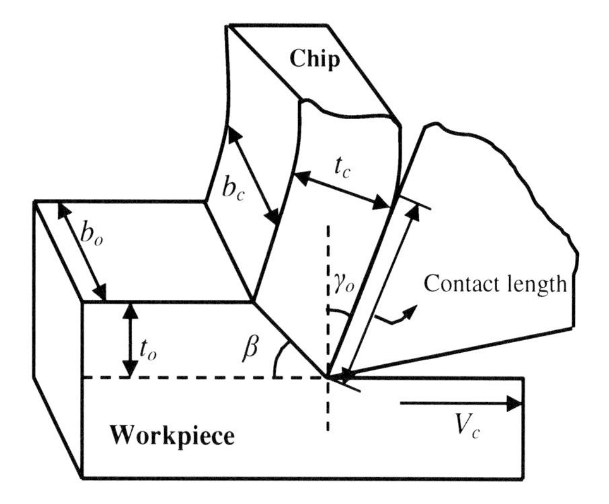

Fig. 3.1. Orthogonal cutting model

β = Shear plane angle which is defined as the angle between the shear plane (plane of separation of chip from the undeformed work material) and the cutting velocity vector

γ_0 = Rake angle of the cutting tool

ϕ_p = Principal cutting edge angle (shown in Fig. 3.2)

λ = Inclination angle

d = Depth of cut (mm)

s = Feed (mm/rev)

V_c = Cutting velocity (m/min)

t_o, b_o, l_o are thickness, width and length (mm) of the uncut chip thickness, respectively, such that

$$t_0 = \frac{s \sin \phi_p}{\cos \lambda}$$
$$b_0 = \frac{d}{\sin \phi_p} \tag{3.1}$$

For pure orthogonal cutting case, $\lambda = 0°$ and $\phi_p = 90°$ giving $t_0 = s$ and $b_0 = d$.

t_c, b_c, l_c are thickness, width and length (mm) of the deformed chip thickness, respectively.

ζ, Chip reduction coefficient is defined as the index of the degree of deformation involved in the chip formation such that

$$\zeta = \frac{t_c}{t_0} \tag{3.2}$$

The degree of chip thickening can also be expressed in terms of cutting ratio $r\left(=\frac{1}{\zeta}\right)$.

Since $t_c > t_0$, ζ is generally greater than one. Larger value of ζ signifies higher value of cutting forces that are required to carry out the machining process. The chip reduction coefficient is affected by the tool rake angle and chip-tool interfacial friction coefficient (μ). This is clear from the relation obtained from velocity analysis based on D'Alemberts' principle applied to chip motion, given as [15]:

$$\zeta = e^{\mu\left(\Pi/2 - \gamma_0\right)} \tag{3.3}$$

The ζ can be reduced considerably by using cutting tools with larger positive rake angles and by using cutting fluids.

Shear angle can be expressed in terms of ζ :

$$\tan\beta = \frac{\cos\gamma_0}{\zeta - \sin\gamma_0} \tag{3.4}$$

It is evident from Eq. (3.4) that decrease in ζ and increase in rake angle tend to increase the shear angle. This suggests that increasing values of shear angle requires lesser forces for cutting resulting in favorable machining.

3.2.2 Cutting Forces

The first scientific studies of metal cutting were conducted in about 1850, which were aimed at establishing the power requirements for various operations. A considerable amount of investigations has been directed toward the prediction and measurement of cutting forces. This is because cutting force is a result of the extreme conditions at the tool-workpiece interface and this interaction can be directly related to many other output variables such as generation of heat and consequently tool wear and quality of machined surface as well as chip formation process and the chip morphology [16, 17]. Measurement of forces becomes mandatory at certain cases say, adequate equations are not available, evaluation of effect of machining parameters cannot be done analytically and theoretical models have to be verified. Several works are available in the literature that makes use of different types of dynamometers to measure the forces. The dynamometers being commonly used nowadays for measuring forces are either strain gauge or piezoelectric type. Though piezoelectric dynamometer is highly expensive, this has almost become standard for recent experimental investigations in metal cutting due to high accuracy, reliability and consistency.

Estimation of forces acting between tool and work material is one of the vital aspects of mechanics of cutting process since it is essential for:

- Determination of the cutting power consumption
- Structural design of machine-fixture-tool system
- Study of the effect of various cutting parameters on cutting forces
- Condition monitoring of both the cutting tools and machine tools

Analysis of force during machining includes magnitude of cutting forces and their components and location of action of those forces as well as the pattern of the forces, say, static or dynamic.

A single cutting force is known to act in case of a single point cutting tool being used in turning operation which can be resolved into three components along three orthogonal directions i.e. X ,Y and Z , as shown in Fig. 3.2.

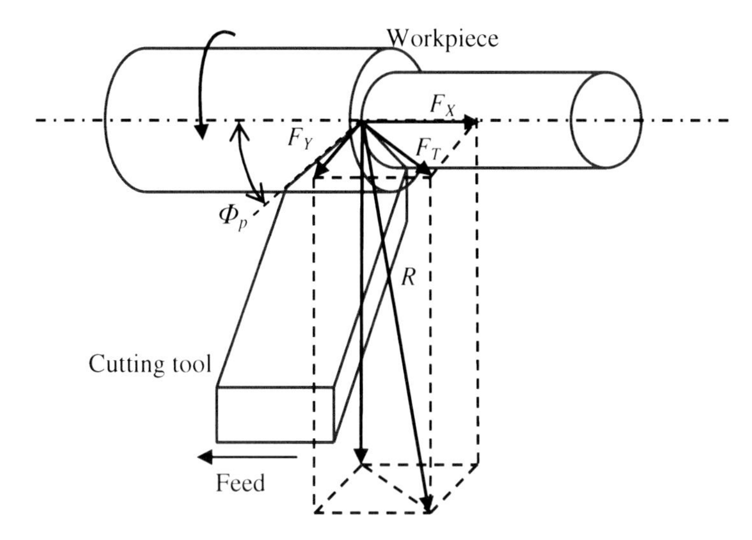

Fig. 3.2. Forces acting on a cutting tool in turning

The resultant cutting force $\overline{R}$ is resolved as:

$$\overline{R} = \overline{F}_c + \overline{F}_T \tag{3.5}$$

$$\overline{F}_T = \overline{F}_X + \overline{F}_Y \tag{3.6}$$

such that $\overline{F}_X = \overline{F}_T \sin\phi_p$ and $\overline{F}_Y = \overline{F}_T \cos\phi_p$.

The force components are described as follows:

F_c – This is the main or tangential cutting force acting in the direction of cutting velocity. This when multiplied with cutting velocity gives the value of cutting power consumption.

F_X – It is feed force acting in the direction of feed.

F_Y – This is radial force acting in the direction perpendicular to the feed force.

F_T – This is known as thrust force.

The above-mentioned forces act on the cutting tool during machining process and can be measured directly using tool force dynamometers. But there exist another set of forces acting on the chip both from the tool side and the workpiece side that cannot be measured directly. Merchant's circle diagram, one of the oldest techniques, is widely used for analytical estimation of these kind of forces for orthogonal cutting process.

Merchant's circle diagram is shown in Fig. 3.3 with schematic representation of forces acting on a portion of continuous chip passing through the shear plane at a constant speed which is in a state of equilibrium.

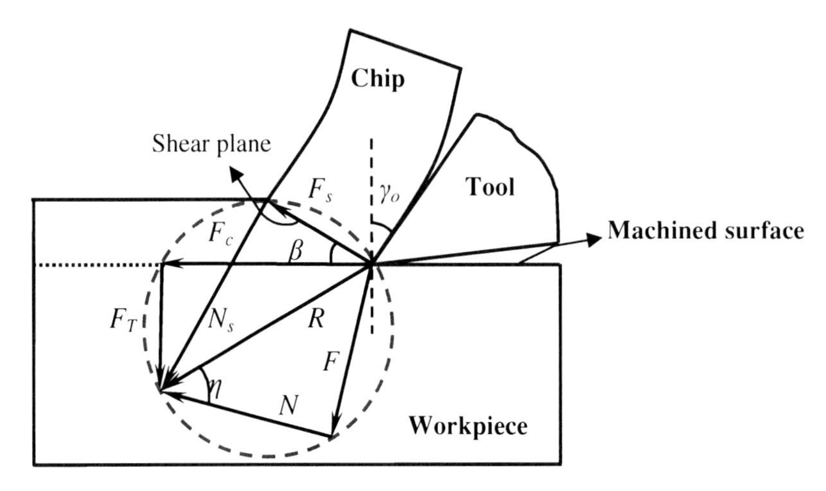

Fig. 3.3. Merchant's circle diagram for 2D orthogonal cutting

F_s and N_s are called shear force and normal force, respectively, that act on the chip from workpiece side i.e. in the shear plane. F and N are friction force at chip-tool interface and force normal rake face, respectively, that act on the chip from the tool side i.e. in the chip-tool interface. These forces can be determined as follows [10]:

$$\begin{aligned} F_s &= F_c \cos\beta - F_T \sin\beta \\ N_s &= F_T \cos\beta + F_c \sin\beta \end{aligned} \tag{3.7}$$

$$\begin{aligned} F &= F_T \cos\gamma_0 + F_c \sin\gamma_0 \\ N &= F_c \cos\gamma_0 - F_T \sin\gamma_0 \end{aligned} \tag{3.8}$$

The average coefficient of friction between chip and tool (μ) can be deduced either in terms of friction angle (η) or F and N , as given:

$$\mu = \tan\eta = \frac{F}{N} \tag{3.9}$$

3.2.3 Cutting Temperature

Heat has a critical influence on machining process. It is found that almost all of the mechanical energy is converted into heat during chip formation process that tends to increase the temperature to very high values in the cutting zone. The heating action is known to pose many of the economic and technical problems of machining. During machining heat is generated in the cutting zone from three distinct zones: Primary shear deformation zone (PSDZ), secondary shear deformation zone (SSDZ) and tertiary shear deformation zone (TSDZ) (see Fig. 3.4). Major part of the heat is generated due to severe plastic deformation in a very narrow zone called PSDZ. Further heating is generated in the SSDZ due to friction between tool and chip as well as shearing at the chip-tool interface. It is considered that both these zones together account for almost 99% of the total energy converted into heat in the cutting process [18]. A very small fraction of heat is generated at the TSDZ due to rubbing between flank of the tool and machined surface. The effect of this zone comes into play only when the flank wear develops. The TSDZ is therefore neglected when studying unworn tools [19]. The heat so generated is shared by the chip, cutting tool and the workpiece, the major portion of heat being carried away by the flowing chip. As chip is meant for disposal only, focus should be on the chip to take as much heat as possible with it so that small amount of heat gets retained in the cutting tool.

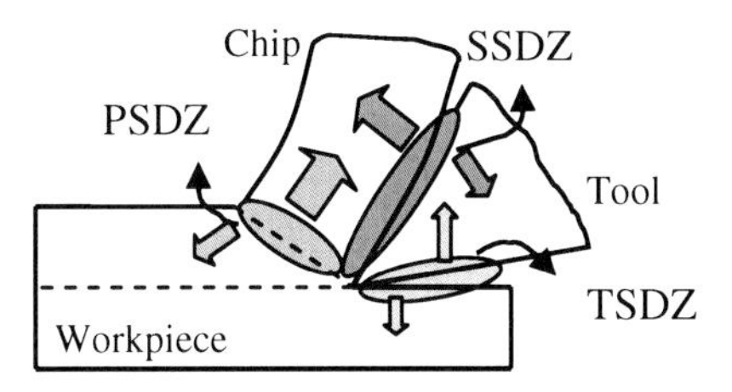

Fig. 3.4. Heat generation zones during metal cutting

Elevated temperature in the cutting zone adversely affects the strength, hardness and wear resistance of the cutting tool, thereby inducing rapid tool wear and reduced tool life. Temperature rise in workpiece material may cause dimensional inaccuracy of the machined surface and can also damage the surface properties of the machined component by oxidation, rapid corrosion, burning, etc. Thus, estimation of cutting temperature is a crucial aspect in the study of metal cutting.

Cutting temperatures are more difficult to measure accurately than cutting forces. No simple analog to the cutting force dynamometer exists for measuring the cutting temperatures; rather numerous methods have been found in the literature to experimentally measure the machining temperature [20]. These methods are thermocouples, radiation methods, metallographic techniques and application of thermal paints, fine powders and Physical Vapor Deposition (PVD) coatings [21]. A particular method generally gives only limited information on the complete temperature distribution.

3.2.4 Chip Morphology

Chip formation and its morphology are the key areas in the study of machining process that provide significant information on the cutting process itself. The process variables such as cutting force, temperature, tool wear, machining power, friction between tool-chip interface and surface finish are greatly affected by the chip formation process and chip morphology. Chip is formed due to deformation of the metal lying ahead of the cutting tool tip by shearing process. The extent of deformation that the work piece material undergoes determines the nature or type of chip produced. The extent of deformation of chips again depends upon cutting tool geometry (positive or negative rake angle), workpiece material (brittle or ductile), cutting conditions (speed, feed and depth cut), machining environment (dry or wet machining) [10].

The main chip morphologies observed in cutting process are described briefly as follows:

Discontinuous chip: These chips are small segments that adhere loosely to each other. The phenomenon can be attributed to the repeated fracturing that limits the amount of deformation the chip undergoes. Hard and brittle materials like gray cast iron, bronze, brass when machined with larger feed and negative rake cutting tool in the absence of cutting fluid produce discontinuous chips.

Continuous chip: Continuous chips are in the form of long coil. These are formed by the continuous plastic deformation of material without fracture ahead of the cutting edge of the tool resulting in a smooth flow of chip up the tool face. Ductile material when machined under low feed and high cutting speed, generally, in the presence of cutting fluid produces such kind of chips.

Continuous chip with built-up edge: This kind of chip is not as smooth as that of continuous chip and affects the surface finish adversely. Built-up edge is mostly found when ductile materials are machined under larger feed and lower cutting speed with inadequate or no cutting fluid.

Cyclic or serrated chips: These chips are continuous, but, possess a saw-tooth appearance that is produced by a cyclical chip formation of alternating high shear strain followed by low shear strain. Machining of certain difficult to machine metals such as titanium alloys, nickel-base super alloys and austenitic stainless steels are generally characterized by formation of segmented chip formation [22].

3.3 Basics of FEM

The considerable rise in computer processing speeds makes it possible to incorporate the FEM for analyzing various aspects of metal cutting process, thus overcoming most of the restrictive assumptions associated with analytical models. Many investigators have adopted FEM to gain better understanding of machining process. One of the first FE models developed for metal cutting process was by Klamecki in the year 1973 [23]. Since then many works have been found to use FEM for gaining better understanding of the machining process. Numerous FE codes such as DEFORM, FORGE2, ABAQUS/Standard, ABAQUS/Explicit, Ansys/LS-DYNA, MSC Marc and Thirdwave AdvantEdge have come up that are being used by the researchers, thus giving results closer to the experimental ones. However, it should be noted that the accuracy of FE modeling is determined by how adequately the characterization of selected input parameters reflects the deformation behavior undergoing during the chip formation in the actual practice.

3.3.1 Generalized Steps in FEM

The finite element method is a numerical method seeking an approximated solution to a wide variety of engineering problems. The engineering problems are basically boundary value problems (or field problems) containing one or more dependent variables that must satisfy a differential equation everywhere within a known domain and satisfy specific conditions on the boundary of the domain. The field is the domain of interest, often representing a physical structure while the dependent variables of interest are called field variables. The specified values of the field variables on the boundaries of the field are called boundary conditions. The field variables may include displacement in solid mechanics problem, temperature in heat flow problem, velocity in fluid flow problem and so on. In formulating an FE model for a physical problem, few basic steps are required to be followed that are common to all kinds of analysis whether structural, heat flow or fluid flow problems. The basic concepts, procedures and formulations of FEM can be found in many available books [24–30].

Discretization: The idea of FEM is to approximate the problem domain by replacing it with a finite number of distinct elements and solving them separately. This is done by subdividing the domain into smaller units of simpler geometry called finite elements. Such a process is called discretization of domain [25]. The number of elements used to mesh the domain (mesh density) depends upon the type of element chosen, accuracy deserved and computer resource available. The elements selected must be small enough to give practical results and yet large enough to reduce computational effort.

Interpolation functions: The interpolation function or shape function is in the form of polynomial which is defined within an element using the values of field variables at the nodes. For problems involving curved boundaries, the construction of shape functions that satisfy consistency requirements for higher order elements becomes increasingly complicated. Moreover, exact evaluation of the integrals that appear in the expressions of the element stiffness matrix is not always feasible because of the algebraic complexity. These two obstacles can be surmounted through the concepts of isoparametric elements and numerical integration, respectively. The numerical integration or numerical quadrature, the most popular of which is Gaussian (Gauss–Legendre) quadrature involves approximation of the integrand by a polynomial of sufficient degree because the integral of a polynomial can be evaluated exactly [30].

***Derivation of element stiffness equation*:** Derivation of basic equations for a finite element is based on governing equations and constitutive laws. This preferably involves weak form of system equation which is often an integral form and requires a weaker requirement on the field variables and the integral operation. A formulation based on weak form usually leads to a set of algebraic system equations when the problem domain, no matter how complex, is discretized into smaller elements [29]. There are three different methods for obtaining weak form, namely, direct method (for simple problems), variational method and weighted residual method [30]. The variational method is generally based on either principle of virtual work, principle of minimum potential energy or Castigliano's theorem to derive the element equation [25]. Whereas, method of weighted residual (MWR) uses trial functions to minimize the error over the problem domain [28]. MWR is particularly useful when functional such as potential energy is not readily available. Galerkin method is one such technique which is widely used for its simplicity and easy adaptability in FEM. The constitutive equations are employed which are nothing but the stress-strain laws giving much of the information about the material behavior. Finally, the element stiffness matrix is derived by substituting the interpolation functions into the weak form.

***Assembly of elements*:** The individual element equations are combined together using a method of superposition called direct stiffness method to obtain the global stiffness matrix.

***Boundary conditions*:** After the global equations are ready, necessary boundary conditions are imposed to the global equations which lead to a set of simultaneous equations

***Solution of system equations*:** The system equations are often solved by using direct method such as Guass elimination method and LU decomposition method for small systems. While iterative methods namely, Gauss–Jacobi method, Gauss-Siedel method, generalized conjugate residual method, etc. are generally employed for large systems. For transient problems (highly dynamic and time-dependent problems), another loop is required meant for time stepping. The widely used method is called direct integration method such as explicit and implicit methods [29].

3.3.2 Modeling Techniques

With the emergence of more and more potential computers as well as powerful general purpose FE software packages, FEM is becoming a popular tool in the analysis of various kinds of manufacturing processes. It is true that there have been remarkable developments in both computer hardware and software but proper background in FEM is a must otherwise using FEM packages for analysis would simply appear as a black box. It is the task of the researcher to make use of such resources tactfully and judiciously to come up with favorable and meaningful results. Thus, in order to develop a good FE model, few critical issues which may be referred as modeling techniques must be taken care of. This not only improves the computational efficiency but also ensures the reliability and accuracy of the obtained results [29].

3.3.2.1 Type of Approach

The selection of an appropriate formulation or approach is of prime importance especially for problems involving nonlinearity and large deformations. There are mainly three ways to formulate the problems involving the motion of deformable materials, given as:

- Lagrangian approach
- Eulerian approach
- Arbitrary Lagrangian Eulerian (ALE) approach

The formulations are distinguished from each other by mesh descriptions, kinetics and kinematics description [31].

Lagrangian and Eulerian formulations are the two classical approaches for description of motion in continuum mechanics that differ mostly in terms of the behavior of nodes. In Lagrangian meshes, the nodes and elements move with the material i.e. the nodes are coincident with material points. But if the mesh is Eulerian, the Eulerian coordinates of nodes are fixed, i.e. the nodes are coincident with spatial points.

Lagrangian approach: It can be seen that in Lagrangian approach, both the material points and nodes change position as the body deforms such that position of the material points relative to the nodes remains fixed. Boundary nodes tend to remain along the element edges throughout the evolution of the problem which in turn simplifies the imposition of boundary conditions and interface conditions. Quadrature points also remain coincident with the material points which signify constitutive equations are always evaluated at the same material points, thus leading to a straightforward treatment of constitutive equations. For these reasons, Lagrangian approach is widely used in solid mechanics. However, this approach may encounter severe mesh distortion because the mesh deforms with the material. This becomes highly critical when the problems involve large deformation as this may affect the performance of FE simulations greatly.

Lagrangian formulation can be again of two types, namely, total Lagrangian formulations and updated Lagrangian formulations [31]. In the total Lagrangian formulation, the weak form involves integrals over the initial (reference) configuration and derivatives are taken with respect to the material coordinates. In the updated Lagrangian formulation, the weak form involves integrals over the current i.e. deformed configuration and the derivatives are with respect to the spatial (Eulerian) coordinates. It has been suggested that the updated and total Lagrangian formulations are two representations of the same mechanical behavior and each can be transformed to the other, thus, implying that the choice of formulation is purely a matter of convenience.

Many of the researchers have used Lagrangian formulation in metal cutting models since it is easy to implement and is computationally efficient because of easy tracking of the material boundaries. But difficulties arise when elements get highly distorted as the mesh deforms with the deformation of the material in front of the tool tip. Since element distortion degrades its element accuracy, the degree of deformation that can be simulated with a Lagrangian mesh is restricted. In order to overcome this, either chip separation criteria or remeshing techniques have been used in the literature to simulate the cutting action at the cutting zone [32].

Eulerian approach: In Eulerian approach, the reference frame is fixed in space that allows for the material to flow through the grid [33]. As the mesh is fixed in space, the numerical difficulties associated with the distortion of elements are eliminated. This approach permits simulation of machining process without the use of any mesh separation criterion. The main drawback of Eulerian formulation is that it is unable to model free boundaries since boundary nodes may not be coincident with the element edge. This may only be used when boundaries of the deformed material are known a priori. Hence, dimension of the chip must be specified in advance to produce a predictive model for chip formation in an Eulerian framework [34]. Furthermore, handling and updating of constitutive equations become complex because of the fact that material flows through the mesh i.e. element quadrature points may not be coincident with material points. These drawbacks often limit the application of Eulerian formulation in modeling the metal cutting process.

Arbitrary Lagrangian Eulerian approach: It is known that Eulerian and Lagrangian approaches have their own advantages and disadvantages. As the name suggests, in ALE formulation the nodes can be programmed to move arbitrarily in such a way that advantages of both Lagrangian and Eulerian formulations can be exploited and combined into one while minimizing their disadvantages. The idea is that the boundary nodes are moved to remain on the material boundaries, while the interior nodes are moved to minimize mesh distortion. While simulating metal cutting process in ALE framework, ALE reduces to a Lagrangian form on free boundaries while maintains an Eulerian form at locations where significant deformations occur, as found during the deformation of material in front of the tool tip. This approach not only enables to reduce mesh distortion without the need of remeshing but also eliminates the need of a priori assumption about the shape of the chip [35].

3.3.2.2 Geometry Modeling

Creating the geometry of the problem domain is the first and foremost step in any analysis. The actual geometries are usually complex. The aim should not be simply to model the exact geometry as that of the actual one, instead focus should be made on how and where to reduce the complexity of the geometry to manageable one such that the problem can be solved and analyzed efficiently without affecting the nature of problem and accuracy of results much. Hence, proper understanding of the mechanics of the problem is certainly required to analyze the problem and examine the geometry of the problem domain. It is generally aimed to make use of 2 D elements rather than 3 D elements since this can drastically reduce the number of degrees of freedom (DOFs).

3.3.2.3 Computation Time

Computation time which is nothing but the time that CPU takes for solving finite element equation is affected markedly by the total number of DOFs in the FE equation as shown [29]:

$$t_{comp} \propto DOF^{\beta} \tag{3.10}$$

Here, β is a constant which generally lies in the range of 2-3 depending on the type of solver used and the structure or bandwidth of the stiffness matrix. A smaller bandwidth yields smaller value of β resulting in faster computation. The Eq. (3.10) suggests that a finer mesh with larger number of DOFs increases the computation time exponentially. Thus, it is always preferred to create FE model with elements possessing lower dimensions so that number of DOFs are reduced as far as possible. In addition, meshing should be done in such a way that critical areas possess finer meshing while others possess coarse meshing. This is one way of reducing the computation time without hampering the accuracy of results.

3.3.2.4 Mesh Attributes

Choice of element type: Based on the shape, elements can be classified as one dimensional (line or beam), two-dimensional or plane (triangular and quadrilateral) and three-dimensional (tetrahedral and hexahedral) elements. Each of these elements can again be either in their simplest forms i.e. linear or higher order forms such as quadratic or cubic depending upon the number of nodes an element possess. As discussed earlier, 2D elements are generally preferred over 3D elements as far as cost of computation is concerned. The linear triangular element was the first type of element developed for defining 2D geometries, the formulation of which is simplest of all. However, quadrilateral elements are mostly preferred nowadays for 2D solids [36]. This is not simply because quadrilateral element contains one node more than that of triangular elements but also gradients of quadrilateral elements are linear functions of the coordinate directions compared to the gradient being constant in triangular elements. Besides, the number of elements is reduced to half that of a mesh consisting of triangular elements for the same number of nodes. In nearly all instances, a mesh consisting of quadrilateral elements is sufficient and usually more accurate than that of triangular elements. Besides for higher order elements, more complex representations are achieved that are increasingly accurate from an approximation point of view. However, care must be taken to evaluate the benefits of increased accuracy against the computational cost associated with more sophisticated elements.

Mesh density: A mesh of varying density is generally preferred. The critical areas need to be finely meshed while rest of the areas may contain coarse mesh. The control of mesh density can be performed by placing the nodes according to the given geometry and required degree of accuracy in FEM packages.

Element distortion: It is not always possible to have regularly shaped elements for irregular geometries. Irregular or distorted elements are acceptable in the FEM, but there are limitations, and one needs to control the degree of element distortion in the process of mesh generation. If these are overlooked then there are chances of termination of simulation program because of element distortion. Few possible forms of distortion error of elements are aspect ratio distortion, angular distortion, curvature distortion and so on. Aspect ratio error arises when the elongation of element occurs beyond a particular limit disturbing the aspect ratio of the element. Angular distortion occurs when the included angle between edges of the element approaches either 0° or 180°. Similarly, when the straight edges of the element are distorted into curves, curvature distortion is most likely to occur [29].

3.3.2.5 Locking and Hourglassing

Approximating the problem domain using simpler elements is relatively easy, but, very often formulation of these elements can give inaccurate results due to some abnormalities namely, shear locking, volumetric locking and hourglassing. Locking, in general, is the phenomenon of exhibiting an undesirable stiff response to deformation in finite elements. This kind of problem may arise when the element interpolation functions are unable to approximate accurately the strain distribution in the solid or when the interpolation functions for the displacement and their derivatives are not consistent. Shear locking predominantly occurs in linear elements with full integration and results in underestimated displacements due to undesirable stiffness in the deformation behavior. This problem can be alleviated by using reduced integration scheme. Volumetric locking is exhibited by incompressible materials or materials showing nearly incompressible behavior resulting in an overly stiff response [37].

It is known that the stress comprises of two components: deviatoric (distortional) and volumetric (dilatational). The volumetric component is a function of bulk modulus and volumetric strain. The bulk modulus is given by:

$$K = \frac{E}{3 - 6\nu} \tag{3.11}$$

where E is Young's modulus and ν is Poisson's ratio. It is understood that when $\nu = 0$, there is no volumetric locking at all. But in the incompressibility limit, when $\nu \to 0.5$ then $\lim_{\nu \to 0.5} K = \infty$ resulting in overly stiff response. In this limit, the finite element displacements tend to zero and thus, the volumetric locking prevails [37]. Hybrid Elements are often used to overcome volumetric locking [38]. The idea is to include the hydrostatic stress distribution as an additional unknown variable, which must be computed at the same time as the displacement field. This allows the stiff terms to be removed from the system of finite element equations. Further details on hybrid elements can be found in existing literatures [39, 40].

It has been also observed that fully integrated elements have higher tendency of volumetric locking. This is because at each of the integration points the volume remains nearly constant and when the integration points are more, as in case of full integration scheme, it results in overconstraining of the kinematically admissible displacement field. Reduced integration scheme is a possible way to avoid locking wherein the element stiffness is integrated using fewer integration points than that are required for full integration scheme [24]. Reduced integration is an effective measure for resolving locking in elements such as quadratic quadrilateral and brick effectively but not in elements such as 4 noded quadrilateral or 8 noded brick elements. The error occurs because the stiffness matrix is nearly singular which implies that the system of equations includes a weakly constrained deformation mode. This phenomenon is known as hourglassing which results in wildly varying displacement field but correcting stress and strain fields. This can be cured either by employing selectively reduced integration or by adding an artificial stiffness to the element that acts to constrain the hourglass mode (Reduced Integration with Hourglass Control) [37].

3.3.2.6 Solution Methods

Time integration methods are employed for time stepping to solve the transient dynamic system of equations. There are two main types of time integration methods [29, 31]:

- Implicit
- Explicit

Both the procedures have their own benefits and limitations from solution point of view. Selection of an appropriate solution method should be done carefully according to the nature of the problem. The implicit method is generally suitable for linear transient problems. In an implicit dynamic analysis, a global stiffness matrix is assembled and the integration operator matrix must be inverted and a set of non-linear equilibrium equations must be solved at each time increment. Suitable implicit integrator, which is unconditionally stable, is employed for non-linear formulations and the time increment is adjusted as the solution progresses in order to obtain a stable, yet time-efficient solution. However, there are situations where implicit method encounters problems finding a converged solution. In analyses such as machining which involves complex contact problems, this algorithm is less efficient due to the simultaneous solving of the equation system in every increment. In such situations, explicit time integration method proves to be robust. The simulation program seldom aborts due to failure of the numerical algorithm since the global mass and stiffness matrices need not be formed and inverted. However, the method is conditionally stable i.e. if the time step exceeds critical time step, the solution may grow unboundedly and give erroneous results. The critical time step for a mesh is given by [31]:

$$\Delta t_{\text{crit}} \leq \min\left(L_e / c_e\right) \tag{3.12}$$

where L_e is the characteristic length of the element and c_e is the current wave speed of an element.

3.3.2.7 Mesh Adaptivity

Generally speaking, mesh adaptivity can be described as the capability of the mesh to adapt according to the nature of problem in order to maintain a high quality mesh throughout the analysis. The need of such technique becomes more evident while simulating problems that involve heavy mesh distortion such as in the case of cutting process undergoing large deformations. This is well illustrated by citing an example which considers cutting process as a purely large deformation problem comprising a rectangular block as workpiece and a portion of cutting tool (see Fig. 3.5). Figure 3.5 (a) shows the initial mesh configuration of the workpiece while Fig. 3.5 (b) shows the deformed mesh configuration as tool comes in contact with the workpiece. Since no mesh adaptivity technique is employed, elements get highly distorted near the tool tip as highlighted in Fig. 3.6 (b) that leads to termination of the simulation program.

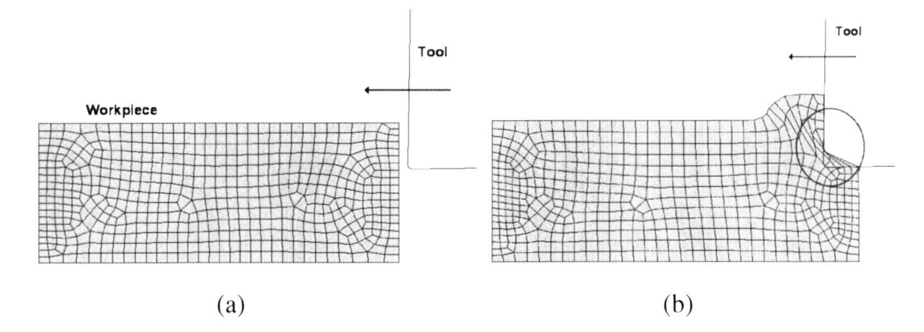

Fig. 3.5. (a) Initial mesh configuration and (b) Deformed mesh configuration

Figures 3.6 (a) and 3.6 (b) show two possible mesh adaptivity techniques that are implemented to control the mesh distortion near the tool tip.

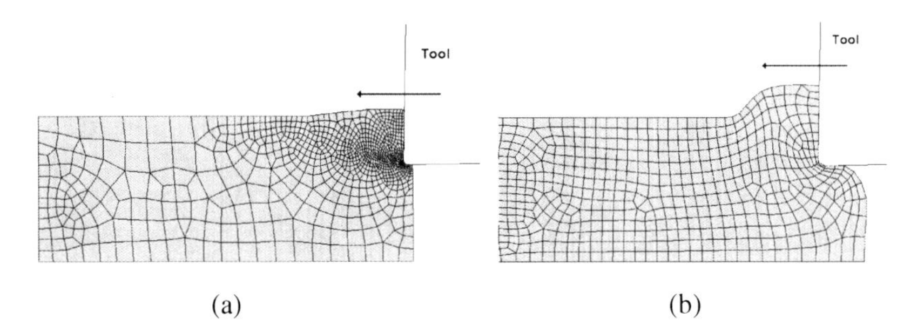

Fig. 3.6. (a) Mesh refinement and (b) Relocation of nodes near the tool tip

In the first case (Fig. 3.6 (a)), it is noted that refinement of mesh takes place at critical areas i.e. finer mesh in the region closer to the tool tip wherein the size of elements is adjusted based on selected error indicators and loading history. Here, the number of elements and connectivity of the nodes are changed. This type of technique is often known as h–adaptivity [41]. In Fig. 3.6 (b), mesh distortion is controlled by relocation of nodes without altering the number of elements and connectivity of nodes. This approach is often termed as r–adaptivity [41]. Since the number of elements remains the same, this approach is computationally less expensive as compared to h–adaptivity. However, both h-refinement and r–refinement are widely implemented in simulating metal cutting process, the former being used in Lagrangian framework while the latter being employed in ALE framework.

3.4 Brief History of FEM in Machining

Pioneering works in FEM include works by Klamecki [23], Tay et al. [42], Muraka et al. [43]. Klamecki [23] was the first one to apply FEM in the analysis of metal cutting process, whereas Tay et al. [42] were the first to calculate the temperature field distribution in orthogonal machining by the application of FEM. Muraka et al. [43] investigated the effect of process variables such as cutting speeds, flank wear, coolant, etc., on the temperature distribution using finite element method. Morikawa et al. [44] developed a rigid-plastic finite element model to examine the effect of the tool edge radius to depth of cut ratio on the micro machining process. Kim and Sin [45] developed a thermo-visco-plastic cutting model by using finite element method to analyze the mechanics of steady state orthogonal cutting process. Liu and Guo [46], on the other hand, proposed a thermoelastic-viscoplastic FE model to investigate the friction at the tool chip interface. Cerreti et al. [47] used DEFORM 2D to estimate the chip and tool temperatures for continuous and segmented chip. Li et al. [48] employed Johnson–Cook's model as constitutive equation of the workpiece material for qualitative understanding of crater-wear from the calculated temperature distribution by using commercial FE code ABAQUS. Arazzola et al. [49] incorporated ALE formulation along with Johnson Cook model as the material model for the workpiece using ABAQUS/Explicit. In many of the works, researchers have attempted to make comparative study of the effect of various material models such as Johnson-Cook model, Zerilli–Armstrong model, Power law rate-dependent model, Oxley model, Litonski–Batra and many more on the predicted results [50–52]. Adibi–Sedeh et al. [51] found that Johnson–Cook model predicts chip thickness better than that of other models when compared with the experimental ones. In few of the works, comparison of some selected friction models has been presented showing their effect on FE simulations for orthogonal cutting of various work materials [53–55].

Mabrouki et al. [56] compared the results obtained from ABAQUS/Explicit and Thirdwave Systems' AdvantEdge with those obtained experimentally. Coelho et al. [57] developed an ALE–FEM model to help understand the wear results found in case of coated PCBN tools for turning hardened AISI 4340. Lorentzon and Jarvstrat [58] investigated the impact of different wear models and friction models on the parameters affecting the wear process for cemented carbide tools while machining alloy 718, using the commercial software MSC Marc. Davim et al. [59] made a comparative study between the performances of PCD (polycrystalline diamond) and K10 (cemented carbide) tools while machining aluminum alloys using AdvantEdge software. Attanasio et al. [60] presented 3D numerical predictions of tool wear based on modified Takeyama and Murata wear model using SFTC Deform 3D software.

3.5 Formulation of Two-Dimensional FE Model for Machining

This section attempts to make understand the basic methodology and the related theory for developing the FE model of orthogonal cutting process. Furthermore, a brief description of various modules contained in ABAQUS program has been presented so that readers become conversant with ABAQUS. It is agreed that ABAQUS software is provided with extensive and excellent manuals [61–63], but our aim is to abridge those details in a very lucid manner that may serve as a quick guide to a beginner. This would facilitate the readers to have a much clearer understanding of the finite element concepts being implemented in the finite element software for simulating machining operation.

3.5.1 Formulation Steps

The present study focuses on developing a 2D FE model of chip formation process based on orthogonal cutting conditions. Although two–dimensional analysis is a restrictive approach from practical point of view, it reduces the computational time considerably and provides satisfactory results regarding the details of the chip formation. A plane strain condition was assumed because the feed value is generally very less as compared to the depth of cut. A fully coupled thermal-stress analysis module of finite element software ABAQUS/Explicit version 6.7–4 has been employed to perform the study of chip formation process. This makes use of elements possessing both temperature and displacement degrees of freedom. It is necessary because metal cutting is considered as a coupled thermo-mechanical process, wherein, the mechanical and thermal solutions affect each other strongly.

3.5.1.1 Geometric Model, Meshing and Boundary Conditions

The 2D model comprises a rectangular block representing the workpiece and only a portion of cutting tool which participates in the cutting. The nose radius is neglected for the simplicity of the problem. Moreover, there is hardly any effect of nose radius once a steady state is reached in cutting. The cutting tool includes the following geometrical angles: inclination angle $\chi = 90^{\circ}$, rake angle $\gamma = -6^{\circ}$ and flank angle $\alpha = 5^{\circ}$. An intermediate layer of elements known as damage zone (highlighted region in Fig. 3.7) has been considered in the workpiece block such that width of the upper surface is equal to the undeformed chip thickness or in other words feed in case of orthogonal cutting process. As mechanical boundary conditions, bottom of work piece is fixed in Y direction and left vertical edge of workpiece is fixed in X direction. The former not only constrains the movement of workpiece in the Y direction but also aids in calculating feed force during machining while the latter, not only constrains the movement of workpiece in the X direction but also aids in calculating cutting force during machining. The reaction force components when added at all the constrained nodes of the left vertical edge of the workpiece in X direction give the cutting force values while at bottom edge of the workpiece in Y direction give the feed force. Tool is given the cutting velocity in negative X direction and top edge of the tool is constrained in Y direction.

Similarly, as thermal boundary conditions the tool and the workpiece are initially considered at the room temperature. Heat transfer from the chip surface to cutting tool is allowed by defining the conductive heat transfer coefficient (h), given as:

$$-k\frac{\partial T}{\partial n} = h\left(T_o - T\right) \tag{3.13}$$

where k is thermal conductivity and T_o is the ambient temperature.

The geometric model taken into consideration is shown in Fig. 3.7.

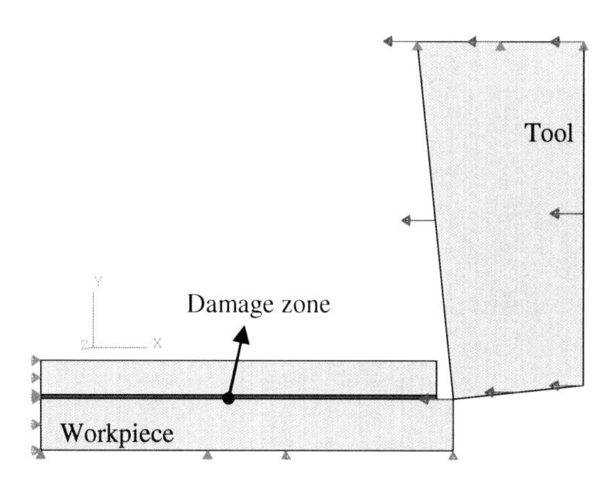

Fig. 3.7. 2D geometric model

Damage zone is a sacrificial layer of elements that defines the path of separation between chip surface and the machined surface which is going to take place as the tool progresses. In actual practice, these two surfaces should be same but such assumption has been taken only for the modeling purpose as a chip separation criterion. The choice of the height of the designated damage zone is purely based on computational efficiency. Generally, a very small value (say, 10–30 μm) which is computationally acceptable has been taken as the width of the damage zone.

Four-node plane strain bilinear quadrilateral (CPE4RT) elements with reduced integration scheme and hourglass control are used for the discretization for both the workpiece and the cutting tool. The workpiece is meshed with CPE4RT-type elements by unstructured grid generation which utilizes advancing front algorithm in ABAQUS/Explicit.

3.5.1.2 Governing Equations

The governing equations of a body undergoing deformation consists of two sets of equations, namely, the conservations laws of physics and the constitutive equations. The conservation laws can be applied to body of any material. But to distinguish between different materials undergoing deformation of varied degrees, one needs the constitutive equations.

Conservation laws: The mass, momentum and energy equations governing the continuum are given as follows:

$$\dot{\rho} + \rho \operatorname{div} \vec{v} = 0 \tag{3.14}$$

$$\rho \dot{\vec{v}} = \vec{f} + \operatorname{div} \sigma \tag{3.15}$$

$$\rho \dot{e} = \sigma : \mathbf{D} - \operatorname{div} \vec{q} + r \tag{3.16}$$

where ρ is the mass density, $\vec{v}$ material velocity, $\vec{f}$ body forces, σ Cauchy stress tensor, e specific internal energy, D strain rate tensor, $\vec{q}$ heat flux vector and r is body heat generation rate. The superposed '•' denotes material derivative in a Lagrangian description and ':' denotes contraction of a pair of repeated indices which appear in the same order such as, $\mathrm{A{:}B} = A_{ij} B_{ij}$.

The basic idea of using FEM is to discretize the above equations and then to seek a solution to the momentum equation.

Workpiece material constitutive equations: Constitutive equations describe the thermo-mechanical properties of a material undergoing deformation. Based on the simplicity or complexity of the material behavior, there could be one constitutive equation or a set of constitutive equations that relate the deformation in the body to the stresses. The elastic response can be described by Hook's law. Whereas, the constitutive equations of plasticity deal with yield criterion, flow rule and strain hardening rule. The yield criterion describes the stress state when yielding occurs, the flow rule defines increment of plastic strain when yielding occurs, and the hardening rule describes how the material is strain hardened as the plastic strain increases. For large deformation problems, as in case of machining, plasticity models based on von Mises yield criterion and Prandtl–Reuss flow rule are generally used to define the isotropic yielding and hardening [64].

It is known that during cutting process, the workpiece material is generally subjected to high levels of, strain, strain rate and temperature which significantly influences flow stress. Thus, accurate and reliable rate-dependent constitutive models are required that can reflect such phenomenon effectively. Johnson–Cook constitutive equation is one such model that considers the flow stress behavior of the work materials as multiplicative effects of strain, strain rate and temperature, given as follows [65]:

$$\bar{\sigma} = \left(A + B\left(\bar{\varepsilon}^{\mathrm{p}}\right)^{n}\right)\left(1 + C\ \ln\dot{\bar{\varepsilon}}^{\mathrm{p}*}\right)\left(1 - \left(\frac{T - T_{\mathrm{room}}}{T_{\mathrm{m}} - T_{\mathrm{room}}}\right)^{m}\right) \tag{3.17}$$

$$\dot{\bar{\varepsilon}}^{*} = \frac{\dot{\bar{\varepsilon}}^{\mathrm{p}}}{\dot{\bar{\varepsilon}}_{0}^{\mathrm{p}}} \text{ for } \dot{\bar{\varepsilon}}_{0}^{\mathrm{p}} = 1\ \mathrm{s}^{-1} \text{ and } T^{*} = \left(\frac{T - T_{\mathrm{room}}}{T_{\mathrm{m}} - T_{\mathrm{room}}}\right) \tag{3.18}$$

where T_{room} is the room temperature taken as 25 ºC, T_{melt} is the melting temperature of the workpiece, A is the initial yield stress (MPa), B the hardening modulus, n the work-hardening exponent, C the strain rate dependency coefficient (MPa), and m the thermal softening coefficient. A, B, C, n and m used in the model are actually the empirical material constants that can be found from different mechanical tests. Johnson-Cook model has been found to be one of the most suitable one for representing the flow stress behavior of work material undergoing cutting. Besides, it is also considered numerically robust as most of the variables are readily acceptable to the computer codes. This has been widely used in modeling of machining process by various researchers. [52, 66–68].

3.5.1.3 Chip Separation Criterion

A damage model should be incorporated in the damage zone along with the material as a chip separation criterion in order to simulate the movement of the cutting tool into workpiece without any mesh distortion near the tool tip. Specification of damage model includes a material response (undamaged), damage initiation criterion, damage evolution and choice of element deletion.

Damage initiation criterion is referred to as the material state at the onset of damage. In the present case, Johnson–Cook damage initiation criterion has been employed. This model makes use of the damage parameter ω_D defined as the sum of the ratio of the increments in the equivalent plastic strain $\Delta\bar{\varepsilon}^{\mathrm{p}}$ to the fracture strain ε^{f}, given as follows [61].

$$\omega_{\mathrm{D}} = \sum \frac{\Delta\bar{\varepsilon}^{\mathrm{p}}}{\varepsilon^{\mathrm{f}}} \tag{3.19}$$

The fracture strain ε^{f} is of the form as follows [69]:

$$\bar{\varepsilon}_{f} = \left[D_{1} + D_{2}\exp\left(D_{3}\sigma^{*}\right)\right] \times \left[1 + D_{4}\ln\dot{\bar{\varepsilon}}^{\mathrm{p}*}\right] \times \left[1 + D_{5}T^{*}\right] \tag{3.20}$$

$$\sigma^{*} = \frac{-\mathrm{P}}{\bar{\sigma}} \tag{3.21}$$

where P is the hydrostatic pressure and D_1 to D_5 are failure parameters determined experimentally. The damage constants can be determined by performing several experiments such as tensile test, torsion test and Hopkinson bar test. On the basis of results obtained, a graph is generally plotted between equivalent plastic strain at fracture and pressure stress ratio such that the constants related to pressure i.e. D_1, D_2, and D3 can be found out at $T^* = 0$ and $\dot{\bar{\varepsilon}}^p = 0$. Similarly, D_4 and D_5 can be obtained by plotting another graph between T^* and ratio of fracture strain at different strain rates. The damage initiation criterion is met when ω_D (Eq. 3.20) reaches one [61].

Once the element satisfies the damage initiation criterion, it is assumed that progressive degradation of the material stiffness occurs, leading to material failure based on the damage evolution. At any given time during the analysis, the stress tensor in the material is given by:

$$\sigma = (1 - D)\bar{\sigma} \tag{3.22}$$

where $\bar{\sigma}$ is the effective (undamaged) stress tensor computed in the current increment. When overall damage variable D reaches a value 1, it indicates that the material has lost its load carrying capacity completely. At this point, failure occurs and the concerned elements are removed from the computation.

The effective plastic displacement ($\bar{u}^p$), after the damage initiation criterion is met can be defined with the evolution law as follows:

$$\bar{u}^p = L_e \bar{\varepsilon}^p \tag{3.23}$$

where L_e is the characteristic length of the element and $\bar{\varepsilon}^p$ equivalent plastic strain.

When a linear evolution of the damage variable with plastic displacement is assumed, the variable increases as per the following [61]:

$$D = \frac{L_e \bar{\varepsilon}^p}{\bar{u}_f^{\,p}} = \frac{\bar{u}^p}{\bar{u}_f^{\,p}} \qquad \bar{u}_f^{\,p} = \frac{2G_f}{\sigma_{y0}} \tag{3.24}$$

where $\bar{u}_f^{\,p}$ is the plastic displacement at failure, G_f is the fracture energy per unit area and σ_{y0} is the value of the yield stress at the time when the failure criterion is reached.

The model ensures that the energy dissipated during the damage evolution process is equal to G_f only if the effective response of the material is perfectly plastic (constant yield stress) beyond the onset of damage. In this study, G_f is provided as an input parameter which is a function of fracture toughness K_C, Young's modulus E and Poisson's ratio v as given in the equation for the plane strain condition [70]:

$$G_f = \left(\frac{1 - v^2}{E} \right) K_C^2 . \tag{3.25}$$

The ELEMENT DELETION = YES module along with the Johnson Cook damage model of the software enables to delete the elements that fail. This produces the chip separation and allows the cutting tool to penetrate further into the workpiece through a predefined path (damage zone).

3.5.1.4 Chip-Tool Interface Model and Heat Generation

A contact is defined between the rake surface and nodes of the workpiece material. Coulomb's law has been assumed in the present study to model the frictional conditions as the chip flows over the rake surface.

During the machining process, heat is generated in the primary shear deformation zone due to severe plastic deformation and in the secondary deformation zone due to both plastic deformation and the friction in the tool–chip interface. The steady state two-dimensional form of the energy equation governing the orthogonal machining process is given as:

$$k\left(\frac{\partial^2 T}{\partial x^2} + \frac{\partial^2 T}{\partial y^2} \right) - \rho C_p \left(u_x \frac{\partial T}{\partial x} + v_y \frac{\partial T}{\partial y} \right) + \dot{q} = 0 \tag{3.26}$$

$$\dot{q} = \dot{q}_p + \dot{q}_f \tag{3.27}$$

$$\dot{q}_p = \eta_p \overline{\sigma} \dot{\overline{\varepsilon}}^p \tag{3.28}$$

$$\dot{q}_f = \eta_f J \overline{\tau} \dot{\gamma} \tag{3.29}$$

where $\dot{q}_p$ is the heat generation rate due to plastic deformation, η_p the fraction of the inelastic heat, $\dot{q}_f$ is the volumetric heat flux due to frictional work, $\dot{\gamma}$ the slip rate, η_f the frictional work conversion factor considered as 1.0, J the fraction of the thermal energy conducted into the chip, and $\overline{\tau}$ is the frictional shear stress. The value of J may vary within a range, say, 0.35 to 1 for carbide cutting tool [71]. In the present work, 0.5 (default value in ABAQUS) has been taken for all the cases. The fraction of the heat generated due to plastic deformation remaining in the chip, η_p , is taken to be 0.9 [56].

3.5.1.5 Solution Method

An ALE approach is incorporated to conduct the FEM simulation. This avoids severe element distortion and entanglement in the cutting zone without the use of remeshing criterion.

***ALE formulation*:** In the ALE approach, the grid points are not constrained to remain fixed in space (as in Eulerian description) or to move with the material points (as in Lagrangian description) and hence have their own motion governing equations. ALE description is given as follows:

$$\tilde{(\ )} = \dot{(\ )} + \vec{c}\nabla(\ ) \tag{3.30}$$

$$\vec{c} = \vec{v} - \vec{\hat{v}} \tag{3.31}$$

where superposed '~' is defined in the ALE description, $\vec{c}$ is the convective velocity, $\vec{v}$ is the material velocity, $\vec{\hat{v}}$ is the grid velocity and ∇ is the gradient operator. The Eqs. (3.14) to (3.16) can now be rewritten in the ALE framework, as given below:

$$\tilde{\rho} + \vec{c}\nabla\rho + \rho\,\mathrm{div}\,\vec{v} = 0 \tag{3.32}$$

$$\rho\tilde{\vec{v}} + \rho\vec{c}\nabla\vec{v} = \vec{f} + \mathrm{div}\,\sigma \tag{3.33}$$

$$\rho\tilde{e} + \rho\vec{c}\nabla e = \sigma : \mathrm{D} - \mathrm{div}\,\vec{q} + r \tag{3.34}$$

***Explicit dynamic analysis*:** The explicit dynamic ALE formulation is primarily used to solve highly non-linear problems involving large deformations and changing contact, as observed in the case of machining. The explicit dynamic procedure in ABAQUS/Explicit is based upon the implementation of an explicit integration scheme for time dicretization of Eqs. (3.32) to (3.34). This analysis calculates the state of a system at a later time from the state of system at the current time.

The explicit formulation advances the solution in time with the central difference method [31]:

$$\dot{v}^i = M^{-1}\left(f^{ext^{(i)}} - f^{\mathrm{int}^{(i)}}\right) \tag{3.35}$$

$$v^{(i+1/2)} = v^{(i-1/2)} + \frac{\Delta t^{(i+1)} - \Delta t^i}{2}\dot{v}^i \tag{3.36}$$

$$x^{(i+1)} = x^i + \frac{\Delta t^{(i+1)} - \Delta t^i}{2}v^{(i+1/2)} \tag{3.37}$$

where x is the displacement, v is the velocity and $\dot{v}$ is the acceleration, M is the lumped mass matrix, f^{ext} is the external force vector and f^{int} is the internal force. The subscript (i) refers to the increment number (i–1/2) and (i+1/2) and refers to mid-increment values. In this algorithm, the accelerations are first integrated to obtain the velocities, integration of which is broken into two half-steps so that the velocities are available at an integer step in the computation of the energy balance. The displacements are then computed in each time step by integrating the velocities.

3.5.2 ABAQUS Platform

The type of software package chosen for the FE analysis of metal cutting process is equally important in determining the quality and scope of analysis that can be performed. There are currently large number of commercial software packages available for solving a wide range of engineering problems that might be static, dynamic, linear or non-linear. Some of the dominant general purpose FE software packages include ABAQUS, ANSYS, MSC/NASTRAN, SRDC-IDEAS, etc. It is obvious that different packages would possess different capabilities. This makes it critical to select the suitable software package with appropriate features required for performing a given analysis successfully. The present study selects ABAQUS as a platform to explore the capabilities of finite element method in analyzing various aspects of metal cutting process. ABAQUS is known to be powerful general purpose FE software that can solve problems ranging from relatively simple linear analyses to the highly complex non-linear simulations. This software does not have any separate module for machining as in the case of Deform or AdvantEdge. As a result, the user has to explicitly define the tool and the workpiece, process parameters, boundary conditions, mesh geometry and simulation controls. This may certainly require more skill, effort and time to set up simulations as no preset controls and assumptions are available. But this is the feature that not only ensures very high level of details from modeling point of view but also a thorough analysis by allowing a precise control on boundary conditions, mesh attribute, element type, solver type and so on.

A complete ABAQUS program can be subdivided into three distinct modules, namely, ABAQUS/CAE, ABAQUS/Standard or ABAQUS/Explicit and ABAQUS/Viewer as shown in Fig. 3.8. These modules are linked with each other by input and output files. ABAQUS/Standard and ABAQUS/Explicit are the two main types of solvers that are available for performing analysis, ABAQUS/Explicit being mainly used for explicit dynamic analysis. It is said that the strength of ABAQUS program greatly lies in the capabilities of these two solvers.

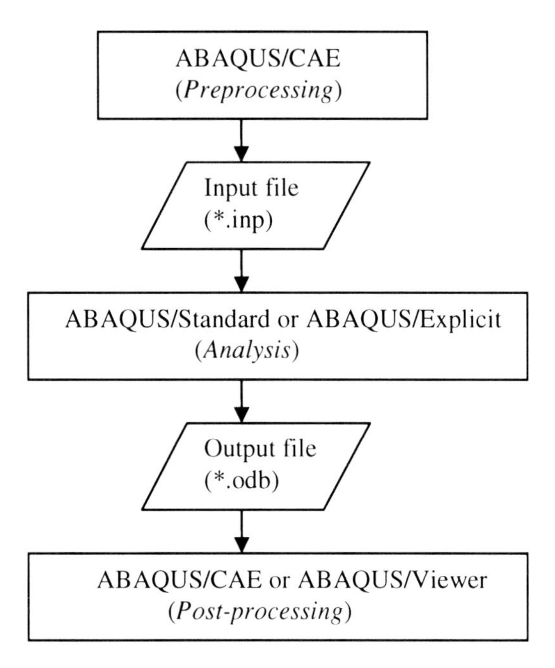

Fig. 3.8. Work flow of ABAQUS program

The model of the physical problem is created in the pre-processing stage, details of which such as discretized geometry, material data, boundary conditions, element type, analysis type and output request are contained in the input file. ABAQUS/CAE is divided into functional units called modules with the help of which the FE model can be created, input file can be generated and results can be extracted from the output file. Each module has been designed to serve a specific portion of the modeling task. The subsequent subsections would discuss about various modules of ABAQUS/CAE in brief.

Part module: Individual parts are created in the part module either by sketching their geometry directly in Abaqus/CAE or by importing their geometry from other geometric modeling programs. Depending upon the analysis the parts can be 2D or 3D deformable, discrete rigid, analytical rigid or Eulerian parts. In the present study, both the cutting tool and the workpiece are considered as 2D deformable bodies. The part tools contained in this module allow editing and manipulating the existing parts defined in the current model.

***Property module*:** Property module allows assigning sections to a part instance or region of a part instance to which various material properties are defined. A material definition specifies all the property data relevant to a particular analysis. For a coupled temperature displacement analysis, both the mechanical strength properties (elastic moduli, yield stress, etc.) and heat transfer properties (conductivity, specific heat) must be given as inputs. Various plasticity models and damage models are also contained in the property module. As an input the material model constants of the selected plastic model and damage model such as Johnson–Cook material model and Johnson–Cook damage model, respectively, are defined in a tabular form.

***Assembly module*:** The individual parts that are created in part module exist in their own coordinate system. It is in the assembly module that these parts are assembled by relative positioning with respect to each other in a global coordinate system.

***Step module*:** The step module is used to create and configure analysis steps, as in the present case is coupled temperature displacement explicit dynamic analysis. The associated output requests can also be created. The sequence of steps provides a convenient way to capture changes that may occur in the model during the course of the analysis such as changes in the loading and boundary conditions of the model or changes in the interaction, etc. In addition, steps allow you to change the analysis procedure, the data output and various controls. An output request contains information regarding which variables will be output during an analysis step, from which region of the model they will be output, and at what rate they will be output. The general solution controls and solver controls can also be customized. Furthermore, adaptive mesh regions and the controls for adaptive meshing in selected regions can be specified in this module. It is noted that implementation of suitable mesh adaptivity (remeshing in Lagrangian framework or repositioning of nodes in ALE framework) depends upon the type of analysis taken into consideration i.e. implicit (ABAQUS/Standard) or explicit (ABAQUS/Explicit). Adaptive remeshing is available in ABAQUS/Standard while ALE adaptive meshing is available in ABAQUS/Explicit.

***Interaction module*:** Interaction module allows to specify mechanical and thermal interactions between regions of a model or between a region of a model and its surroundings. Surface-to-surface contact interaction has been used to describe contact between tool and workpiece in the present study. The interaction between contacting bodies is defined by assigning a contact property model to a contact interaction which defines tangential behavior (friction) and normal behavior. Tangential behavior includes a friction model that defines the force resisting the relative tangential motion of the surfaces. While normal behavior includes a definition for the contact pressure–overclosure relationship that governs the motion of the surfaces. In addition, a contact property can contain information about thermal conductance, thermal radiation and heat generation due to friction. ABAQUS/ Explicit uses two different methods to enforce contact constraints, namely, kinematic contact algorithm and penalty contact algorithm. The former uses a kinematic predictor/corrector contact algorithm to strictly enforce contact constraints such that no penetrations are allowed, while the latter has a weaker enforcement of contact constraints. In this study, a kinematic contact algorithm has been used to enforce contact constraints in master–slave contact pair, where rake surface of the cutting tool is defined as the master surface and chip as the slave (node-based) surface. Both the frictional conditions and the friction-generated heat are included in the kinematic contact algorithm through TANGENTIAL BEHAVIOUR and GAP HEAT GENERATION modules of the software.

Load module: The Load module is used to specify loads, boundary conditions and predefined fields.

Mesh module: The Mesh module allows generating meshes on parts and assemblies created within ABAQUS/CAE as well as allows selecting the correct element depending on the type of analysis performed (as in the present case is CPE4RT) for discretization. Variety of mesh controls are available that help in selecting the element shape (tri, quad or quad dominated), meshing technique (structured or unstructured) or meshing algorithm (medial axis or advancing front). The structured meshing technique generates structured meshes using simple predefined mesh topologies and is more efficient for meshing regular shapes. Free meshing, however, allows more flexibility than structured meshing. Two commonly used free mesh algorithms while meshing with quadrilateral elements are medial axis and advancing front. The advancing front is generally preferable because it generates elements of more uniform size (area–wise) with more consistent aspect ratio. Since in ABAQUS/Explicit small elements control the size of the time step, avoidance of large differences in element size reduces solution stiffness, i.e., makes the numerical procedure more efficient.

Job: The Job module allows to create a job, to submit it to ABAQUS/Standard or ABAQUS/Explicit for analysis and to monitor its progress.

Visualization: The Visualization module provides graphical display of finite element models and results. It extracts the model and result information from the output database.

3.6 Case Studies in ABAQUS Platform

This section demonstrates the efficiency of finite element models to replicate the actual phenomena occurring during the cutting process under varied conditions as well as to understand basic mechanism of chip formation process in terms of various numerical results. Two different work materials are considered, namely AISI 1050 and Ti6Al4V, the former producing continuous chips and the latter producing segmented chips. In the first case study, not only the simulation results for machining AISI 1050 are presented but the predicted results are confirmed with the experimental ones. In the second case study, Ti6Al4V is considered for the study and the predicted results are compared with those of AISI 1050 under similar conditions, thus showing the effect of work material, being one of the machining inputs, on various output variables. Apart from machining inputs, being a numerical model, it is obvious that various FE inputs such as mesh size, material models, friction models and so on also have a typical impact on the output results. Hence, effect of FE inputs, namely, mesh size and Johnson-Cook material model constants have also been studied while simulating segmented chip formation during machining Ti6Al4V.

3.6.1 *Case Study I: FE Simulation of Continuous Chip Formation While Machining AISI 1050 and Its Experimental Validation*

AISI 1050 is a standard grade carbon steel. This particular grade of steel is both relatively easy to machine as well simulate. Therefore, AISI 1050 was chosen as a reference material in the present study. The basic aim of the present section is to develop the FE model for simulating the continuous chip formation process during machining of AISI 1050 as well as to explain the basic mechanism governing the continuous chip formation process in terms of numerical results viz. distributions stress, strain and temperature. The simulated results are then validated with the experimental ones at varied cutting conditions. Table 3.1 and Table 3.2 show the thermo-mechanical properties of the workpiece material as well as the values of Johnson-Cook material and damage constants used as input, respectively. The tool-chip interaction is based on the commonly used Coulomb's law with the mean coefficient of friction 0.2.

Table 3.1. Thermo-mechanical properties of tungsten carbide tool and AISI 1050 [72, 73]

Properties	Carbide tool	AISI 1050
Density, ρ (Kg/m^3)	11900	8030
Elastic modulus, E (GPa)	534	210
Poission's ratio, μ	0.22	0.3
Specific heat, C_p (J/Kg^oC)	400	472
Thermal conductivity, λ (W/m^oC)	50	51.9
Expansion, ($\mu m/m^oC$)	-	13.7
T_{melt} (^oC)	-	1460
T_{room} (^oC)	-	25
Fracture toughness, K_c ($MPa\ m^{½}$)	-	60
Conductive heat transfer coefficient *h*, ($kW\ m^{-2}\ K^{-1}$)	-	10

Table 3.2. Johnson-Cook material and damage constants for AISI 1050 [72]

Material Constants				
A (MPa)	*B* (MPa)	*n*	*C*	*m*
880	500	0.234	0.013	1.00
Damage Constants				
D_1	D_2	D_3	D_4	D_5
0.06	3.31	−1.96	0.018	0.58

3.6.1.1 Simulation Results

Figure 3.9 shows the distributions of stress, strain and temperature while machining of AISI 1050 using tungsten carbide tool for a cutting speed of 120 m/min and feed of 0.2 mm/rev.

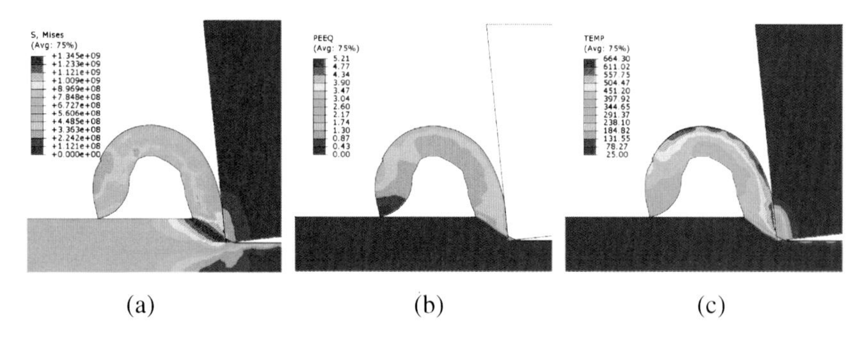

Fig. 3.9. Distribution of (a) stress (b) strain and (c) temperature while machining of AISI 1050 at Vc =120 m/min and f =0.2 mm

As the tool touches the workpiece, compression occurs within the workpiece. With the further advancement of the tool into the workpiece, stresses start developing near the tool tip and attaining high localized values in a confined region called primary shear deformation zone or shear plane as shown in Fig. 3.9 (a). The stresses in these regions are as high as 1.3 GPa. Consequently, such high values of stresses cause high strains to occur in the shear zone. This allows the workpiece material to deform plastically and shear continuously in the form of chip that flows over tool rake face. The type of chip, thus, formed depends largely on the distribution of strain and temperature within the chip surface. As the cutting continues the effective strains (especially, around the tool tip) increase and spread over a wider area of the chip surface with a maximum value not exceeding 1.7 in the shear plane. Consequently, this phenomenon tends to make the temperature distribution uniform in the chip region, thereby resulting in a steady state continuous chip having unvarying chip thickness. It is interesting to note that the maximum equivalent plastic strain and temperature are found along the tool–chip interface.

This can be attributed to the fact that the chip which is entering into the secondary shear deformation zone already possesses accumulated plastic strain and heat. The instant it begins to flow over the rake surface, further plastic straining and local heating occur because of severe contact and friction in the contact zone [74], thus, attaining higher temperature values at the tool-chip interface, specifically in the sliding region.

3.6.1.2 Model Verification

In order to validate the developed model, both the cutting speed and feed or uncut chip thickness values are varied and their effect on predicted cutting forces is studied and compared with the experimental ones. The cutting speed is varied in the range of 72–164 m/min for feed values of 0.1 and 0.2 mm.

It is known that the cutting force and thrust force increase with increasing feed rate almost linearly [75] where as decrease with the increasing cutting velocity. The reason for this can be explained from the expressions of the cutting force and thrust forces which are given as follows:

$$\text{Cutting force, } F_c = ts\tau_s(\zeta - \tan\gamma + 1) \tag{3.38}$$

$$\text{Thrust force, } F_t = ts\tau_s(\zeta - \tan\gamma - 1) \tag{3.39}$$

where t is the depth of cut (mm), s is the feed rate (mm/rev), τ_s is the dynamic shear strength of the workpiece, γ is the rake angle and ζ is the chip reduction coefficient, i.e., the ratio of deformed chip thickness to undeformed chip thickness. From the equations, correlation between feed rate and forces is straightforward. As far as the variation of cutting velocity is concerned, as the former increases, temperature of the shear zone increases. This causes softening of the work piece, which means the value of τ_s decreases and thereby reduces the value of cutting and thrust force.

Predicted cutting forces also show similar kind of trend with variation of speed and feed as shown in Fig. 3.10. Cutting forces measured during experimental tests under conditions similar to that of simulations have also been presented in Fig. 3.10 for comparison. It is noted that at lower feed, predicted results closely match with experimental ones. At higher feed, a flatter curve is observed for predicted values of forces implying that there is, no doubt, a decrease in cutting force values with the increasing speed but not as pronounced as in case of the experimental one. However, the maximum deviation between predicted and measured values of cutting forces remains within an acceptable range of 15%.

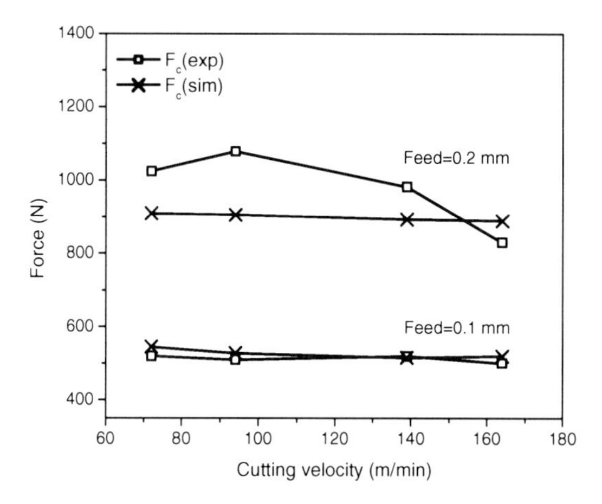

Fig. 3.10. Predicted and measured forces while machining AISI 1050

3.6.2 Case Study II: FE Simulation of Segmented Chip Formation While Machining Ti6Al4V

The present study deals with the simulation of chip formation while machining Ti6Al4V. It is a known fact that machining of Ti6Al4V yields segmented chips. Formation of segmented chips makes not only the machining of Ti6Al4V challenging but also simulating the chip formation process of the same very difficult. In this study, simulation of segmented chip formation is performed based on adiabatic shearing phenomenon. Thermal softening becomes highly pronounced when the rate of heat generation exceeds the rate of heat dissipation which comes into picture when thermal conductivity of the workpiece is very low and cutting speed is high. During cutting process, high stresses and strains along the primary shear zone accumulate and plastic deformation occurs, thereby increasing cutting temperature. As the tool proceeds further, temperature increases in a highly localized area in the shear zone in the form of bands called adiabatic shear bands that extend from tool tip toward the chip free surface. It has been found that temperature at the back of the chip reaches such a high value that thermal softening occurs in a highly localized region on the free side of the chip surface. This induces wave like irregularities on the back of the chip.

Firstly, an attempt is made to simulate the desired chip by fine tuning two of the FE inputs that seem to be critical as far as reliability of the model is concerned. Then a comparative study is done between the predicted results of AISI 1050 and Ti6Al4V to explain the significance of simulating the right kind of chip formation and its correlation with other output variables.

3.6.2.1 Role of FE Inputs

***Meshing*:** Selection of a suitable mesh size is a critical factor from both accuracy and computational time points of view [76]. As discussed earlier, finer mesh leads to greater accuracy but at the cost of higher computational time. It is important to mesh the model in such a way that the model gives results closer to the experimental ones on one hand and consumes fairly less time on the other hand. Moreover, it is illogical to start with a very fine mesh, instead a mesh refinement study should be performed wherein the degree of meshing is gradually changed from coarser mesh to a finer mesh and the corresponding results are compared among each other. There exists a limit beyond which if the mesh is refined further, the CPU time would, no doubt, increase but there would not be any significant changes in the numerical results. It is the duty of the analyst to consider an optimum meshing that comes as a fair compromise between accuracy and computational time. This shows the need of mesh refinement study to prove the reliability of the developed model.

Researchers have pointed that with the decrease in element size, the temperature at the integration point increases. Since this study primarily deals with the mechanism of adiabatic shearing during the formation of segmented chip, it is logical to carry out a mesh refinement study. The meshing is mainly varied in the chip region by considering 5, 10 and 15 elements (on an average) in the chip thickness direction as shown in Fig. 3.11. The average element size for 5, 10 and 15 elements are: 50x50 μm, 20x20 μm and 12x12 μm, respectively. Since advancing front algorithm (free mesh algorithm) is employed for meshing the workpiece, slight skewness has been observed at certain mesh regions of the workpiece with finer meshing at the right edge of the workpiece.

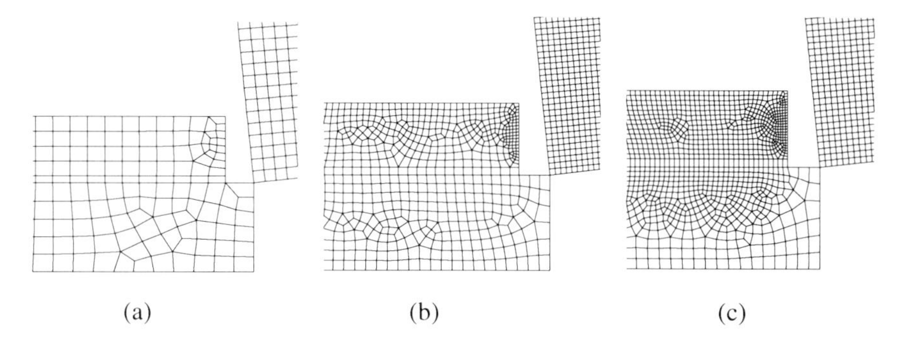

(a) (b) (c)

Fig. 3.11. Mesh configurations for (a) 5 (b) 10 and (c) 15 elements

Figure 3.12 shows the predicted chip morphology and the distribution of temperature within the chip surface for all the three cases for a cutting speed of 210 m/min and uncut chip thickness 0.2 mm.

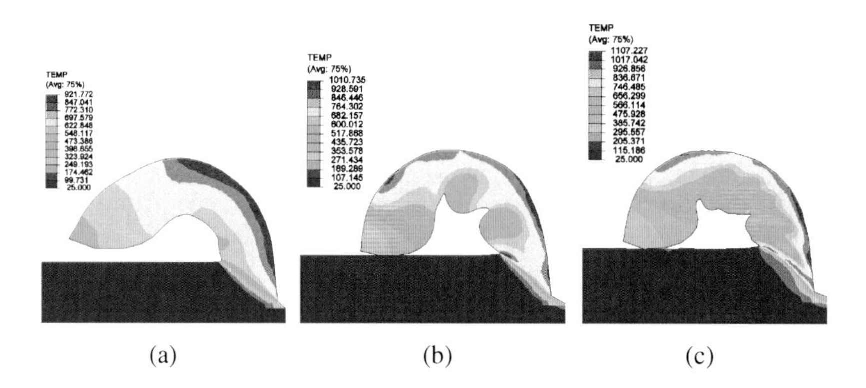

(a) (b) (c)

Fig. 3.12. Temperature distributions for (a) 5 (b) 10 and (c) 15 elements

It is interesting to note that there is a gross difference in the predicted chip morphologies for the first two cases. The difference can be attributed to the variation in stress, strain and temperature distributions within the chip region. Though the maximum value of stress in both the cases is nearly same but the distribution pattern varies. As the meshing gets finer, the stresses become more concentrated along the shear plane, as a result of which the strain and temperature not only get localized in a very narrow zone but also attain very high values. Since an increase in the maximum temperature is found with finer meshing, the tendency to invoke chip segmentation due to thermal softening by adiabatic shearing becomes higher, thus producing segmented chips in the second case. However, with further refinement of mesh as in the case of 15 elements, not much variation in chip morphology is observed when compared with that of 10 elements. The adiabatic shear band, no doubt, appears relatively more distinct (as very thin strip) in the former case, but the average temperature values in the shear band (T_{adia}) increases not even by 1% (see Table 3.3). This implies that further mesh refinement may not be required as far as consistency of the results is concerned. Moreover, there is a limit to the reduction of element size from the software point of view. Since the time step in ABAQUS/Explicit is controlled by the size of the smallest element, reducing the element size beyond this may not allow the simulation to run properly.

Table 3.3. Mesh refinement study

Mesh type	Stress (GPa)	Strain	T_{adia} (°C)
5 elements	1.48	1.32	570
10 elements	1.50	2.0	956
15 elements	1.54	2.5	963

It is observed that finer meshing increases CPU time considerably. Therefore, fixed mass scaling is used to expedite the analysis in the last two cases. Scaling the material density of the small elements throughout the analysis proves to be very useful as it increases the stable time increment allowing the analysis to take less time, while retaining the accuracy to a great extent [77]. Table 3.4 presents a comparative study of the actual CPU time with respect to simulation time for 5, 10 and 15 elements.

Table 3.4. CPU time for 5, 10 and 15 elements

Simulation time (ms)	CPU time (h:min:s)		
	5 elements	10 elements	15 elements
0.4	0:09:53	1:05:20	1:50:02
0.6	0:21:37	1:44:55	---
1	0:35:07	2:55:52	---

From Table 3.4, it can be inferred that the model considering 10 elements along the uncut chip thickness is the best compromise between accuracy and computation time. In case of 15 elements, though adiabatic shearing is very prominent, it was very difficult to run the simulation beyond 0.4 ms.

***Johnson–Cook material model constants*:** Several classical plasticity models have been widely employed that represent, with varying degrees of accuracy, the material flow stresses as a function of strain, strain rate and temperature. These models include the Litonski-Batra model [78, 79], Usui model [80], Maekawa model [81], the Johnson–Cook model [65], Zerilli–Armstrong model [82], Oxley model [83], Mechanical Threshold Stress (MTS) model [84] etc. It is very important to carefully select the appropriate material model that satisfactorily predicts the desired chip morphology and other output variables. Johnson-Cook model, being most widely used, is employed to describe the flow stress property of workpiece material Ti6Al4V in this study (see Eq. 3.17). This material model defines the flow stress as a function of strain, strain-rate and temperature such that, it not only considers the strain rates over a large range but also temperature changes due to thermal softening by large plastic deformation. The work material constants contained in this constitutive equation have been found by various researchers by applications of several methods, thus producing different values of data sets for a specific material. As a result, selection of suitable data sets along with appropriate material model becomes equally important [85]. The present study selects two sets of Johnson-Cook material constants from the available literature namely, *M1* and *M2* as listed in Table 3.4 [86, 87]. Lee and Lin [87] obtained the Johnson–Cook material constants from a high strain rate mechanical testing using Split Hopkinson Pressure Bar (SHPB) method under a constant strain rate of 2000 s^{-1} within the temperature range of 700–1000 ºC and maximum true plastic strain of 0.3. While Meyer and Kleponis [86] obtained the material constants by considering strain rate levels of 0.0001, 0.1 and 2150 $s^{-1}$and a maximum plastic strain of 0.57.

Table 3.5. Johnson-Cook material parameters for Ti6Al4V [86, 87]

Johnson-Cook constants	*A* (MPa)	*B* (MPa)	*C*	*n*	*m*
M1 model	862.5	331.2	0.012	0.34	0.8
M2 model	782.7	498.4	0.028	0.28	1.0

Besides the material model constants, all other parameters are kept constant so that the results can be compared on the same conditions. Meshing in both the cases is similar to the one that was considered as optimum in the previous case study.

Figures 3.13 and 3.14 show the distribution of stress, strain and temperature for the models *M1* and *M2* at a cutting speed of 210 m/min and uncut chip thickness of 0.2 mm.

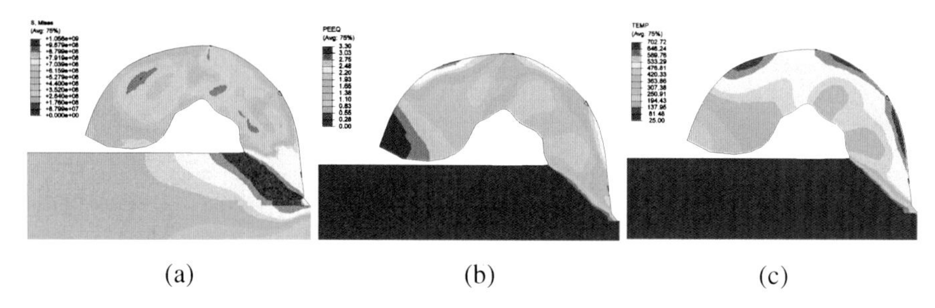

(a) (b) (c)

Fig. 3.13. Distribution of (a) Stress (Pa) (b) Strain and (c) Temperature (ºC) for *M1* model

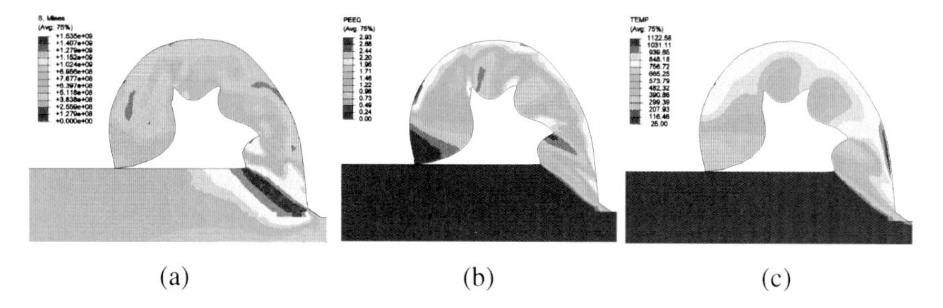

(a) (b) (c)

Fig. 3.14. Distribution of (a) Stress (Pa) (b) Strain and (c) Temperature (ºC) for *M2* model

It is observed that model *M1* is unable to simulate the formation of segmented chips. The distribution of stress, strain and temperature when compared with that of model *M2* explains much about the variation in chip morphology. This suggests that material model constants used in the constitutive equations also affect the predicted results. Thermal softening coefficient m and work hardening exponent n are often considered as the critical parameters that influence the segmentation of chip [88]. Baker [88] suggested increase in m leads to increase in chip segmentation while increase in n causes lesser tendency to undergo chip segmentation. In the present study, similar kind of observation has been found. Greater value of m and lower value of n can be attributed for the formation of well-defined segmented chips in case of model *M2*.

3.6.2.2 Significance of FE Outputs

Various difficult to measure output variables such as stresses, strains, temperatures and many more can be determined easily from the FE simulations of chip formation process. These predicted outputs greatly help in understanding the basic mechanism of chip formation under different cutting conditions. Chip morphology, considered as an important index in study of machining, is dependent on various input variables like workpiece material, cutting conditions, tool material and so on. The type of chip produced affects the cutting force, the residual stresses and the cutting temperatures to a great extent. This section, thus, demonstrates the same by a comparative study of predicted forces and cutting tool temperature while machining AISI 1050 and Ti6Al4V.

Cutting forces: Figure 3.15 shows the variation of predicted cutting force with time while machining AISI 1050 and Ti6Al4V.

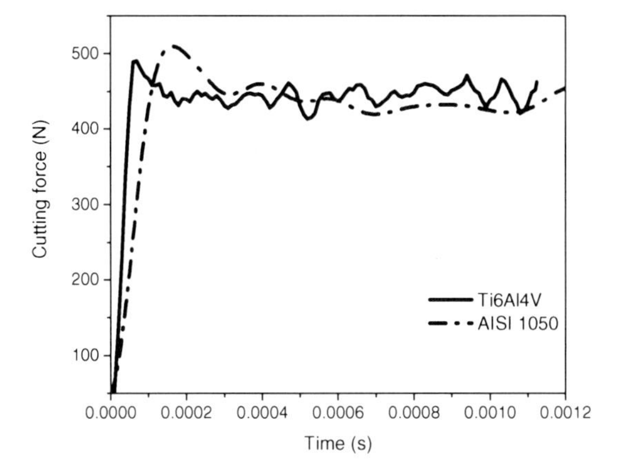

Fig. 3.15. Time signature of cutting forces for Ti6Al4V and AISI 1050

The force signatures obtained for both the materials vary remarkably not only in magnitude but also in nature from each other. As expected, the cutting force profile closely resembles the predicted chip morphology i.e. continuous for AISI 1050 and wavy profile for Ti6Al4V. While machining Ti6Al4V, chip formation process begins with the bulging of the workpiece material in front of the tool as a result of which the forces increase gradually but drop suddenly when the shear band begins to form due to thermal softening. Obviously, magnitude of the waviness (difference between the peak and lower values) affects the surface finish quality and thermal load that the cutting tool undergoes during cutting process.

Cutting temperature: In cutting operation, cutting tool is subjected to highly adverse conditions such as high stresses and high temperatures due to which tool wears out. As the tool wear occurs, tool geometry changes affecting residual stresses, surface finish and dimensional accuracy of the machined surface. Hence, cutting temperature is a critical parameter which is again a function of various machining inputs such as cutting conditions, workpiece material, tool material, tool geometry, etc. Tool condition monitoring and optimization of process parameters are of paramount importance that can lead to higher productivity and accuracy.

Figures 3.16 (a) and (b) present the temperature distributions over the cutting tool that has developed during machining of AISI 1050 and Ti6Al4V, respectively.

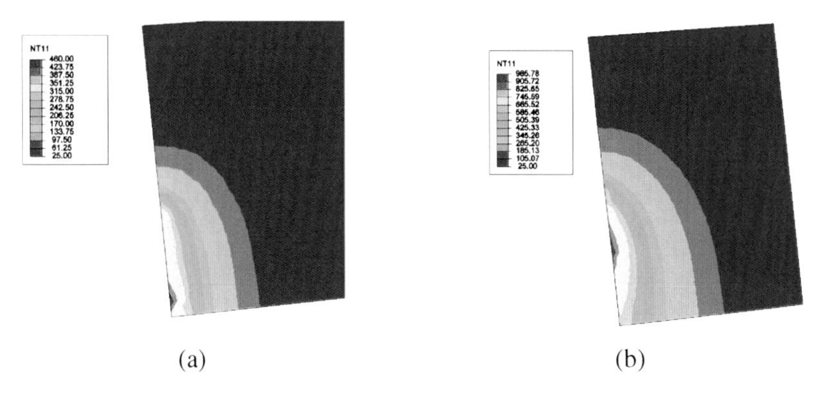

Fig. 3.16. Temperature contours on cutting tool while machining (a) AISI 1050 and (b) Ti6Al4V

While machining Ti6Al4V the maximum temperature attained by carbide tool is as high as 1000ºC as compared to that of AISI 1050 with maximum value not even exceeding 400ºC. It can be seen that not only the maximum value of temperature attained by cutting tool is different but also the distribution pattern over the cutting tool varies widely depending upon the type of worpiece material machined. This gives a basic idea of how the tool wear progress takes place. Furthermore, the local variables such as temperature, stress, strain, etc., at different nodes that are predicted prove helpful in developing a tool wear prediction method by integrating FEM simulation of the cutting process with a suitable tool wear rate model.

In general, it can be concluded that FEM can prove to be a very effective tool in analyzing various aspects of metal cutting process, if used with caution. The stresses, strains, temperatures and many other outputs that are obtained from FE simulations greatly help in understanding the basic mechanism of chip formation under different cutting conditions. This consequently helps in the cutting tool design, optimization of cutting parameters for higher tool life and good surface finish, prediction of tool wear growth, etc., thus leading to higher productivity.

References

[1] Astakhov, V.P.: Tribology of Metal Cutting. Elsevier (2006) ISBN: 978-0-444-52881-0

[2] Shaw, M.C.: Metal Cutting principles, 2nd edn. Oxford University Press, Oxford (2005)

[3] Hahn, R.S.: On the temperature developed at the shear plane in the metal cutting process. In: Proceedings of First U.S. National Congress Appl. Mech. ASME 661 (1951)

[4] Chao, B.T., Trigger, K.J.: An analytical evaluation of metal cutting temperature. Trans. ASME 73, 57–68 (1951)

[5] Leone, W.C.: Distribution of shear-zone heat in metal cutting. Trans. ASME 76, 121–125 (1954)

[6] Loewen, E.G., Shaw, M.C.: On the analysis of cutting tool temperatures. Transactions of the ASME 71, 217–231 (1954)
[7] Weiner, J.H.: Shear plane temperature distribution in orthogonal machining. Trans. ASME 77, 1331–1341 (1955)
[8] Rapier, A.C.: A theoretical investigation of the temperature distribution in the metal cutting process. Br. J. Appl. Phys. 5, 400–405 (1954)
[9] Bagci, E.: 3-D numerical analysis of orthogonal cutting process via mesh-free method. Int. J. the Physical Sciences 6(6), 1267–1282 (2011)
[10] ASM Handbook, Volume 16-Machining, ASM International Handbook Committee, ASM International, Electronic (1989) ISBN: 978-1-61503-145-0
[11] Juneja, B.L., Sekhon, G.S., Seth, N.: Fundamentals of metal cutting and machine tools, 2nd edn. New Age International Publishers, New Delhi (2003)
[12] Merchant, M.E.: Mechanics of the metal cutting process. J. Appl. Phys. 16, 318–324 (1945)
[13] Lee, E.H., Shaffer, B.W.: The theory of plasticity applied to a problem of machining. Trans. ASME, J. Appl. Mech. 18, 405–413 (1951)
[14] Oxley, P.L.B.: Shear angle solutions in orthogonal machining. Int. J. Mach. Tool. Des. Res. 2, 219–229 (1962)
[15] Bhattacharya, A.: Metal cutting theory and practice. Central book publishers, Kolkata (1984)
[16] Lacale, L.N., Guttierrez, A., Llorente, J.I., Sanchez, J.A., Aboniga, J.: Using high pressure coolant in the drilling and turning of low machinability alloys. Int. J. of Adv. Tech. 16, 85–91 (2000)
[17] Tobias, S.A.: Machine tool Vibration. Blackie and Sons Ltd, Scotland (1965)
[18] Ghosh, A., Mallik, A.K.: Manufacturing science (1985) ISBN: 81-85095-85-X
[19] Jacobson, S., Wallen, P.: A new classification system for dead zones in metal cutting. Int. J. Mach. Tool. Manufact. 28, 529–538 (1988)
[20] Trent, E.M., Wright, P.K.: Metal cutting, 4th edn. Butterworth-Heinemann (2000)
[21] Komanduri, R., Hou, Z.B.: A review of the experimental techniques for the measurement of heat and temperatures generated in some manufacturing processes and tribology. Tribol. Int. 34, 653–682 (2001)
[22] Groover, M.P.: Fundamentals of modern manufacturing: materials processes, and systems, 2nd edn. Wiley, India (2002)
[23] Klamecki, B.E.: Incipient chip formation in metal cutting- A three dimensional finite element analysis. Ph.D. Thesis. University of Illinois, Urbana (1973)
[24] Hughes, J.R.T.: The Finite Element Method. Prentice-Hall International, Inc. (1987)
[25] Reddy, J.N.: An introduction to the finite element method, 2nd edn. McGraw-Hill Inc. (1993)
[26] Bathe, K.J.: Finite Element Procedures. Prentice Hall, Englewood Cliffs (1996)
[27] Rao, S.S.: The Finite Element in Engineering, 3rd edn. Butterworth-Heinemann (1999)
[28] Zienkiewicz, O.C., Taylor, R.L.: The Finite Element Method, 5th edn. Butterworth-Heinemann (2000)
[29] Liu, G.R., Quek, S.S.: The Finite Element Method: A Practical Course. Butterworth Hienemann (2003)
[30] Hutton, D.V.: Fundamentals of finite element analysis, 1st edn. Mc Graw Hill (2004)
[31] Belytschko, T., Liu, W.K., Moran, B.: Nonlinear Finite Elements for Continua and Structures. John Wiley and Sons, New York (2000)

[32] Strenkowski, J.S., Moon, K.-J.: Finite element prediction of chip geometry and tool/workpiece temperature distributions in orthogonal metal cutting. ASME J. Eng. Ind. 112, 313–318 (1990)
[33] Raczy, A., Elmadagli, M., Altenhof, W.J., Alpas, A.T.: An Eulerian finite-element model for determination of deformation state of a copper subjected to orthogonal cutting. Metall Mater. Trans. 35A, 2393–2400 (2004)
[34] Mackerle, J.: Finite element methods and material processing technology, an addendum (1994–1996). Eng. Comp. 15, 616–690 (1962)
[35] Rakotomalala, R., Joyot, P., Touratier, M.: Arbitrary Lagrangian-Eulerian thermomechanical finite element model of material cutting. Comm. Numer. Meth. Eng. 9, 975–987 (1993)
[36] Pepper, D.W., Heinrich, J.C.: The Finite Element Method: Basic Concepts and Applications. Hemisphere Publishing Corporation, United States of America (1992)
[37] Bower, A.F.: Applied mechanics of solid. CRC Press, Taylor and Francis Group, New York (2010)
[38] Dhondt, G.: The finite element method for three-dimensional thermomechanical applications. John Wiley and Sons Inc., Germany (2004)
[39] Pian, T.H.H.: Derivation of element stiffness matrices by assumed stress distributions. AIAA J. 2, 1333–1336 (1964)
[40] Zienkiewicz, O.C., Taylor, R.L.: The Finite Element Method. Basic formulations and linear problems, vol. 1. McGraw-Hill, London (1989)
[41] Liapis, S.: A review of error estimation and adaptivity in the boundary element method. Eng. Anal. Bound. Elem. 14, 315–323 (1994)
[42] Tay, A.O., Stevenson, M.G., de Vahl Davis, G.: Using the finite element method to determine temperature distribution in orthogonal machining. Proc. Inst. Mech. Eng. 188(55), 627–638 (1974)
[43] Muraka, P.D., Barrow, G., Hinduja, S.: Influence of the process variables on the temperature distribution in orthogonal machining using the finite element method. Int. J. Mech. Sci. 21(8), 445–456 (1979)
[44] Moriwaki, T., Sugimura, N., Luan, S.: Combined stress, material flow and heat analysis of orthogonal micromachining of copper. CIRP Annals - Manufact. Tech. 42(1), 75–78 (1993)
[45] Kim, K.W., Sin, H.C.: Development of a thermo-viscoplastic cutting model using finite element method. Int. J. Mach. Tool Manufact. 36(3), 379–397 (1996)
[46] Liu, C.R., Guo, Y.B.: Finite element analysis of the effect of sequential cuts and tool-chip friction on residual stresses in a machined layer. Int. J. Mech. Sci. 42(6), 1069–1086 (2000)
[47] Ceretti, E., Falbohmer, P., Wu, W.T., Altan, T.: Application of 2D FEM to chip formation in orthogonal cutting. J. Mater Process Tech. 59, 169–180 (1996)
[48] Li, K., Gao, X.-L., Sutherland, J.W.: Finite element simulation of the orthogonal metal cutting process for qualitative understanding of the effects of crater wear on the chip formation. J. Mater Process Tech. 127, 309–324 (2002)
[49] Arrazola, P.J., Ugarte, D., Montoya, J., Villar, A., Marya, S.: Finite element modeling of chip formation process with abaqus/explicit. VII Int. Conference Comp., Barcelona (2005)
[50] Davies, M.A., Cao, Q., Cooke, A.L., Ivester, R.: On the measurement and prediction of temperature fields in machining AISI 1045 steel. Annals of the CIRP 52, 77–80 (2003)

[51] Adibi-Sedeh, A.H., Vaziri, M., Pednekar, V., Madhavan, V., Ivester, R.: Investigation of the effect of using different material models on finite element simulations of metal cutting. In: 8th CIRP Int Workshop Modeling Mach Operations, Chemnitz, Germany (2005)
[52] Shi, J., Liu, C.R.: The influence of material models on finite element simulation of machining. J. Manufact. Sci. Eng. 126, 849–857 (2004)
[53] Ozel, T.: Influence of Friction Models on Finite Element Simulations of Machining. Int. J. Mach. Tool Manufact. 46(5), 518–530 (2006)
[54] Filice, L., Micari, F., Rizzuti, S., Umbrello, D.: A critical analysis on the friction modeling in orthogonal machining. International Journal of Machine Tools and Manufacture 47, 709–714 (2007)
[55] Haglund, A.J., Kishawy, H.A., Rogers, R.J.: An exploration of friction models for the chip-tool interface using an Arbitrary Lagrangian-Eulerian finite element model. Wear 265(3-4), 452–460 (2008)
[56] Mabrouki, T., Deshayes, L., Ivester, R., Regal, J.-F., Jurrens, K.: Material modeling and experimental study of serrated chip morphology. In: Proceedings of 7th CIRP Int Workshop Model. Machin, France, April 4-5 (2004)
[57] Coelho, R.T., Ng, E.-G., Elbestawi, M.A.: Tool wear when turning AISI 4340 with coated PCBN tools using finishing cutting conditions. J. Mach. Tool Manufact. 47, 263–272 (2006)
[58] Lorentzon, J., Jarvstrat, N.: Modelling tool wear in cemented carbide machining alloy 718. J. Mach. Tool Manufact. 48, 1072–1080 (2008)
[59] Davim, J.P., Maranhao, C., Jackson, M.J., Cabral, G., Gracio, J.: FEM analysis in high speed machining of aluminium alloy (Al7075-0) using polycrystalline diamond (PCD) and cemented carbide (K10) cutting tools. Int. J. Adv. Manufact. Tech. 39, 1093–1100 (2008)
[60] Attanasio, A., Cerretti, E., Rizzuti, S., Umbrello, D., Micari, F.: 3D finite element analysis of tool wear in machining. CIRP Annals – Manufact. Tech. 57, 61–64 (2008)
[61] ABAQUS Analysis User's manual. Version 6.7-4 Hibbitt, Karlsson & Sorensen, Inc. (2007)
[62] ABAQUS Theory manual, Version 6.7-4 Hibbitt, Karlsson & Sorenson, Inc. (2007)
[63] ABAQUS/CAE User's manual. Version 6.7-4 Hibbitt, Karlsson & Sorensen, Inc. (2007)
[64] Wu, H.-C.: Continuum Mechanics and Plasticity. Chapman and Hall/CRC (2004)
[65] Johnson, G.R., Cook, W.H.: A constitutive model and data for metals subjected to large strains, high strain rates and high temperatures. In: Proceedings of 7th Int Symp Ballistics, the Hague, The Netherlands, pp. 541–547 (1983)
[66] Umbrello, D., M'Saoubi, R., Outeiro, J.C.: The influence of Johnson–Cook material constants on finite element simulation of machining of AISI 316L steel. Int. J. Mach. Tool Manufact. 47, 462–470 (2007)
[67] Davim, J.P., Maranhao, C.: A study of plastic strain and plastic strain rate in machining of steel AISI 1045 using FEM analysis. Mater Des. 30, 160–165 (2009)
[68] Vaziri, M.R., Salimi, M., Mashayekhi, M.: A new calibration method for ductile fracture models as chip separation criteria in machining. Simulat Model Pract. Theor. 18, 1286–1296 (2010)
[69] Johnson, G.R., Cook, W.H.: Fracture characteristics of three metals subjected to various strains, strains rates, temperatures and pressures. Eng. Fract. Mech. 21(1), 31–48 (1985)

[70] Mabrouki, T., Girardin, F., Asad, M., Regal, J.-F.: Numerical and experimental study of dry cutting for an aeronautic aluminium alloy. Int. J. Mach. Tool Manufact. 48, 1187–1197 (2008)
[71] Mabrouki, T., Rigal, J.: -F A contribution to a qualitative understanding of thermo-mechanical effects during chip formation in hard turning. J. Mater. Process Tech. 176, 214–221 (2006)
[72] Duan, C.Z., Dou, T., Cai, Y.J., Li, Y.Y.: Finite element simulation & experiment of chip formation process during high speed machining of AISI 1045 hardened steel. Int. J. Recent Trend Eng. 1(5), 46–50 (2009)
[73] Priyadarshini, A., Pal, S.K., Samantaray, A.K.: A Finite Element Study of Chip Formation Process in Orthogonal Machining. Int . J. Manufact., Mater. Mech. Eng. IGI Global(accepted, in Press, 2011)
[74] Shi, G., Deng, X., Shet, C.: A finite element study of the effect of friction in orthogonal metal cutting. Finite Elem. Anal. Des. 38, 863–883 (2002)
[75] Lima, J.G., Avila, R.F., Abrao, A.M., Faustino, M., Davim, J.P.: Hard turning: AISI 4340 high strength alloy steel and AISI D2 cold work tool steel. J. Mater. Process. Tech. 169, 388–395 (2005)
[76] Priyadarshini, A., Pal, S.K., Samantaray, A.K.: Finite element study of serrated chip formation and temperature distribution in orthogonal machining. J. Mechatron Intell. Manufact. 2(1-2), 53–72 (2010)
[77] Wang, M., Yang, H., Sun, Z.-C., Guo, L.-G.: Dynamic explicit FE modeling of hot ring rolling process. Trans. Nonferrous Met. Soc. China 16(6), 1274–1280 (2006)
[78] Litonski, J.: Plastic flow of a tube under adiabatic torsion. Bulletin of Academy of Pol. Science, Ser. Sci. Tech. XXV, 7 (1977)
[79] Batra, R.C.: Steady state penetration of thermo-visoplastic targets. Comput Mech. 3, 1–12 (1988)
[80] Usui, E., Shirakashi, T.: Mechanics of machining–from descriptive to predictive theory: On the art of cutting metals-75 Years Later. ASME PED 7, 13–55 (1982)
[81] Maekawa, K., Shirakashi, T., Usui, E.: Flow stress of low carbon steel at high temperature and strain rate (Part 2)–Flow stress under variable temperature and variable strain rate. Bulletin Japan Soc Precision Eng 17, 167–172 (1983)
[82] Zerilli, F.J., Armstrong, R.W.: Dislocation-mechanics-based constitutive relations for material dynamics calculations. J. Appl. Phys. 61, 1816–1825 (1987)
[83] Oxley, P.L.B.: The mechanics of machining: An analytical approach to assessing machinability. Ellis Horwood Limited, Chichester (1989)
[84] Banerjee, B.: The mechanical threshold stress model for various tempers of AISI 4340 steel. Int. J. Solid Struct. 44, 834–859 (2007)
[85] Priyadarshini, A., Pal, S.K., Samantaray, A.K.: On the Influence of the Material and Friction Models on Simulation of Chip Formation Process. J. Mach. Forming Tech. Nova Science (accepted, 2011)
[86] Meyer, H.W., Kleponis, D.S.: Modeling the high strain rate behavior of titanium undergoing ballistic impact and penetration. Int. J. Impact Eng. 26(1-10), 509–521 (2001)
[87] Lee, W.S., Lin, C.F.: High temperature deformation behaviour of Ti6Al4V alloy evluated by high strain rate compression tests. J. Mater. Process. Tech. 75, 127–136 (1998)
[88] Baker, M.: The influence of plastic properties on chip formation. Comp. Mater. Sci. 28, 556–562 (2003)

4

GA-Fuzzy Approaches: Application to Modeling of Manufacturing Process

Arup Kumar Nandi

Central Mechanical Engineering Research Institute (CSIR-CMERI)
Durgapur-713209, West Bengal, India
nandiarup@yahoo.com, nandi@cmeri.res.in

This chapter presents various techniques using the combination of fuzzy logic and genetic algorithm (GA) to construct model of a physical process including manufacturing process. First, an overview on the fundamentals of fuzzy logic and fuzzy inferences systems toward formulating a rule-based model (called fuzzy rule based model, FRBM) is presented. After that, the working principle of a GA is discussed and later, how GA can be combined with fuzzy logic to design the optimal knowledge base of FRBM of a process is presented. Results of few case studies of modeling various manufacturing processes using GA-fuzzy approaches conducted by the author are presented.

4.1 Introduction

Optimal selection of machining parameters is an imperative issue to obtain a better performance of machining, cost effectiveness as well as to achieve a desired accuracy of the attributes of size, shape and surface roughness of the finished product. Selection of these parameters is traditionally carried out on the basis of the experience of process planners with the help of past data available in machining handbooks and tool catalogs. Practitioners continue to experience great difficulties due to the lack of sufficient data on the numerous new cutting tools with different materials. Specific data on relevant machining performance measures such as tool life, surface roughness, chip form, etc. are very difficult to find due to the lack of reliable information or predictive models for these measures. In automated manufacturing processes, it is required to control the machining process by determining the optimum values of machining parameters online during machining. Therefore, it is important to develop a technique to predict the attribute of a product before machining to evaluate the robustness of machining parameters for keeping a desired attribute and increasing product quality. Construction of suitable machining process model and evaluation of the optimal values of machining parameters using this model as predictor are essential and challenging tasks.

The model of a machining process represents a mapping of input and output variables in specific machining conditions. The input variables differ corresponding to the type of machining process and the desired output. For example, in turning, the surface roughness (output variable) is dependent on a number of variables that can be broadly divided into four groups: major variables which include cutting speed, feed rate, depth of cut and tool wear; flow of coolant, utilization of chip breaker, work-holding devices and selection of tool type belong to the second group. The third group includes machine repeatability, machine vibration and damping, cutting temperature, chip formation and chip exit speed, thermal expansion of machine tool and power consumption; room temperature, humidity, dust content in the air and fluctuation in the power source are involved in the fourth group. Among these four groups of input variable, the major variable can be measured and controlled during machining process. Though the other variables are not directly involved and uncontrolled during machining but their effect cannot be neglected to obtain a desired surface roughness. In a specific machining condition, these variables are assumed to be fixed at a particular state.

Various approaches have been proposed to model and simulate the machining processes. Analytical methods, which are generally based on the established science principles, are probably the first modeling approach. Experimental or empirical approaches use experimental data as the basis for formulating the models. Mechanistic and numerical methods integrate the analytical and empirical methods, generally by the use of modern computer techniques.

Due to the complex and nonlinear relationship among the input–output variables, influence of uncontrollable parameters and involvement of random aspect, prediction of an output of machining processes using mathematical/analytical approaches is not accurate. This leads to the development of empirical equations for a particular machine tool, machining parameters and work piece-cutting tool material combination. Empirical models do not consider the underlying principles and mathematical relationships. These are usually obtained by performing statistical analysis or through the training of data-driven models to fit the experimental data [1].

The significant drawback of empirical models is their sensitivity to process variation though they have the advantages of accuracy due to the use of experimental data. The accuracy of a model degenerates rapidly due to the variation of experimental data as the machining conditions deviate from the experimental settings. In addition, quality characteristics of machined parts exhibit stochastic variations over time due to changes in a machine tool structure and the environment. Therefore, the modeling techniques should have enough capability to adapt the variations in machining process. Most of the statistical process control models do not account for time-varying changes. Involvement of uncertainty and imprecision in machining processes is another aspect affecting the variation of machining output. In such cases, techniques of modeling using fuzzy logic are most useful because fuzzy logic is a powerful tool for dealing with imprecision and uncertainty [2]. The basic concept of fuzzy logic is to categorize the variables into fuzzy sets with a degree of certainty in the numerical interval (0 and 1), so that ambiguity and vagueness in the data structure and human knowledge can be handled without constructing complex mathematical models. Moreover, fuzzy logic-based control system has the capability to adapt the variations of a process by learning and adjusting itself to the environmental changes by observing current system behavior.

Fuzzy logic is an application of fuzzy set theory, and was first proposed by Prof. L.A. Zadeh [3]. Fuzzy logic rules, which are derived based on fuzzy set theory, are used in fuzzy inference system toward formulating a rule-based model (called fuzzy rule-based model, FRBM). The performance of a FRBM mainly depends on two different aspects: structure of fuzzy logic rules and the type/shape of associated fuzzy subsets (membership function distributions, MFDs) those constitute the knowledge base (KB) of FRBM. Manually constructed KB of a FRBM may not be optimal in many cases since it strongly demands a thorough knowledge of the process which is difficult to acquire, particularly in a short period of time. Therefore, design of an optimal KB of a fuzzy model needs the help of other optimization/learning techniques. Genetic algorithm (GA), a population-based search and optimization technique is used by many researchers to design the optimal KB of FRBM for various processes. The systems of combining Fuzzy logic and genetic algorithm are called genetic-fuzzy systems.

4.2 Fuzzy Logic

4.2.1 Crisp Set and Fuzzy Set

A set (A) is a collection of any objects (a_1, a_2, a_3, ….., a_n), which according to some law can be considered as a whole and it is usually written as

$A=\{a_1, a_2,\ldots..,a_n\}$ or

$A=\{x \mid P(x)\}$, means A is the set of all elements (x) of universal set X for which the proposition P(x) is true (e.g. P(x) > 3).

In crisp set, the function $X_A(x)$ (so-called characteristic function) assigns a value of either 1 or 0 to each individual object, x in the universal set, thereby discriminating between members and non-members of the crisp set, A under consideration. That means there exists no uncertainty or vagueness in the fact that the object belongs to the set or does not belong to the set. Set A is defined by its characteristic function, $X_A(x)$ as follows

$$X_A(x) = \begin{cases} 1 \text{ for } x \in A \\ 0 \text{ for } x \notin A \end{cases}$$

In the year, 1965, Lotfi A. Zadeh [3] proposed a completely new and elegant approach to vagueness and uncertainty in a seminal paper, called *fuzzy set theory*. In his approach an element, x can belong to a set to a degree, k ($0 \leq k \leq 1$) in contrast to classical set theory where an element must definitely belong or not to a set. A fuzzy set, Ã is usually written as

$\tilde{A}=\{x, \mu_{\tilde{A}}(x) \mid x \in X\}$

The function $\mu_{\tilde{A}}(x)$ is called membership function (MF) of the (fuzzy) set $\tilde{A}$ and defined as $\mu_{\tilde{A}}(x) \rightarrow [0,1]$. The value of $\mu_{\tilde{A}}(x)$ is called the degree of membership of x in the fuzzy set $\tilde{A}$.

4.2.2 *Fuzzy Membership Function*

Graphically, a membership function is represented as a curve (as shown in Figure 4.1) that defines how each element in the set is mapped to a membership value (or degree of membership) between 0 and 1. There are many ways to assign membership values or functions to fuzzy variables compared to that of assigning probability density function to random variables. The membership function assignment process can be intuitive or based on some algorithmic or logical operations.

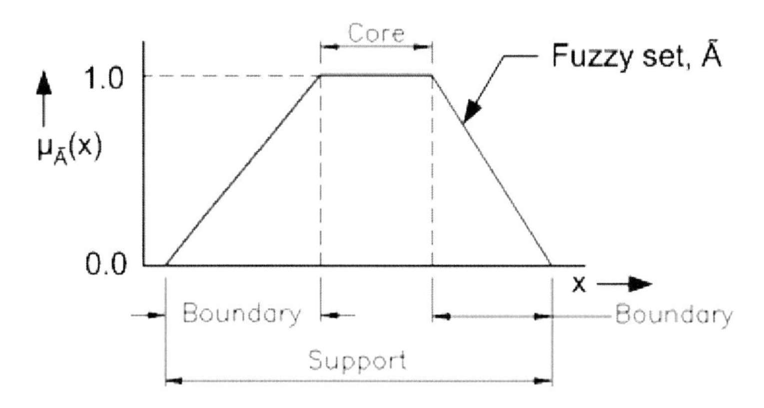

Fig. 4.1. A graphical representation of fuzzy set

4.2.2.1 Some Key Properties of Fuzzy Set

i) Having two fuzzy sets $\tilde{A}$ and $\tilde{B}$ based on X, then both are *equal* if their membership functions are equal, i.e. $\tilde{A} = \tilde{B} \Leftrightarrow \mu_{\tilde{A}}(x) = \mu_{\tilde{B}}(x)$ for all $x \in X$

ii) Given a fuzzy set $\tilde{A}$ defined on X and any number $\alpha \in [0,1]$, the *α-cut*, ${}^{\alpha}\tilde{A}$, and the *strong α-cut*, ${}^{\alpha+}\tilde{A}$, are the crisp sets: ${}^{\alpha}\tilde{A} = \{x \mid \tilde{A}(x) \geq \alpha\}$ and ${}^{\alpha+}\tilde{A} = \{x \mid \tilde{A}(x) > \alpha\}$

iii) The *height* of a fuzzy set is the largest membership grade obtained by any element in that set i.e., $\text{height}(\tilde{A}) = \max_{x \in X} \mu_{\tilde{A}}(x)$

iv) The *crossover points* of a membership function are defined as the elements in the universe for which a particular fuzzy set A has values equal to 0.5, i.e., for which $\mu_{\tilde{A}}(x) = 0.5$

v) A fuzzy set $\tilde{A}$ is called *normal* when $\text{height}(\tilde{A}) = 1$ and *subnormal* when $\text{height}(\tilde{A}) < 1$.

vi) The *support* of a fuzzy set $\tilde{A}$ is the crisp set that contains all the elements of X that have non-zero membership grades, i.e. $\text{support}(\tilde{A}) = \{x \in X \mid \mu_{\tilde{A}}(x) > 0\}$, refer to Figure 4.1.

vii) The *core* of a normal fuzzy set $\tilde{A}$ is the crisp set that contains all the elements of X that have the membership grades of one in $\tilde{A}$, i.e. $\text{core}(\tilde{A}) = \{x \in X \mid \mu_{\tilde{A}}(x) = 1\}$, refer to Figure 4.1.

viii) The *boundary* is the crisp set that contains all the elements of X that have the membership grades of $0 < \mu_{\tilde{A}}(x) < 1$ in $\tilde{A}$, i.e. $\text{boundary}(\tilde{A}) = \{x \in X \mid 0 < \mu_{\tilde{A}}(x) < 1\}$, refer to Figure 1.

ix) Having two fuzzy sets $\tilde{A}$ and $\tilde{B}$ based on X, then both are *similar* if $\text{core}(\tilde{A}) = \text{core}(\tilde{B})$ and $\text{support}(\tilde{A}) = \text{support}(\tilde{B})$.

x) If the support of a normal fuzzy set consists of a single element x_o of X, which has the property $\text{support}(\tilde{A}) = \text{core}(\tilde{A}) = \{x_0\}$, this set is called a *singleton*.

xi) A fuzzy set $\tilde{A}$ is said to be a *convex* fuzzy set if for any elements x, y and z in fuzzy set $\tilde{A}$, the relation x<y<z implies that $\mu_{\tilde{A}}(y) \geq \min[\mu_{\tilde{A}}(x), \mu_{\tilde{A}}(z)]$. The intersection of two convex fuzzy sets is also a convex fuzzy set, i.e., if $\tilde{A}$ and $\tilde{B}$ are *convex* fuzzy sets, then $\tilde{A} \cap \tilde{B}$ is also convex fuzzy set.

xii) If $\tilde{A}$ is convex single-point normal fuzzy set defined on the real line, then $\tilde{A}$ is often termed as a *fuzzy number*.

xiii) Any fuzzy set $\tilde{A}$ defined on a universe X is a *subset* of that universe.

4.2.2.2 Various Types of Fuzzy Membership Function and Its Mathematical Representation

The common types of membership function (MF) used in FRBM are triangular, (higher order) polynomial, trapezoidal, Gaussian, etc.

Trapezoidal MF: Mathematically a trapezoidal MF can be represented as shown in Figure 4.2(a).

$$\mu_{\tilde{A}}(x,a,b,c,d) = \begin{matrix} 0 \\ \frac{x-a}{b-a} \\ 1 \\ \frac{d-x}{d-c} \end{matrix} \quad \text{for} \quad \begin{cases} x < a, x > d \\ a \leq x \leq b \\ b < x < c \\ c \leq x \leq d \end{cases}$$

The controlling parameters toward the configuration of trapezoidal MF (as shown in Figure 4.2(a)) are b_1=b-a, b_2=c-b and b_3=d-c.

Polynomial MF: A polynomial MF can be expressed mathematically as shown in Figure 4.2(b):

$$\mu_{\tilde{A}}(x,a,b,c,d) = \begin{array}{c} 0 \\ f_1(x,b_1) \\ 1 \\ f_2(x,b_2) \end{array} \quad \text{for} \begin{cases} x < a, x > c \\ a \le x < b \\ x = b \\ b < x \le c \end{cases}$$

where the functions f_1 and f_2 are the polynomial type. Polynomial MF is treated as triangular type MF when the functions, f_1 and f_2 of the above empirical expression are linear. f_1 and f_2 may be also exponential or any other kind of functions.

The controlling parameters toward the configuration of polynomial MF (as shown in Figure 4.2(b)) are b_1=b-a and b_2=c-b. Mathematically, the second order polynomial function can be represented as $\mu_{\tilde{A}}(x) = c_0 x + c_1 x^2$, where x is the distance measured along the base-width of membership function distributions, $\mu_{\tilde{A}}(x)$ is the fuzzy membership function value and, c_0 and c_1 are the coefficients which can be determined based on some specified conditions, such as

$$\mu_{\tilde{A}} = \begin{cases} 1 \text{ at } x = b_1 \\ 0 \text{ at } x = 0 \end{cases} \text{ and } \frac{\partial \mu_{\tilde{A}}}{\partial x} = 0, \text{at } x = b_1,$$

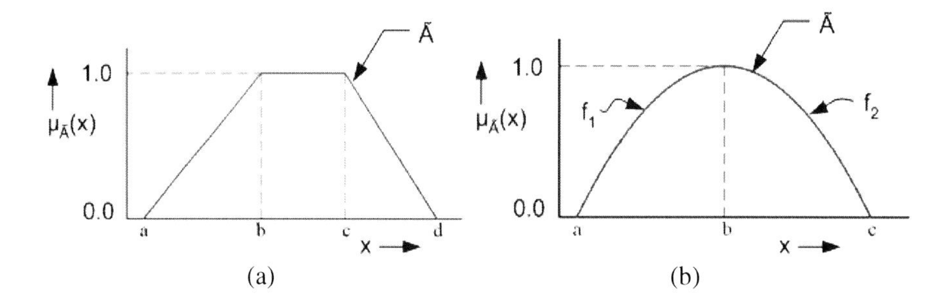

Fig. 4.2. Membership function configuration (a) Trapezoidal (b) Polynomial

Finally the coefficients of the 2nd order polynomial function become $c_0 = \frac{2}{b_1}$ and $c_1 = -\frac{1}{b_1^2}$.

From Figure 4.2, it has been seen that for a given value of support of the membership function of a fuzzy set, only one parameter, b_1 is required to describe triangular and polynomial type MFDs (membership function distributions) with 3 fuzzy sub-sets, whereas two parameters, b_1 and b_2 are required to explain the (semi) trapezoidal MFDs with two fuzzy subsets. The number of controlling parameter increases with increasing the number of fuzzy sub-sets involved in the MFDs.

4.2.3 Fuzzy Set Operators

Defining the fuzzy sets $\tilde{A}$ and $\tilde{B}$ on the universe X, for a given element x of the universe, the fuzzy set operations, intersection (t-norm), union (t-conorm) and complement are expressed as follows and the corresponding Venn diagrams are shown in Figure 4.3.

Intersection: $\mu_{\tilde{A}\cap\tilde{B}}(x) = \mu_{\tilde{A}}(x) \wedge \mu_{\tilde{B}}(x)$

Union: $\mu_{\tilde{A}\cup\tilde{B}}(x) = \mu_{\tilde{A}}(x) \vee \mu_{\tilde{B}}(x)$

Complement: $\mu_{\bar{\tilde{A}}}(x) = 1 - \mu_{\tilde{A}}(x)$

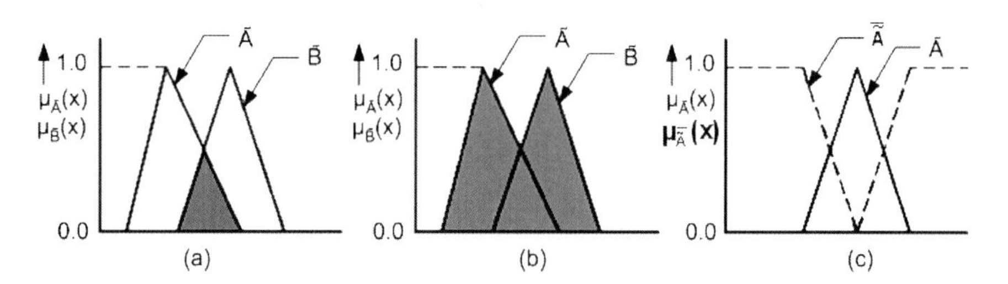

Fig. 4.3. (a) Intersection of fuzzy sets $\tilde{A}$ and $\tilde{B}$ (b) Union of fuzzy sets $\tilde{A}$ and $\tilde{B}$ (c) Complement of fuzzy set $\tilde{A}$

4.2.4 Classical Logical Operations and Fuzzy Logical Operations

Classical logic deals with classical proposition (P) which is a collection of elements, that is, a set, where all the truth values, T(P) for all elements in the set are either all true (1) or all false (0), and follows the two-valued logical operations and Boolean algebra.

Let the sets A and B are defined from universe X and the proposition, P measures the truth of the statement that an element, x from the universe X is contained in set, A and Q measures the truth of the statement that this element, x is contained in set,

B, i.e., if $x \in A, T(P)=1$; otherwiase $T(P)=0$ and if $x \in B, T(Q)=1$; otherwise $T(Q)=0$. There are five logical connectives (defined as follows) used to combine multiple simple propositions and to form new propositions:

Disjunction (OR):

$$P \vee Q : x \in A \text{ or } x \in B, \text{Hence}, T(P \vee Q) = \max(T(P), T(Q))$$

Conjunction (AND):

$$P \wedge Q : x \in A \text{ or } x \in B, \text{Hence}, T(P \wedge Q) = \min(T(P), T(Q))$$

Negation: $\text{If } T(P) = 1, \text{ then } T(\overline{P}) = 0; \text{ if } T(P) = 0, \text{ then } T(\overline{P}) = 1$

Implication: $P \rightarrow Q : x \notin A \text{ or } x \in B, \text{Hence}, T(P \rightarrow Q) = T(\overline{P} \cup Q)$

Equivalence: $(P \leftrightarrow Q) : T(P \leftrightarrow Q) = \begin{cases} 1 \text{ for } T(P) = T(Q) \\ 0, \text{ for } T(P) \neq T(Q) \end{cases}$

For two different universes of discourse where P is a proposition described by set A, which is defined on universe X, and Q is a proposition described by set B, which is defined on universe Y. Then the implication P→Q (which is also equivalent to the linguistic rule form, IF A THEN B) can be represented in set-theoretic terms by the relation, R which is defined by

$$R = (A \times B) \cup (\overline{A} \times Y) \equiv \text{IF A, THEN B}$$

$$\text{IF } x \in A, \text{ where } x \in X \text{ and } A \subset X$$

$$\text{THEN } y \in B, \text{ where } y \in Y \text{ and } B \subset Y$$

The other connectives are applicable to two different universes of discourse as usual. Classical logical compound propositions that are always true irrespective of the truth values of the individual simple propositions are called tautologies.

Fuzzy propositional logic generalizes the classical propositional operations by using the truth set [0, 1] instead of either 1 or 0. The above logical connectives are also defined for a fuzzy logic. Like classical logic, the implication connective in fuzzy logic can be modeled in rule-based form: $\tilde{P} \rightarrow \tilde{Q}$ is, IF x is $\tilde{A}$ THEN y is $\tilde{B}$ (where IF part is called *antecedent* and THEN part is called *consequent*) and it is equivalent to the fuzzy relation $\tilde{R} = (\tilde{A} \times \tilde{B}) \cup (\overline{\tilde{A}} \times Y)$ where the fuzzy proposition $\tilde{P}$ is assigned to fuzzy set $\tilde{A}$ which is defined on universe X, and the fuzzy proposition $\tilde{Q}$ is described by fuzzy set $\tilde{B}$, which is defined on universe Y. The membership function of $\tilde{R}$ is expressed by $\mu_{\tilde{R}}(x,y) = \max[(\mu_{\tilde{A}}(x) \wedge \mu_{\tilde{B}}(y)), (1 - \mu_{\tilde{A}}(x))]$. The implication connective can be defined in several distinct forms. While these forms of implication are equivalent in classical logic, their extensions to fuzzy logic are not equivalent and result in distinct classes of fuzzy implications.

4.2.5 Fuzzy Implication Methods

The fuzzy implication operation is used to find the fuzzy relation $\tilde{R}$ between two fuzzy sets $\tilde{A}$ and $\tilde{B}$ which are defined on the universes of discourse X and Y, respectively based on the rule IF x is $\tilde{A}$, THEN y is $\tilde{B}$, where $x \in X$ and $y \in Y$. Mathematically, the fuzzy relation, $\tilde{R}$ is defined as $\tilde{R}(x, y) = \wp[A(x), B(y)]$, where $\wp$ is called the implication operator. Besides the implication method as presented in Section 4.2.4, there are other forms of implication operator, among them *min* and *product* implication operators are mostly used in fuzzy inference system for practical applications. The membership function values of fuzzy relation $\tilde{R}$ defined on the Cartesian product space $X \times Y$ using min and product implication operators are obtained by the Equation (4.1) and Equation (4.2), respectively and graphically represented in Figure 4.4.

min:
$$\mu_{\tilde{R}}(x, y) = \min\left[\mu_{\tilde{A}}(x), \mu_{\tilde{B}}(y)\right] \tag{4.1}$$

product:
$$\mu_{\tilde{R}}(x, y) = \mu_{\tilde{A}}(x) \bullet \mu_{\tilde{B}}(y) \tag{4.2}$$

In Figure 4.4, the MF value, 0.7 of $\mu_{\tilde{B}}(x)$ corresponds to rule weight obtained after decomposition of the IF part of rule.

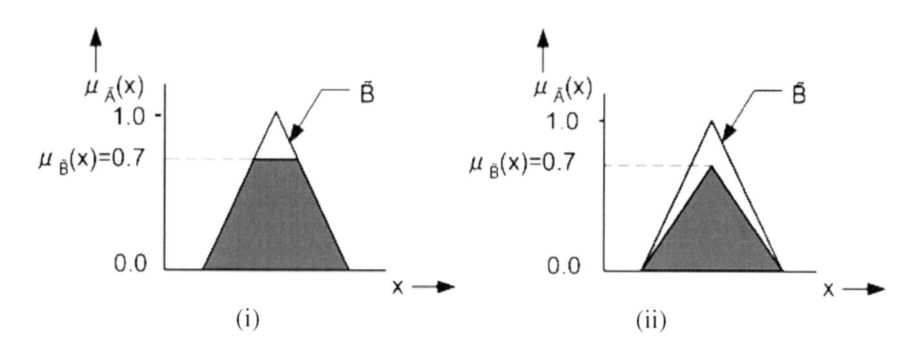

Fig. 4.4. Graphical representation of fuzzy implication (i) min (ii) product

4.2.6 Decomposition of Compound Rules

The most common techniques for decomposition of compound linguistic rules into simple canonical forms are described as follows:

Multiple conjunctive antecedents: IF x is $\tilde{A}_1$ AND x is $\tilde{A}_2$ .... AND x is $\tilde{A}_L$ THEN y is $\tilde{B}_S$

This fuzzy rule can be written in canonical form as: IF $\tilde{A}_S$ THEN $\tilde{B}_S$, where the fuzzy subset $\tilde{A}_S$ is defined as $\tilde{A}_S = \tilde{A}_1 \cap \tilde{A}_2 \cap ...... \cap \tilde{A}_L$) with the membership function $\mu_{\tilde{A}_S}(x) = \min[\mu_{\tilde{A}_1}(x), \mu_{\tilde{A}_2}(x), ..., \mu_{\tilde{A}_L}(x)]$ obtained by the definition of the fuzzy intersection operation.

Multiple disjunctive antecedents: IF x is $\tilde{A}_1$ OR x is $\tilde{A}_2$ .... OR x is $\tilde{A}_L$ THEN y is $\tilde{B}_S$

Assuming a new fuzzy subset $\tilde{A}_S$ (as $\tilde{A}_S = \tilde{A}_1 \cup \tilde{A}_2 \cup ...... \cup \tilde{A}_L$) expressed by means of membership function $\mu_{\tilde{A}_S}(x) = \max[\mu_{\tilde{A}_1}(x), \mu_{\tilde{A}_2}(x), ..., \mu_{\tilde{A}_L}(x)]$ based on the definition of the fuzzy union operation, the compound rule may be written in canonical form as: IF $\tilde{A}_S$ THEN $\tilde{B}_S$.

4.2.7 Aggregation of Rule

The technique of obtaining the overall rule consequent by combining the individual consequents contributed by each rule in the rule base (which comprises multiple rules) is known as aggregation of rules. The most popular aggregation techniques of fuzzy rules are as follows:

Conjunctive system of rules (MIN): In the case of a system of rules that must be jointly satisfied, the rules are connected by AND connectives. In this case, the aggregated consequent, y is obtained by fuzzy intersection of all individual rule consequents (y_1, y_2, ..., y_r), as $y = y_1 \cap y_2 \cap ...... \cap y_r$ which is defined by the membership function: $\mu_y(y) = \min[\mu_{y_1}(y), \mu_{y_2}(y), ..., \mu_{y_r}(y)]$ for $y \in Y$.

Disjunctive system of rules (MAX): For the case of a disjunctive system of rules where the satisfaction of at least one rule is required, the rules are connected by the OR connectives. In this case, the aggregated consequent, y is obtained by fuzzy union of the individual rule consequents as $y = y_1 \cup y_2 \cup ...... \cup y_r$ which is defined by the membership function: $\mu_y(y) = \max[\mu_{y_1}(y), \mu_{y_2}(y), ..., \mu_{y_r}(y)]$ for $y \in Y$.

4.2.8 Composition Technique of Fuzzy Relation

Let R and S be the relations that relate elements from universe X to universe Y, and elements from universe Y to universe Z, respectively. Now, composition is an operation to find another relation, T that relates the same elements in universe X that R contains to the same elements in universe Z that S contains. The composition operation of fuzzy relation reflects the inference of a fuzzy rule-based system and is expressed by $\tilde{B} = \tilde{A} \circ \tilde{R}$, where $\tilde{A}$ is the input, or antecedent defined on the universe X, $\tilde{B}$ is the output or consequent defined on universe Y and $\tilde{R}$ is a fuzzy relation characterizing the relationship between specific input(s), x and specific output(s), y. Among various methods of composition of fuzzy relation, max-min and max-product are the most commonly used techniques and defined by membership function-theoretic expressions as follows.

$$\text{max-min: } \mu_{\tilde{B}}(y) = \max_{x \in X} \{\min[\mu_{\tilde{A}}(x), \mu_{\tilde{R}}(x, y)]\} \tag{4.3}$$

$$\text{max-product: } \mu_{\tilde{B}}(y) = \max_{x \in X} [\mu_{\tilde{A}}(x) \bullet \mu_{\tilde{R}}(x, y)] \tag{4.4}$$

The method of composition of fuzzy relation basically includes the implication method and technique of aggregation of fuzzy rule. In the above methods of composition of fuzzy relation, max is the aggregation technique of rule, whereas min and product are the implication methods used in Equation (4.3) and Equation (4.4), respectively.

4.2.9 Fuzzy Inferences

Inference is a process of combining the measurement of input(s)/antecedent(s) with one or more relevant fuzzy rules in a proper manner to infer the output(s)/consequent(s). In order to demonstrate the inference method, consider a system of one input (antecedent) and a single output (antecedent) described by two IF-THEN rules as follows:

Rule 1: if input1 is $\tilde{A}_1$, then output1 is $\tilde{B}_1$
Rule 1: if input1 is $\tilde{A}_2$, then output1 is $\tilde{B}_2$

Now given the IF-THEN rules (Rule 1 and Rule 2) and a fact/measurement "Input1 is $\tilde{A}$" it is inferred that "output1 is $\tilde{B}$", where $\tilde{A}, \tilde{A}_1 \in F(\chi)$ and $\tilde{B}, \tilde{B}_1 \in F(\Upsilon)$, where $F(\chi)$ and $F(\Upsilon)$ denote the sets of all ordinary fuzzy sets that can be defined within the X and Y, respectively. χ and Υ are the sets of values (x and y) of variables, input1 (condition variable) and output1 (action variable), respectively. In order to determine B, method of interpolation is used which consists of two steps explained as follows:

Step 1: Calculate the *degree of consistence,* $r_j(x)$ between the given fact/measurement and the antecedent of each rule, j in terms of the *height* of intersection of the associated sets $\tilde{A}_1$ and $\tilde{A}$. $r_j(x)$ is expressed using the standard fuzzy intersection and the definition of height of a fuzzy set (described in Section 4.2.2.1) as follows:

$$r_j(x) = \max_{x \in X} \min\left[\mu_{\tilde{A}}(x), \mu_{\tilde{A}_j}(x)\right], \; j=1, 2 \tag{4.5}$$

Step 2: Calculate the conclusion $\tilde{B}$ by truncating each set $\tilde{B}_j$ by the value of $r_j(x)$ (i.e., *min* implication method) which expresses the degree to which the antecedent $\tilde{A}_j$ is compatible with the given fact $\tilde{A}_1$, and taking the union of the truncated sets as the rules are satisfied independently (i.e., *max* aggregation method) .

$$\mu_{\tilde{B}}(y) = \max_{j=1,2} \min\left[r_j(x), \mu_{\tilde{B}_j}(y)\right] \text{ for all } y \in Y \tag{4.6}$$

The above steps are graphically presented in Figure 4.5.

$$\text{Now, } \mu_{\tilde{B}}(y) = \max_{j=1,2} \min\left[r_j(x), \mu_{\tilde{B}_j}(y)\right]$$

$$= \max_{j=1,2} \min\left[\max_{x \in X} \min\left(\mu_{\tilde{A}}(x), \mu_{\tilde{A}_j}(x)\right), \mu_{\tilde{B}_j}(y)\right]$$

$$= \max_{j=1,2} \max_{x \in X}\left[\min\left(\mu_{\tilde{A}}(x), \mu_{\tilde{A}_j}(x)\right), \mu_{\tilde{B}_j}(y)\right]$$

$$= \max_{x \in X} \max_{y=1,2} \min\left[\mu_{\tilde{A}}(x), \min\left(\mu_{\tilde{A}_j}(x), \mu_{\tilde{B}_j}(y)\right)\right]$$

$$= \max_{x \in X} \min\left[\mu_{\tilde{A}}(x), \max_{j=1,2} \min\left(\mu_{\tilde{A}_j}(x), \mu_{\tilde{B}_j}(y)\right)\right]$$

$$= \max_{x \in X} \min\left[\mu_{\tilde{A}}(x), \mu_{\tilde{R}}(x, y)\right], \tag{4.7}$$

which is equivalent to the expression of max-min composition presented in Equation (4.3) and accordingly, this inference method is also called max-min inference method. In the above step 2, if *product* implication technique and *max* aggregation method are used, we can obtain the expression (defined in Equation 4.7) similar to max-product composition (Equation (4.4)). Moreover, the above inference method may also be applicable to the system with any number of fuzzy rule (j) and inputs (Input1, input2). In case of multiple inputs/antecedents of fuzzy rule, the (effective) *degree of consistence*, $r_j(x)$ between the given facts/measurements and the antecedents of each rule is obtained by first finding the *degree of consistence* between each fact/measurement and the related antecedent of the rule (using Equation (4.5)), then adopting the technique of decomposition of compound rules according to the type of logical connectives (AND or OR) as explained graphically in Figure 4.6 for the following two IF-THEN rules as follows:

Rule 1: if input1 is $\tilde{A}_{11}$ AND input2 is $\tilde{A}_{12}$ then output1 is B_1
Rule 1: if input1 is $\tilde{A}_{21}$ AND input2 is $\tilde{A}_{22}$ then output1 is B_2
Fact: input1 is $\tilde{A}_1$ AND input2 is $\tilde{A}_2$
Conclusion: output1 is B

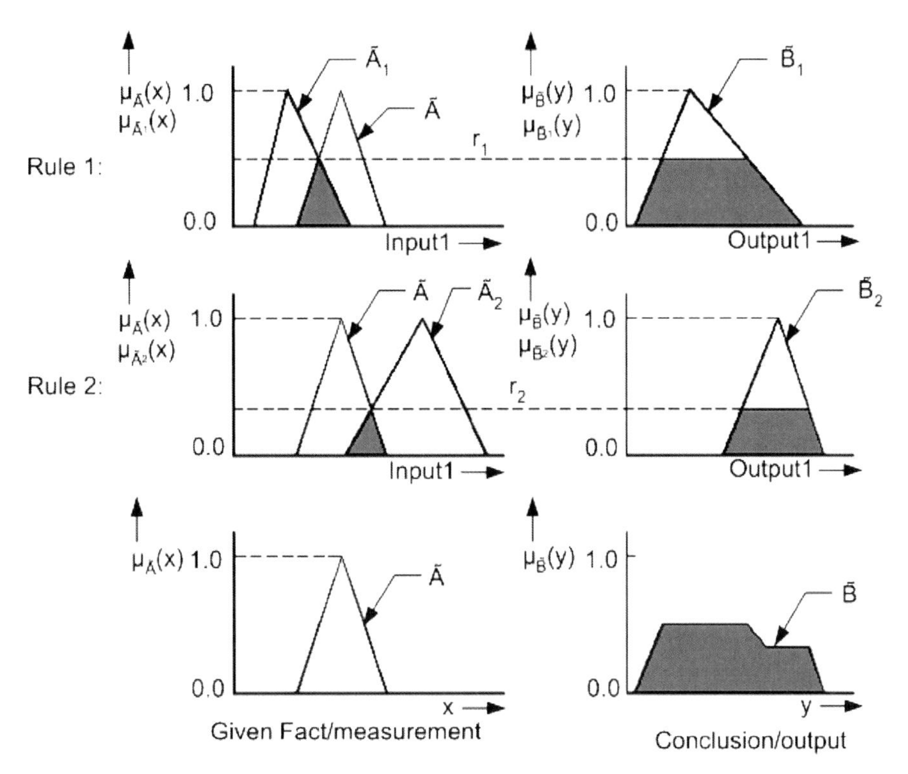

Fig. 4.5. Illustration of the method of interpolation in fuzzy inferences

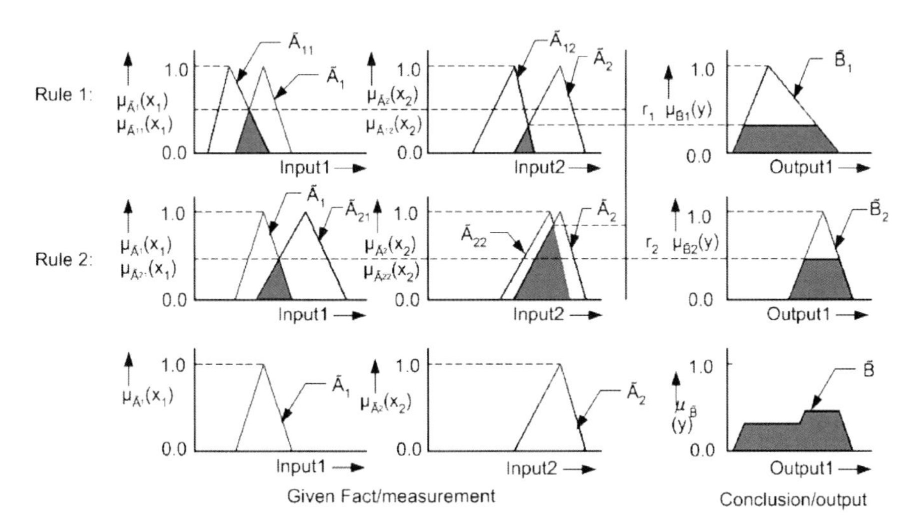

Fig. 4.6. Illustration of the method of interpolation in fuzzy inferences with multiple inputs

In the above illustration of fuzzy inference, we have considered fuzzy value ($\tilde{A}$) for the variable input1. Figure 4.7 demonstrates the above (max-min) inference method for a two input and a single output system where the values of input variables are considered as crisp type (for instance, Fact: input1 is x_1 AND input2 is x_2) and the (max-product) inference method for the same system is demonstrated in Figure 4.8.

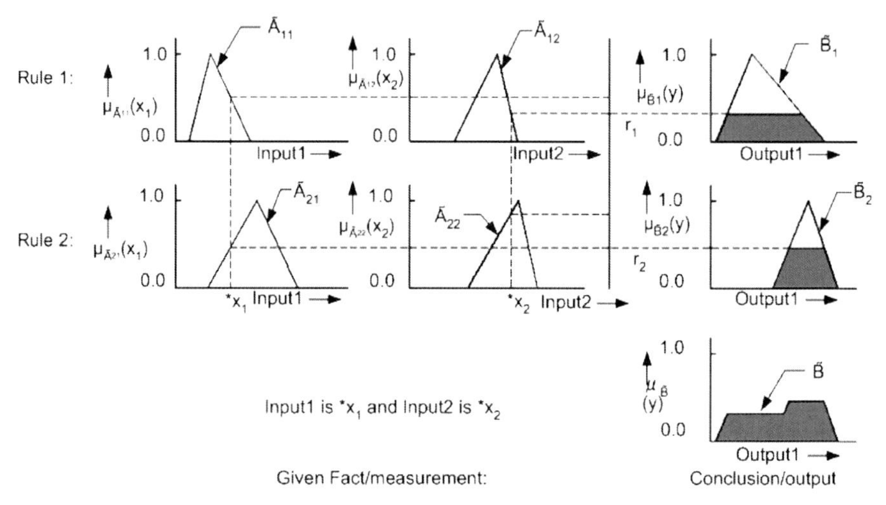

Fig. 4.7. Graphical representation of max-min inference method with crisp type of input values

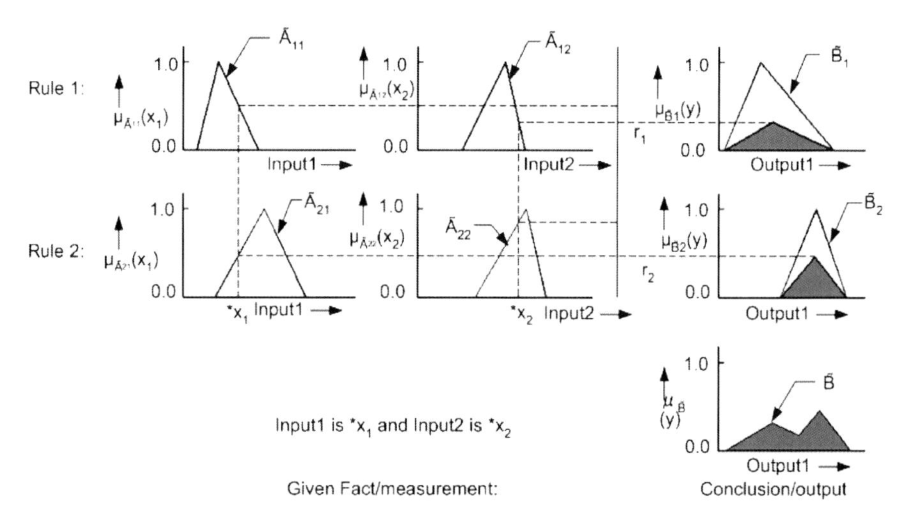

Fig. 4.8. Graphical representation of max-product inference method with crisp type of input values

4.2.10 Fuzzification and De-fuzzification

The fuzzification is a process of transforming the crisp value into a grade of membership using a membership function of the associated fuzzy set as shown in Figure 4.9. Figure 4.9 demonstrates that the given (crisp) value x_0 of variable x belongs to a grade of $\mu_{\tilde{A}_1}(x_0) = 0.8$ to the fuzzy set $\tilde{A}_1$ and with a grade of $\mu_{\tilde{A}_2}(x_0) = 0.2$ to the fuzzy set $\tilde{A}_2$. Fuzzification is required in the fuzzy inference system when the values of input variables to system are considered as crisp type (as described in Figure 4.7/Figure 4.8).

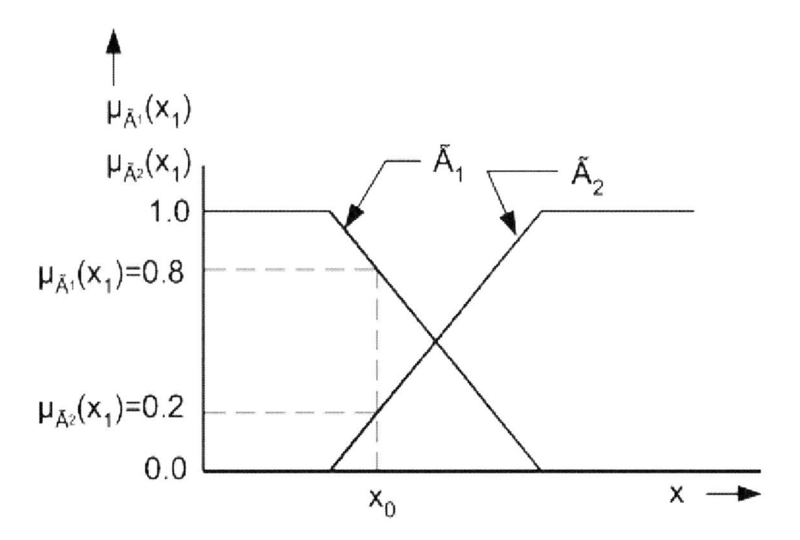

Fig. 4.9. Fuzzification method

Defuzzification is the conversion of a fuzzy quantity to a crisp quantity analogous to fuzzification. Defuzzification is used in fuzzy inference system to convert the fuzzy value of the output (i.e., $\tilde{B}$ of output1 in Section 4.2.9) to a crisp value (*y). Among various defuzzification methods available in the literature, centroid method (also called center of area (COA) or center of gravity) is most popular, which is mathematically expressed by Equation (4.8) and graphically expressed in Figure 4.10:

$$y_{COA} = \frac{\int \mu_{\tilde{B}}(y) \cdot y dy}{\int \mu_{\tilde{B}}(y) dy} \tag{4.8}$$

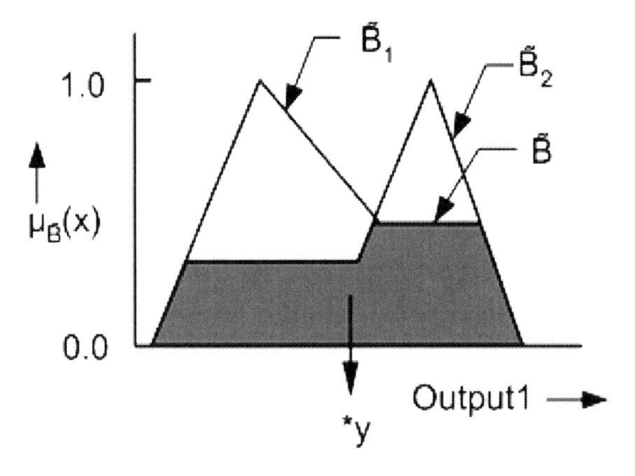

Fig. 4.10. Centroid defuzzification method

4.2.11 Fuzzy Rule-Based Model

4.2.11.1 Working Principle of a Fuzzy Rule-Based Model

A fuzzy rule base, a fuzzy inference engine, fuzzification and de-fuzzification: these are the four modules involved in a FRBM (Fuzzy rule-based model). Figure 4.11 shows a schematic diagram explaining the working cycle of a FRBM. The following steps are involved in the working cycle of a FRBM:

- The output(s) (action) and input(s) (condition) variables needed to control a particular process are chosen and measurements are taken for all the condition variables.
- The measurements taken in the previous steps are converted into appropriate fuzzy sets to express measurement uncertainties (called fuzzification as described in Section 4.2.10).
- The fuzzified measurements are then used by the interference engine to evaluate the control rules stored in the rule base and a fuzzified output is determined (as discussed in Section 4.2.9).
- The fuzzified output is then converted into a single (crisp) value (called a de-fuzzification as illustrated in Section 4.2.10). The de-fuzzified value(s) represent action(s)/prediction(s) to be made by the FRBM in controlling a process.

The kernel of a FRBM that is the knowledge base (KB) is constituted by rule base, RB (a set of fuzzy logic rules) and membership functions/membership function distributions, MFDs (fuzzy subsets). Two types of fuzzy logic rules (FLRs) that are commonly used for constructing the RB of a FRBM are Mamdani-type and TSK-type. In both types of FLRs, the input variables are expressed by linguistic terms (fuzzy subsets) in the rule antecedent part. But the main difference between these two types of fuzzy rules lies in the rule consequent part. The output variable in Mamdani-type FLR is defined by linguistic term also, whereas in TSK-type FLR, it is not defined by linguistic term rather it is defined by a linear combination of the input variables. The shape of fuzzy subsets (MFDs of input-output variables) is also an important factor that is to be decided appropriately to achieve the best performance of a FRBM for a typical process.

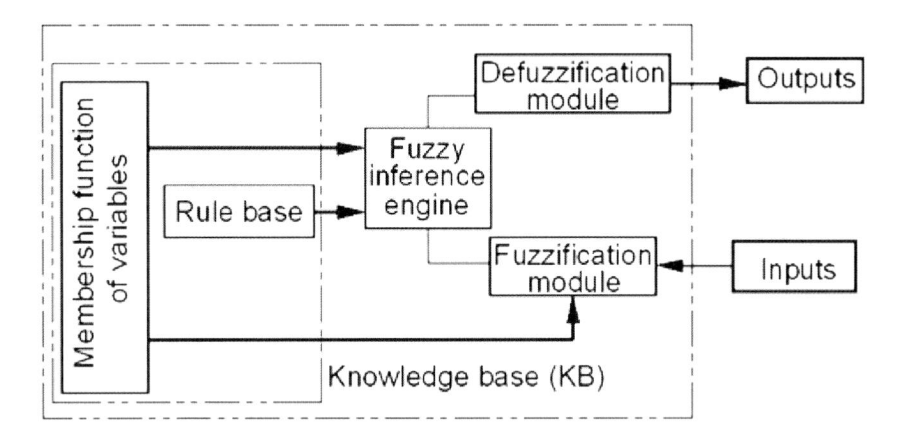

Fig. 4.11. A schematic showing the working cycle of a FRBM

4.2.11.2 Various Types of Fuzzy Rule-Based Model

Mamdani-type [4]:
The structure of Mamdani-type fuzzy logic rule is expressed as follows:

IF x_1 is A_1 AND x_2 is A_2 AND........AND x_n is A_n THEN y is B

where x_i (i=1, 2, ..., n) are input variables and y is the output variable. A_1, A_2, ..., A_n and B are the linguistic terms (say, Low, Medium, High, etc.) used for the fuzzy subsets (membership function distributions) of the corresponding input and output variables, respectively.

Sugeno-type [5]:
The Sugeno-type fuzzy rule is defined as follows:

IF x_1 is A_1 AND x_2 is A_2 AND........AND x_n is A_n THEN $y = f(x_1, x_2, .., x_n)$

Unlike Mamdani-type, the rule consequent/output is expressed by a function of the input variables.

Tsukamoto [6]:
The Tsukamoto -type fuzzy rule is defined as follows:

IF x_1 is A_1 AND x_2 is A_2 AND........AND x_n is A_n THEN y =z

where z is a monotonical membership function.

4.2.11.2.1 Mamdani-Type Fuzzy Rule Based Model

The output of a Mamdani-type FRBM whose rule base (RB) is constructed using Mamdani-type fuzzy logic rule is obtained as follows (when centroid method is considered for defuzzification):

$$Y = \frac{\sum_{r=1}^{R_f^1} A^{\alpha r} C_{A^{\alpha r}}}{\sum_{r=1}^{R_f^1} A^{\alpha r}} \tag{4.9}$$

where $A^{\alpha r}$ is the area of fuzzy subset of output variable, covered by α membership value (equivalent to *degree of consistence* as obtained in Step 1 in Section 4.2.9) that is obtained by r^{th} rule after fuzzy inference method. $C_{A^{\alpha r}}$ is the center distance of the area, $A^{\alpha r}$. R_f^1 ($R_f^1 \subseteq R_f$) is the number of rules fired out of a total of R_f rules present in the rule base for a set of input values.

In order to determine the output in Equation 4.9, all the four modules (fuzzy rule base, a fuzzy inference engine, fuzzification and de-fuzzification) as mentioned in Section 4.2.11.1 are involved in a Mamdani-type FRBM. The performance of a Mamdani-type fuzzy model is relied on the appropriate fuzzy subsets of rule consequents and antecedent, and the type of fuzzy subsets (membership function distributions) considered for input-output variables. Therefore, the tasks of designing FRBMs with Mamdani-type FLRs are:

- construction of an optimal set of rules (R_f) with its appropriate outputs (B),
- selection of shape of fuzzy subsets/MFDs for both the input and output variables
- tuning of MFDs.

4.2.11.2.2 TSK-Type Fuzzy Rule-Based Model

The TSK-type fuzzy logic rule is defined as follows [5, 7]:

If x_1 is A_1 and x_2 is A_2 and......................and x_n is A_n, then $y = \sum_{j=1}^{K} c_j f_j(x_{1,\dots,n})$

where $A_1, \dots, A_n$ are the fuzzy subsets of the respective input variables, $x_1, \dots, x_n$. The output function of fuzzy rule is a linear function (say, polynomial) in the form of

$$y = \sum_{j=1}^{K} c_j f_j(x_{1,\dots,n}) = c_1 f_1(x_{1,\dots,n}) + c_2 f_2(x_{1,\dots,n}) + \dots + c_k f_K(x_{1,\dots,n}) \tag{4.10}$$

The overall output of the TSK-type fuzzy model can be obtained for a set of inputs ($x_1, x_2, \dots, x_n$) using the following empirical expression.

$$Y = \frac{\sum_{r=1}^{R_f^l} \left(\prod_{v=1}^{n} \mu_v(x_v) \right) \sum_{j=1}^{K} c_j^r f_j^r (x_{1,\dots,n})}{\sum_{r=1}^{R_f^l} \left(\prod_{v=1}^{n} \mu_v^r (x_{1,\dots,n}) \right)} \quad (4.11)$$

$\prod$ is the product representing a conjunction decomposition method. $\sum_{j=1}^{K} c_j^r f_j^r (x_{1,\dots,n})$ is the output function of rth rule and c_j^r are the function coefficients of the corresponding rule consequent, where K is the number of coefficients present in the consequent function of each rule.

Unlike Mamdani-type FRBM, TSK-type FRBM includes only the fuzzy rule base, a fuzzy inference engine, and fuzzification module to determine the output in Equation (4.11). The performance of a TSK-type fuzzy model is mainly depended on the optimal values of the rule output (consequent) functions which are depended on the coefficients (c_j), the exponential parameters of the input variables (not shown in the Equation (4.10)) and choice of the fuzzy subsets (membership function distributions). Thus, the steps of developing FRBM with TSK-type FLRs are:

- construction of an optimal set of rules (R_f) with the appropriate structures of rule output/consequent functions
- selection of shapes of fuzzy subsets/MFDs of input variables
- determination of optimal values of coefficients and power terms of rule consequent functions
- tuning of MFDs of the input variables.

4.3 Genetic Algorithm

Genetic algorithm is a search and optimization technique which mimics the principle of natural selection and natural genetics [8] to find the best solution for a specific problem. The genetic algorithm is an approach to solve problems which are not yet fully characterized or too complex to allow full characterization, but for which some analytical evaluation or physical interpretation to evolutes the performance of a solution, is available. GA is a stochastic global search method that mimics the metaphor of natural biological evolution. Genetic algorithms operate on a population of feasible solutions by applying the principle of survival of the fittest to produce better approximations to a solution. At each generation, a new set of approximations is created by the process of selecting individuals according to their level of fitness in the problem domain and breeding them together using operators borrowed from natural genetics. This process leads to the evolution of populations of individuals that are better suited to their environment than the individuals that they were created from, just as in natural adaptation.

The basic concept of a genetic algorithm is to encode a potential solution to a problem as a series of parameter strings, *chromosomes*, composed over some alphabet, so that chromosome values (*genotypic*) are uniquely mapped onto the decision variable (*phenotypic*). A single set of parameter string or chromosome is treated as the genetic material of an individual solution. Initially a large population of candidate solutions is created with random parameter values. These solutions are essentially bred with each other for several simulated generations under the principle of survival of the fittest, meaning that the probability that an individual solution will pass on some of its parameter values to subsequent children is directly related to the fitness of individual i.e. how good that solution is relative to the others in the population.

Breeding takes place through use of recombination operators such as crossover, which simulates basic biological cross-fertilization, and mutation, essentially the introduction of noise. The simple application of these operators with a reasonable selection mechanism has produced surprisingly good results over a wide range of problems. After recombination and mutation, the individual strings are then, if necessary, decoded, the objective function evaluated, a fitness value assigned to each individual and individuals selected for mating according to their fitness, and so the process continues through subsequent generations. In this way, the average performance of individuals in a population is expected to increase, as good individuals are preserved and bred with one another and the less fit individuals die out. The GA is terminated when some criteria are satisfied, e.g. a certain number of generations, a mean deviation in the population, or when a particular point in the search space is encountered.

4.3.1 Genetic Algorithms and the Traditional Methods

The working principle of GA clearly indicates that the GA significantly differs in some very fundamental way from other traditional search and optimization methods. The four major significant differences are:

- GAs search a population of solutions instead of a single point solution as in traditional search methods.
- GAs do not use derivative-based algorithm. It does not use any derivative information or other auxiliary knowledge; only the objective function and corresponding fitness levels influence the directions of search.
- GAs use probabilistic transition rules instead of deterministic principle.

It is important to note that the GA provides a number of potential solutions to a given problem and the choice of final solution is left to the user. In cases where a particular problem does not have one individual solution, for example a family of Pareto-optimal solutions, as is the case in multi-objective optimization and scheduling problems, then the GA is potentially useful for identifying these alternative solutions simultaneously.

4.3.2 Simple Genetic Algorithm

The schematic shown in Figure 4.12 illustrates the structure of a simple genetic algorithm (SGA) as described by Goldberg [8]. GA starts with initial random population consisted of potential solution points called individuals. The decision is made whether the individual is good or bad for the given problem based on the fitness obtained from the evaluation of objective function. Once the fitness value is evaluated and assigned to each individual, then initial population meets the first genetic operator, selection process. This operator provides more chances of survival for the strong individuals and to decay the weakest ones according to their fitness.

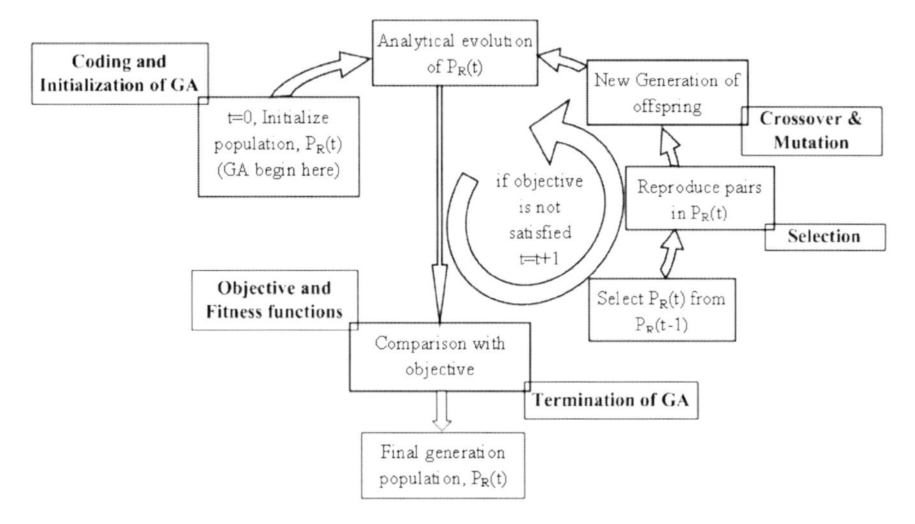

Fig. 4.12. A schematic representation of simple genetic algorithm outline population representation and initialization

Next, crossover operator is performed on selected individuals to build the new individuals by combining the existing ones. Crossover follows reproduction and allows two individuals to swap their structures depending on the probability factor. This result in the creation of a pair of offspring solution containing characteristics of their parents. Then the mutation operator is applied to supply diversity in the population. As the fitness of a population may remain static for a number of generations before a superior individual is found, the application of conventional termination criteria becomes problematic. A common practice is to terminate the GA after a pre-defined number of generations and then test the quality of the best members of the population against the problem definition. If no acceptable solutions are found, the GA may be restarted or a fresh search initiated with more number of generations.

4.3.2.1 Coding and Initialization of GA

4.3.2.1.1 Binary Coding

The most commonly used representation of chromosomes in the GA is that of the single-level binary string. Here, each decision variable (say, d and h) in the parameter set is encoded as a binary string (s_d and s_h, respectively) and these are concatenated to form a chromosome as shown in Figure 4.13. Binary-coded GAs are not restricted to use only integer and for the given lower bound (d_{min}) and upper bound (d_{max}) of a variable (say, d), the value of the variable (d) is calculated from the GA-string using the decoding scheme represented by Equation (4.12).

$$d = d^{min} + \frac{d^{max} - d^{min}}{2^{l_d} - 1} DV(s_d) \tag{4.12}$$

where l_d is the string length used to code the d variable and $DV(s_d)$ is the decoded value of the string s_d. This mapping function allows

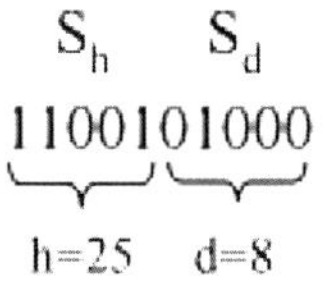

$$\overset{S_h}{\underbrace{11001}_{h=25}}\overset{S_d}{\underbrace{01000}_{d=8}}$$

Fig. 4.13. A schematic representation of a chromosome with 5 bits for S_d and 5 bits for S_h

4.3.2.1.2 Real Coding

For a continuous search space, binary-coded GA faces many problems such as

- Hamming cliffs associated with certain strings (e.g., 01111 and 10000) from which a transition to a neighboring solution requires the alteration of many bits.
- Inability to achieve any arbitrary precision in the optimal solution. The more the required precision, the larger is the string length, results in more computational complexity.

In real coding, the variables are directly represented in real type as $\underbrace{25}_{h}\ \underbrace{8}_{d}$.

4.3.2.1.3 Initialization of GA

Initial population of a GA is normally determined at random. With a binary population of N_{ind} individuals whose chromosomes are L_{ind} bits long, $N_{ind} \times L_{ind}$ random numbers uniformly distributed from the set {0, 1} would be produced.

4.3.2.2 Objective Function and Fitness Function

The objective function is used to provide an analytical measure of how individuals have performed in the problem domain. The objective function, f is a function of decision variable(s). The fitness value of an individual/solution in the population is determined based on the fitness function which consists of the objective function value and the penalty value for constraint violation which is determined by a penalty function, $f_{costraint}$. Thus, the fitness value is calculated as follows:

$$\text{Fitness value} = f + f_{costraint},$$

Since GAs mimic the survival-of-the-fittest of the nature to make a search process, GAs are suitable for maximization problem where objective function is directly used as fitness function F(x). In the case of a minimization problem, the fit individuals will have the lowest numerical value of the associated objective function. This situation is handled using a conversion of the objective function into an equivalent maximization problem and used as fitness function so that the optimum point remain unchanged.

4.3.2.3 Selection

Selection guides the tool to find the optimized solution by preferring individuals/members of the population with higher fitness over one with lower fitness. It is the operator which generates the mating pool. This operator determines that the number of times a particular individual will be used for reproduction and the number of offspring that an individual will produce. Some of the popularly used selection methods are as follows:

Roulette Wheel Selection Methods: Roulette wheel selection scheme chooses a certain individual with a probability proportional to its fitness.

$$p\left[I_{j,t}\right] = \frac{f\left(I_{j,t}\right)}{\sum_{k=1}^{n} f\left(I_{k,t}\right)} \tag{4.13}$$

where $p\left[I_{j,t}\right]$ is the probability of getting selected of any j^{th} individual at a generation t, $f\left(I_{j,t}\right)$ and $\sum_{k=1}^{n} f\left(I_{k,t}\right)$ are corresponding individual fitness and the sum of the fitness of the population with size n, respectively.

The property as represented by Equation (4.13) is satisfied by applying a random experiment that has some similarity with a generalized roulette game. In the roulette game, the slots are not equally wide that is, why different outcomes occur with different probabilities. Figure 4.14 gives a graphical representation of how this roulette wheel game works.

Linear rank selection: In this plan, a small group of individuals is taken from the population and the individual with best fitness is chosen for reproduction. The size of the group chosen is called the tournament size. A tournament size of two is called binary tournament.

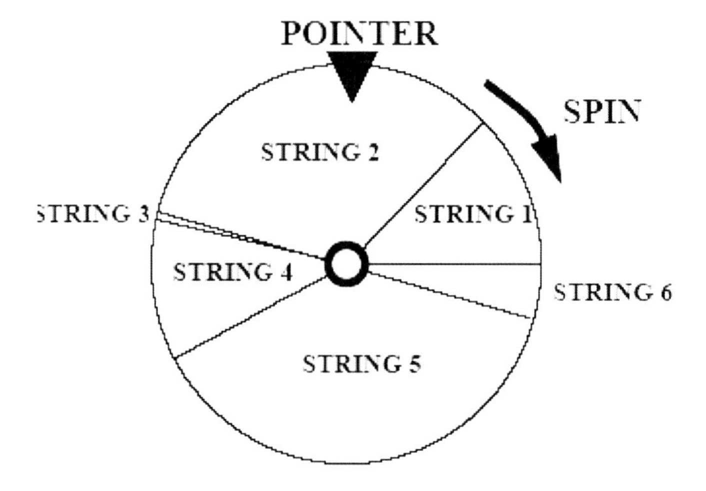

Fig. 4.14. Graphical representation of the Roulette wheel selection mechanism

In addition another scheme for selection is applied along with all three selection schemes discussed above which is called ‘elitism’. The idea of elitism is to avoid the observed best-fitted individual dies out just by selecting it for the next generation without any random experiment. Elitism significantly influences the speed of the convergence of a GA. But it can lead to premature convergence also.

4.3.2.4 Crossover (Recombination)

The basic operator for producing new chromosomes in the GA is that of crossover. Like natural reproduction, crossover produces new individuals so that some genes of a new child come from one individual while others come from the other individual. In essence, crossover is the exchange of genes between the chromosomes of the two parents. The process may be described as cutting two strings at a randomly chosen position and swapping the two tails. It is known as the single-point crossover, and the mechanism is visualized in Figure 4.15. An integer position, i is selected at random with a uniform probability between one and the string length, l, minus one (i.e., $i \in [1, l-1]$). When the genetic information is exchanged among the parent individuals (represented by the strings, P_1 and P_2) about this point, two new offspring (represented by the strings, O_1 and O_2) are produced. The two offspring in Figure 4.15 are produced when the crossover point, i=4 is selected.

P_1=1 0 0 1 | 0 1 1 0 → O_1=1 0 0 1 *1 0 0 0*
P_2=1 0 1 1 | 1 0 0 0 → O_2=1 0 1 1 *0 1 1 0*

Fig. 4.15. A typical example of single point crossover

For multi-point crossover, multiple crossover positions (m) are chosen at random with no duplicates and sorted into ascending order. Then the bits between two successive crossover points are exchanged between the two parents to produce two new offspring. The process of multi-point crossover is illustrated in Figure 4.16 with shaded color.

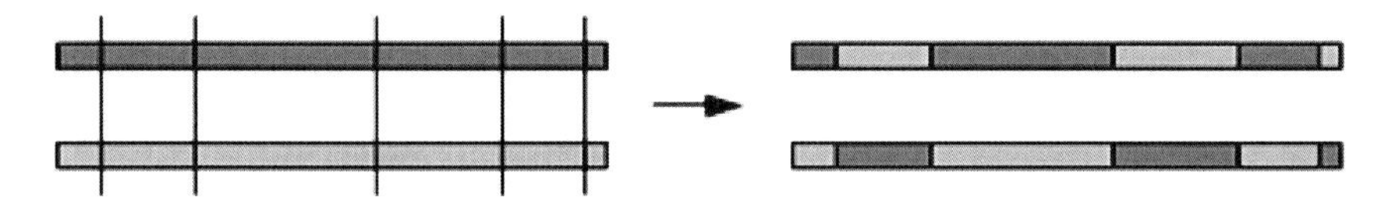

Fig. 4.16. Typical example of multi-point crossover (with m=5)

The idea behind multi-point crossover is that the parts of the chromosome representation that contributes to the most to the performance of a particular individual may not necessarily be contained in adjacent substrings. Further, multi-point crossover appears to encourage the exploration of the search space, thus making the search more robust.

4.3.2.5 Mutation

Mutation is nothing but deformation of the genetic information of an individual (solution) by means of some external influences. The bit-wise mutation operator changes a bit, 1 to 0, and vice versa, with a prescribed probability (called, mutation probability) as shown in Figure 4.17. In real reproduction, the probability that a certain gene is mutated is almost equal for all genes. So, it is near at hand to use the mutation technique for a given binary string, where there is a given probability that a single gene is modified. The probability should be rather low in order to avoid chaotic behavior of the GA.

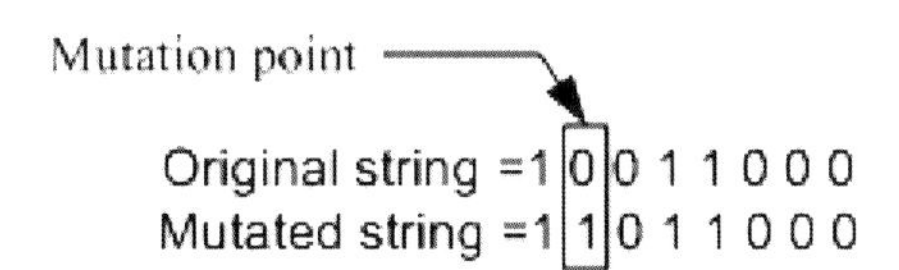

Fig. 4.17. Mutation effect on offspring's strings

4.3.2.6 Termination of the GA

Because the GA is a stochastic search method, it is difficult to formally specify convergence criteria. As the fitness of a population may remain static for a number of generations before a superior individual is found, the application of conventional termination criteria becomes problematic. A common practice is to terminate the GA after a pre-specified number of generations and then test the quality of the best members of the population against the problem definition. If no acceptable solutions are found, the GA may be restarted or a fresh search can be initiated. Attaining a pre-specified fitness function value or when best fitness of the population does not change appreciably over successive iterations may be also considered as a termination criteria.

4.3.3 Description of Working Principle of GA

The working principle of a (binary-coded) GA is described here with an engineering optimization problem, namely designing a can [9] as shown in Figure 4.18. The objective of the problem is to determine the optimum values of the diameter (d, in cm) and height (h, in cm) of the can in order to minimize its cost, f(d, h) subject to some constraints and the problem is defined as follows:

Minimize $$f(d,h) = c\left(\frac{\pi d^2}{2} + \pi dh\right),$$

Subject to

$$g(d,h) \equiv \frac{\pi dh^2}{4} \geq 300,$$
$$d_{min} \leq d \leq d_{max}$$
$$h_{min} \leq h \leq h_{max},$$

where c is the cost of the can material per square cm, which is taken as 0.005 and the minimum and maximum values of d and h are taken as, $d_{min}=h_{min}=0$ and $d_{max}=h_{max}=31$.

Therefore, in this problem the number of decision variables is two (d and h). The population size of GA is considered as six and it is kept constant throughout the GA-operation. GA operates in a number of iterations until a specified termination criterion is satisfied and in the Figure 4.18, the maximum number of iteration/generation (max_gen) is treated as the termination criteria.

GA iteration starts with the creation of six random solutions which are treated as the parent solutions. The chromosome structure of each solution is the same as presented in Figure 4.13. Then, the fitness values of all the parent solutions in the population are calculated using following fitness function (discussed in Section 4.3.2.2).

$$\text{Fitness value} = \begin{cases} f(d,h) + P_p \times [300 - g(d,h)] & \text{if } g(d,h) \leq 300 \\ f(d,h) & \text{if } g(d,h) > 300 \end{cases} \tag{4.14}$$

where P_p is the penalty parameter and the value of P_p is taken as 25. After that selection/reproduction (discussed in Section 4.3.2.3) operation is performed on the parent solutions based on the fitness values of solutions (as depicted in the binary tournament selection table in Figure 4.18) to form the mating pool. The number of solutions contained in the mating pool is to be equal to the number of parent

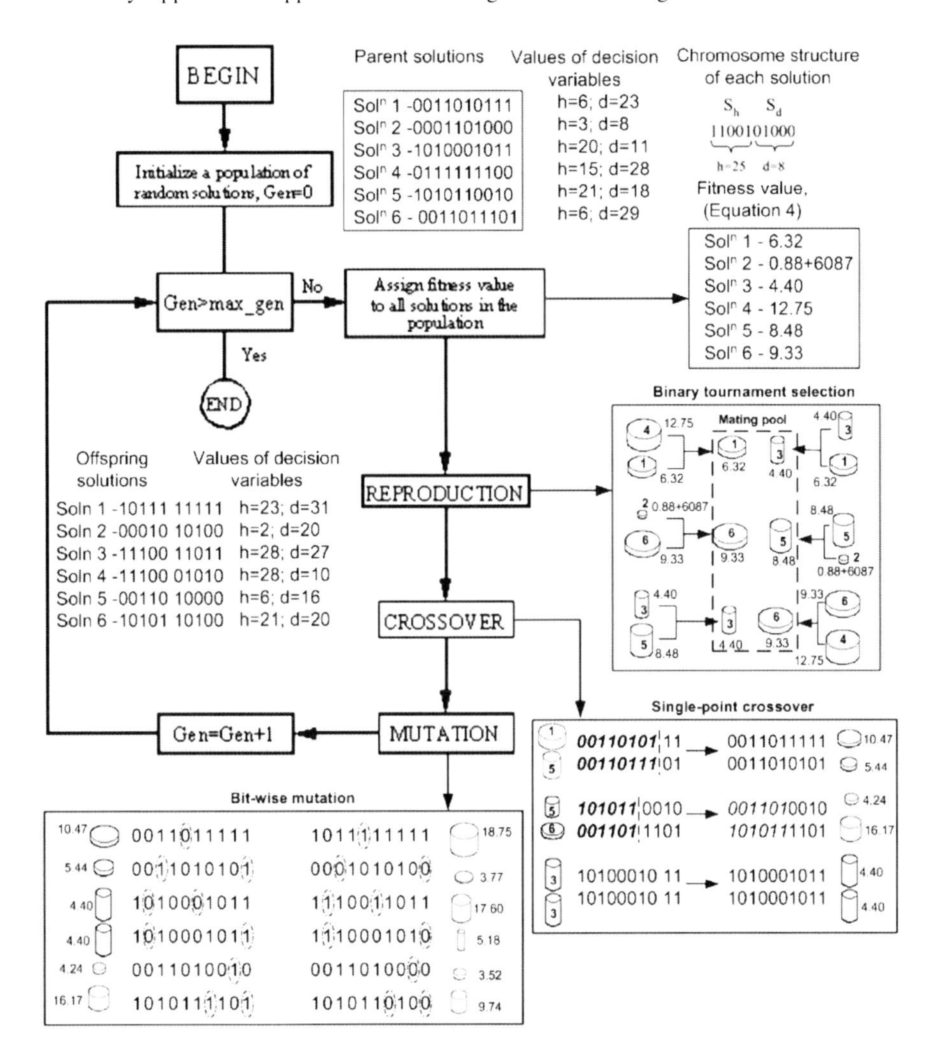

Fig. 4.18. Working principle of a genetic algorithm

solution in order to maintain a constant population size through out the GA-iteration. After that, crossover operator (discussed in Section 4.3.2.4) is applied among the two randomly chosen solutions from the mating pool based on a given probability (crossover-probability, say 0.9). Then, bit-wise mutation (discussed in Section 4.3.2.5) is carried out on each of the six solutions obtained after employing the crossover operator using a given mutation probability (say 0.02) and produces six new (offspring) solutions. The objective function values corresponding to each solution are depicted in the single-point crossover and bit-wise mutation tables in Figure 4.18. It completes one iteration/generation of GA. Now, the checking of GA-termination criterion is performed and if it satisfies the termination criteria, stop the GA-iteration process, otherwise start another generation/iteration with treating the six offspring solutions obtained in the previous iteration as the parent solutions and then assigning fitness value to all the solutions in the population and continue the same iteration procedure as described above.

4.4 Genetic Fuzzy Approaches

The performance of a FRBM depends on its knowledge base (KB) which consists of database (DB) (that is, information regarding membership functions) and rule base (RB). It is important to mention that determination of an appropriate knowledge base for a FRBM is not an easy task. The genetic algorithms (GAs) have been used by several investigators to design database and/or rule base of a FRBM. The fuzzy systems making use of a GA in their design process are called genetic-fuzzy systems (GFS). Figure 4.19 shows the schematic diagram of a genetic-fuzzy system, in which a GA-based learning/tuning based on example data (training data) is adopted, off-line to design the KB a FRBM. GA can also be used to tune the existing KB of a FRBM which may be designed based on the some common knowledge or expert information, to improve the performance of the existing FRBM. As the GA is found to be computationally expensive due to the nature of population-based optimization, the GA-based tuning is normally carried out off-line.

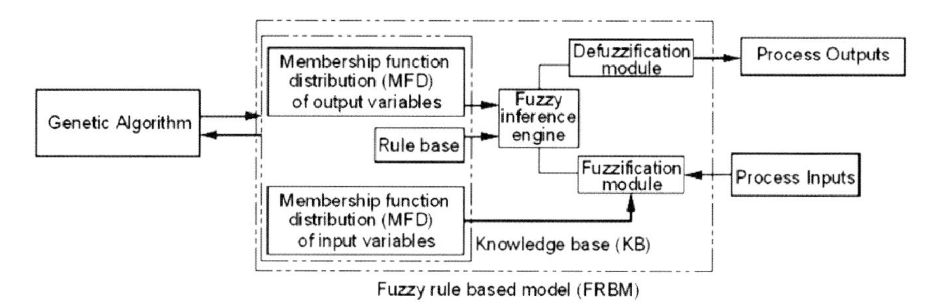

Fig. 4.19. Schematic layout of genetic-fuzzy system

A Mamdani-type fuzzy logic rule for a particular process having say, two input variables (x_1 and x_2) and one output variable (y) (each having triangular-type MFDs with 3 fuzzy subsets) may be expressed as

$$\text{IF } x_1 \text{ is } A_1 \text{ AND } x_2 \text{ is } A_2 \text{ THEN } y \text{ is } B,$$

where A_1, A_2 and B are the fuzzy subsets (those can be expressed by suitable linguistic expression, such as LOW, MEDIUM, HIGH, etc.) of triangular-type membership function. In Section 4.2.2.2, it was discussed that in order to describe triangular-type MFDS with 3 fuzzy subsets one controlling parameter is required. A typical binary coded GA-string for optimizing the KB will look as shown in Figure 4.20.

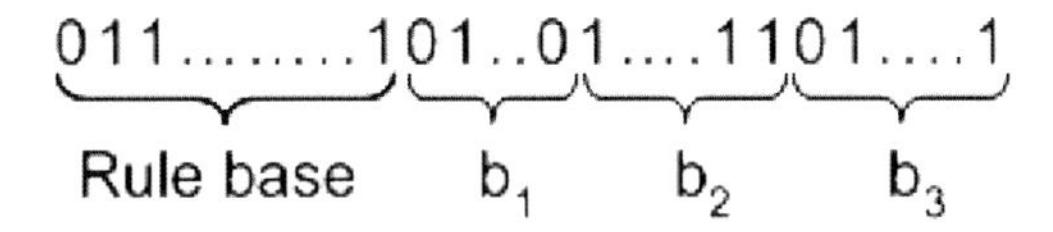

Fig. 4.20. A GA-string representing the rule base and the parameters related to membership functions of input-output variables a FRBM

where b_1, b_2 and b_3 are the (continuous) control (GA) variables related to MFDs corresponding to the two inputs and a single output variables x_1, x_2 and y, respectively. The number of bits used for optimizing the RB is equal to the number of maximum possible rules present in the RB. In this case, the number of rules will be $3 \times 3 = 9$, since each of the two input variables comprise of 3 fuzzy subsets. The information of b_1, b_2 and b_3 are coded by the next bits in GA-string.

There are, in fact, three different approaches of designing genetic-fuzzy system (GFS), according to the KB components including in the learning process. These are as follows

Genetic Learning /Tuning of the Fuzzy Logic Controller Data Base
Here, the GA is used to optimize the appropriate value(s) of the controlling parameter(s) that define the typical type of MFDs. In other words, for examples in case of triangular type MFDs, it is used to move and to expand or shrink the base width(s) ($b_1/b_2/b_3$) of each interior isosceles triangle. The extreme triangles will be right triangles and the GA will make it either bigger or smaller. During GA-based optimization the parameters (b_1, b_2 and b_3) are allowed to vary in a range specified by the designer. Thus, in this approach through evolution, the GA will find a good database for the FRBM but the RB will be kept as same what was initially considered by the designer. In this case, the GA-string shown in Figure 4.21 will look as follows:

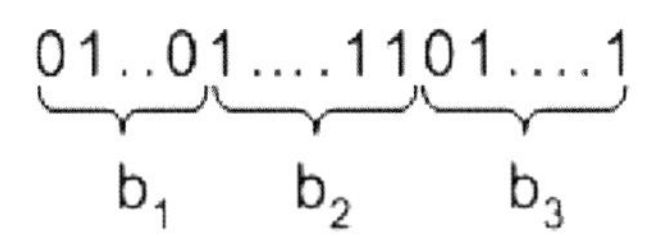

Fig. 4.21. A GA-string representing the parameters related to membership functions of input-output variables of a FRBM

Genetic Learning/Tuning of the Fuzzy Rule Base
All methods belonging to this family assume the existence of a pre-defined DB for the FRBM. In one method, an initial user-defined rule base (assignment of fuzzy subset to the output variable of each rule) constructed based on designer's experience of the process to be controlled is tuned using GA. In GA-sting (as shown in Figure 4.22), each rule is represented by a single bit (1 or 0), where 1 and 0 respectively indicate the presence or absence of the rule in the optimum rule base.

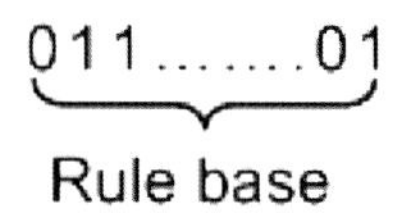

Fig. 4.22. A GA-string representing the rule base of a FRBM

In the method where the RB is designed using GA automatically, additional bits will be included in the GA-string demonstrated in Figure 4.22 (as shown in Figure 4.23) [10]. For the case where output variable is considered to have 2 fuzzy subsets, one bit will be required in order to determine the fuzzy subset to the output variable for each rule (e.g., 0 for LOW and 1 for MEDIUM). Two bits required for the case where output variable is considered to have 3/4 fuzzy subsets (e.g., 00 for LOW, 01 for MEDIUM, 10 for HIGH and 11 for VERY HIGH), and so on.

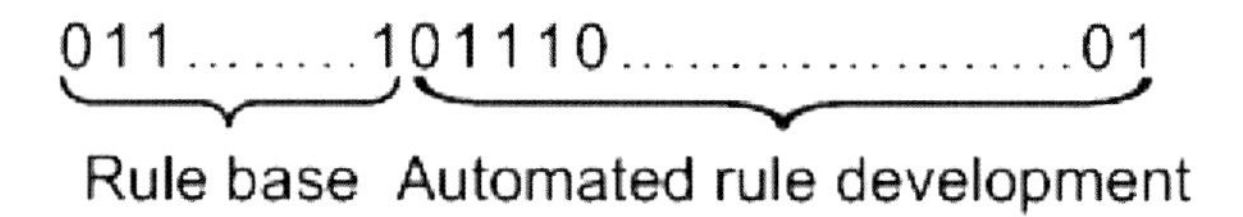

Fig. 4.23. A GA-string representing the rule base and automated rule development of a FRBM

Genetic Learning/Tuning of the Fuzzy Knowledge Base
In this approach both the RB and DB are designed/optimized using GA as stated above simultaneously and the corresponding GA-string will look as shown in Figure 4.24.

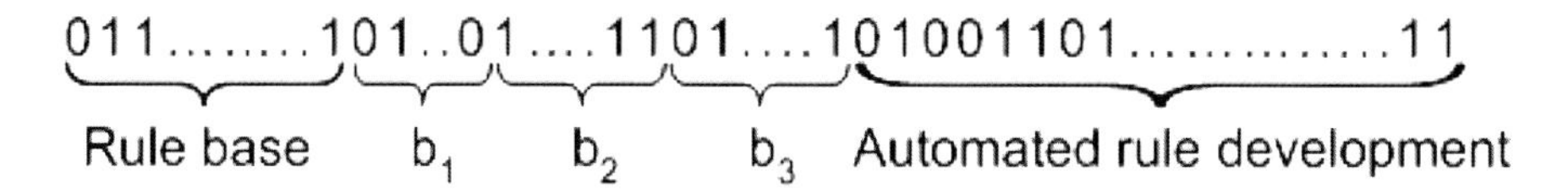

Fig. 4.24. A GA-string representing the rule base with automated rule development and the parameters related to membership functions of input-output variables of a FRBM

Besides the way how to construct/design the KB of FRBM, selection of the appropriate shape of fuzzy subsets/membership function distributions (MFDs) for both the input and output variables in case of Mamdani-type FRBM and selection of shapes of fuzzy subsets of input variables as well as the appropriate structure(s) of rule output/consequent function(s) and determination of optimal values of coefficients and power terms of rule consequent functions are important issues. In order to overcome these problems, a rigorous study with different choices is required in order to obtain a good model for a manufacturing process.

GA is also used in the genetic-fuzzy system where the TSK-type fuzzy logic rules (as defined in Section 4.2.11.2.1) are employed. Genetic Linear Regression (GLR) approach [11] is one of the most popular approaches in designing the KB of TSK-type FRBM. In genetic linear regression (GLR) approach, the GA is introduced partly in the multiple regression method. The GLR method will take the advantages of both the regression technique and GA, and have a capability to find global optimum and good convergence properties. In this approach, the KB of the FRBM is optimized using the combined method of linear regression (LR) approach and genetic algorithm. In this method the coefficients of output function of each rule are determined using a linear regression approach whereas the input variables' exponential parameters are simultaneously optimized using a GA. In addition to that, the MFDs also tune using a GA. In order to accomplish this, besides the exponential parameters of input variables, the controlling parameter(s) describing the MFDs (as discussed in Section 4.2.2.2) are considered as GA-variables. The working principle of GLR approach (as illustrated in Figure 4.25) consists of following five major steps:

- Step-I: Set an initial set (population) of values of power terms of a given regression function at random
- Step-II: Evaluate the function coefficients based on least square method
- Step-III: Checking of fitness value (if satisfied terminate the iteration procedure)
- Step-IV: Update the values of power terms of regression function using GA-operators
- Step-V: Repeat Step-II

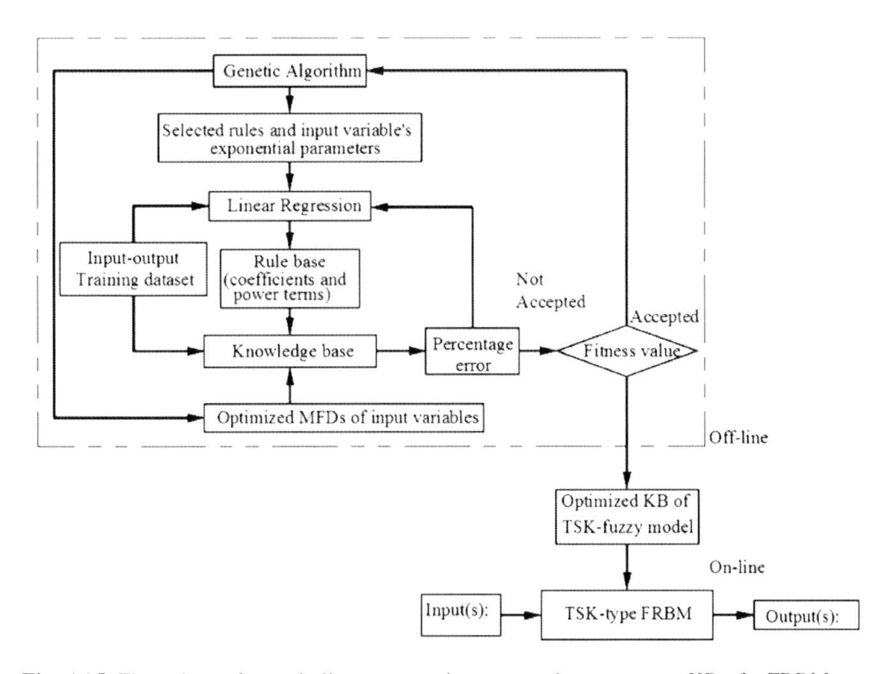

Fig. 4.25. Flow chart of genetic linear regression approach to construct KB of a FRBM

The GA-string of GLR approach will look as presented in Figure 4.26.

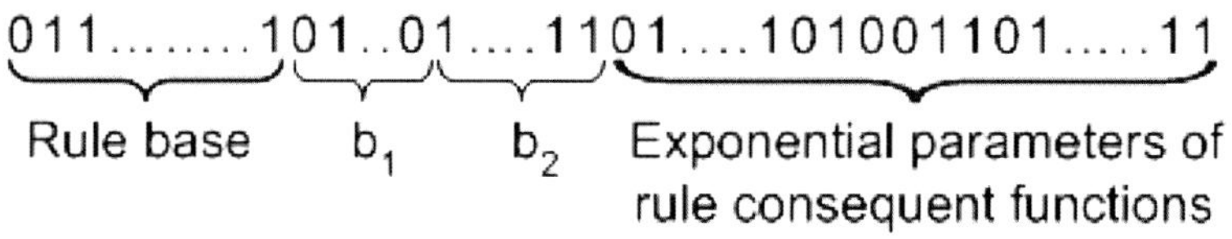

011........101..01....1101....101001101.....11

Rule base b_1 b_2 Exponential parameters of rule consequent functions

Fig. 4.26. A GA-string representing the rule base, parameters related to membership functions of input variables and exponential parameters of rule consequent functions of a TSK-type FRBM

The proposed GLR has an added facility to carryout the task of tuning MFDs of input variables simultaneously in the same framework of GA.

In order to determine the coefficients of the output functions of TSK-type fuzzy rules, a general expression of multiple linear regression system with TSK-fuzzy model is derived as follows. The Equation (4.11) may be rewritten by denoting $\prod_{v=1}^{n} \mu_v(x_{1,\dots,n}) = \eta_r$ for simplicity, in the following form:

$$Y = F(x_1, .. x_n)$$

$$= \frac{\sum_{r=1}^{R_f^l} \eta_r \left(a_1^r f_1^r(x_{1,\dots,n}) + a_2^r f_2^r(x_{1,\dots,n}) + \dots\dots\dots\dots + a_k^r f_k^r(x_{1,\dots,n})\right)}{\sum_{r=1}^{R_f^l} \eta_r}$$

$$= \frac{\begin{array}{l}\eta_1\left(a_1^1 f_1^1(x_{1,..n}) + .. + a_k^1 f_k^1(x_{1,..n})\right) + .. \\ + \eta_{t_r}\left(a_1^{t_r} f_1^{t_r}(x_{1,..n}) + .. + a_{t_j}^{t_r} f_{t_j}^{t_r}(x_{1,..n}) + .. + a_k^2 f_k^2(x_{1,..n})\right) + . \\ . + \eta_{R_f^l}\left(a_1^{R_f^l} f_1^{R_f^l}(x_{1,..n}) + .. + a_k^{R_f^l} f_k^{R_f^l}(x_{1,..n})\right)\end{array}}{\eta_1 + \eta_2 + \dots\dots + \eta_{R_f^l}} \tag{4.15}$$

Let us assume we have a set of input-output tuple (D) of S number of sample data where the output $y^{(i)}$ is assigned to the input $(x_1^{(i)}, x_2^{(i)}, \dots\dots\dots, x_n^{(s)})$

$$D = \left\{\left(x_1^{(1)}, \dots\dots, x_n^{(1)}, y^{(1)}\right), \left(x_1^{(2)}, \dots\dots, x_n^{(2)}, y^{(2)}\right) \dots\dots\dots . \left(x_1^{(s)}, \dots\dots, x_n^{(s)}, y^{(s)}\right)\right\}$$

Now, the total quadratic error that is caused by the TSK-type FRBM with respect to the given data set:

$$E = \sum_{l=1}^{S} \left(f\left(x_1^{(l)}, x_2^{(l)}, \dots . x_n^{(l)}\right) - y^{(l)}\right)^2 \tag{4.16}$$

In order to minimize E, we have to choose the parameters, $\left\{\left(a_1^1, \dots ., a_k^1\right), \left(a_1^2, \dots .. a_k^2\right), \dots\dots, \left(a_1^{R_f}, \dots ., a_k^{R_f}\right)\right\}$ appropriately.

where the parameter a_j^r indicates the j_{th} coefficient of the output function of r_{th} rule.

To determine the above parameters, we take the partial derivatives of E with respect to each parameter and require them to be zero, i.e., $\frac{\partial E}{\partial a_j^r} = 0$, where $j = \{1,2,....,k\}$ and $r = \{1,2,.....,R_f\}$.

Now, we obtain the partial derivation of E with respect to the parameter $a_{t_j}^{t_r}$,

$$\frac{\partial E}{\partial a_{t_j}^{t_r}} = \sum_{l=1}^{S} 2 \cdot \left(f\left(x_1^{(l)},........,x_n^{(l)}\right) - y^{(l)}\right) \cdot \frac{\partial f\left(x_1^{(l)},........,x_n^{(l)}\right)}{\partial a_{t_j}^{t_r}}$$

$$= 2 \cdot \sum_{l=1}^{S} \left(\frac{\sum_{r=1}^{R_f^l} \eta_r \left(a_1^r f_1^r \left(x_{1,..,n}^l\right) + ... + a_k^r f_k^r \left(x_{1,..,n}^l\right)\right)}{\sum_{r=1}^{R_f^l} \eta_r} - y^l \right) \cdot \frac{\eta_{t_r} \cdot f_{t_j}^{t_r}\left(x_{1,..,n}^l\right)}{\sum_{r=1}^{R_f^l} \eta_r}$$

$$= 2\left(\left(\frac{\sum_{l=1}^{S}\sum_{r=1}^{R_f^l} \eta_r \left(a_1^r\right) f_1^r \left(x_{1,..,n}^l\right) \eta_{t_r} f_{t_j}^{t_r}\left(x_{1,..,n}^l\right)}{\left(\sum_{r=1}^{R_f^l} \eta_r\right)^2}\right) + .. + \left(\frac{\sum_{l=1}^{S}\sum_{r=1}^{R_f^l} \eta_r \left(a_k^r\right) f_k^r \left(x_{1,..,n}^l\right) \eta_{t_r} f_{t_j}^{t_r}\left(x_{1,..,n}^l\right)}{\left(\sum_{r=1}^{R_f^l} \eta_r\right)^2}\right)\right)$$

$$- 2\left(\frac{\sum_{l=1}^{S} y^l \cdot \eta_{t_r} f_{t_j}^{t_r}\left(x_{1,..,n}^l\right)}{\sum_{r=1}^{R_f^l} \eta_r}\right) = 0, \tag{4.17}$$

Thus, the Equation (4.16) provides the following system of linear equations from which we can compute the coefficients $\left\{\left(a_1^1,....,\ a_k^1\right), \left(a_1^2,.....\ a_k^2\right),......., \left(a_1^{R_f},....,\ a_k^{R_f}\right)\right\}$:

$$\sum_{r=1}^{R_f}\sum_{j=1}^{K} a_j^r \sum_{l=1}^{S} \frac{\prod_{v=1}^{n} \mu_v^r\left(x_{1,..,n}^l\right)}{\left(\sum_{r=1}^{R_f^l}\prod_{v=1}^{n} \mu_v^r\left(x_{1,..,n}^l\right)\right)^2} f_j^r\left(x_{1,..,n}^l\right) \cdot f_{t_j}^{t_r}\left(x_{1,..,n}^l\right) \cdot \prod_{v=1}^{n} \mu_v^{t_r}\left(x_{1,..,n}^l\right)$$

$$= \sum_{l=1}^{S} \frac{y^l \prod_{v=1}^{n} \mu_v^{t_r}\left(x_{1,..,n}^l\right)}{\sum_{r=1}^{R_f^l}\prod_{v=1}^{n} \mu_v^r\left(x_{1,..,n}^l\right)} f_{t_j}^{t_r}\left(x_{1,..,n}^l\right) \tag{4.18}$$

In matrix form, Equation (4.18) will be written as:

$$\begin{bmatrix} \alpha^r_{11} & \alpha^r_{12} & \cdot & \alpha^r_{1K} \\ \alpha^r_{21} & \alpha^r_{22} & \cdot & \alpha^r_{2K} \\ \cdot & \cdot & \cdot & \cdot \\ \alpha^r_{K1} & \alpha^r_{K2} & \cdot & \alpha^r_{KK} \end{bmatrix} \begin{bmatrix} a^r_1 \\ a^r_{21} \\ \cdot \\ a^r_K \end{bmatrix} = \begin{bmatrix} \beta^r_1 \\ \beta^r_2 \\ \cdot \\ \beta^r_K \end{bmatrix} \tag{4.19}$$

where $\alpha^r_{tj} = \sum_{l=1}^{S} f^r_j(x^l_1, x^l_2 ..., x^l_n) f^r_t(x^l_1, x^l_2 ..., x^l_n)$; $\beta^r_t = \sum_{l=1}^{S} y^l f^r_t$,

where s is the number of training (input-output) sample data. Thus, Equation (4.18) provides a solution of the function coefficients (a^r_j) of the TSK-type fuzzy rule consequents for a given value of the input variable's exponential terms. To solve the Equation (4.18) in order to find the values of function coefficients, the Gauss Algorithm with Column Pivot Search Method is used here. More generally, any conventional numerical method, which provides a representative solution of Equation (4.18), may be adopted.

4.5 Application to Modeling of Machining Process

Numerous works on modeling of machining processes using soft computing tools including different GA-fuzzy approaches can be found in the review paper [12]. In this chapter, some of the previous research works of the author on modeling of manufactuing processes using GA-fuzzy approaches are presented.

4.5.1 Modeling Power Requirement and Surface Roughness in Plunge Grinding Process

In [13], a study was carried out to model power requirement and surface roughness in plunge grinding process using a GA-fuzzy approach. The model considers the input variables such as wheel speed, work speed and feed rate those mainly influence the power requirement and surface roughness obtained on the grind surface. In this study the main objective was to find the effect of various types of MFs considered for input-output variables on the performances of Mamdani-type FRBMs. The approach of tuned rule base and MFs simultaneously using GA was adopted. The performance of the models was tested (as shown in Figure 4.27) with experimental results considering 52100 steel as work material and D126K5V as grinding wheel specification. A digital clamp power meter is used to take the measurements of power requirement in grinding and the surface roughness is measured with the help of a perthometer (S6R). In Figure 4.27, Error 1, Error 2 and Error 3 are the percentage deviations of results predicted by FRBMs with triangular, 2nd order polynomial and 3rd order polynomial-type MFDs, respectively from that obtained in experimentation. It is observed that higher order polynomial type MFs showed better results. It may happen because the input-output relationship in grinding is highly nonlinear and the linear MFDs (triangular type) may not be sufficient.

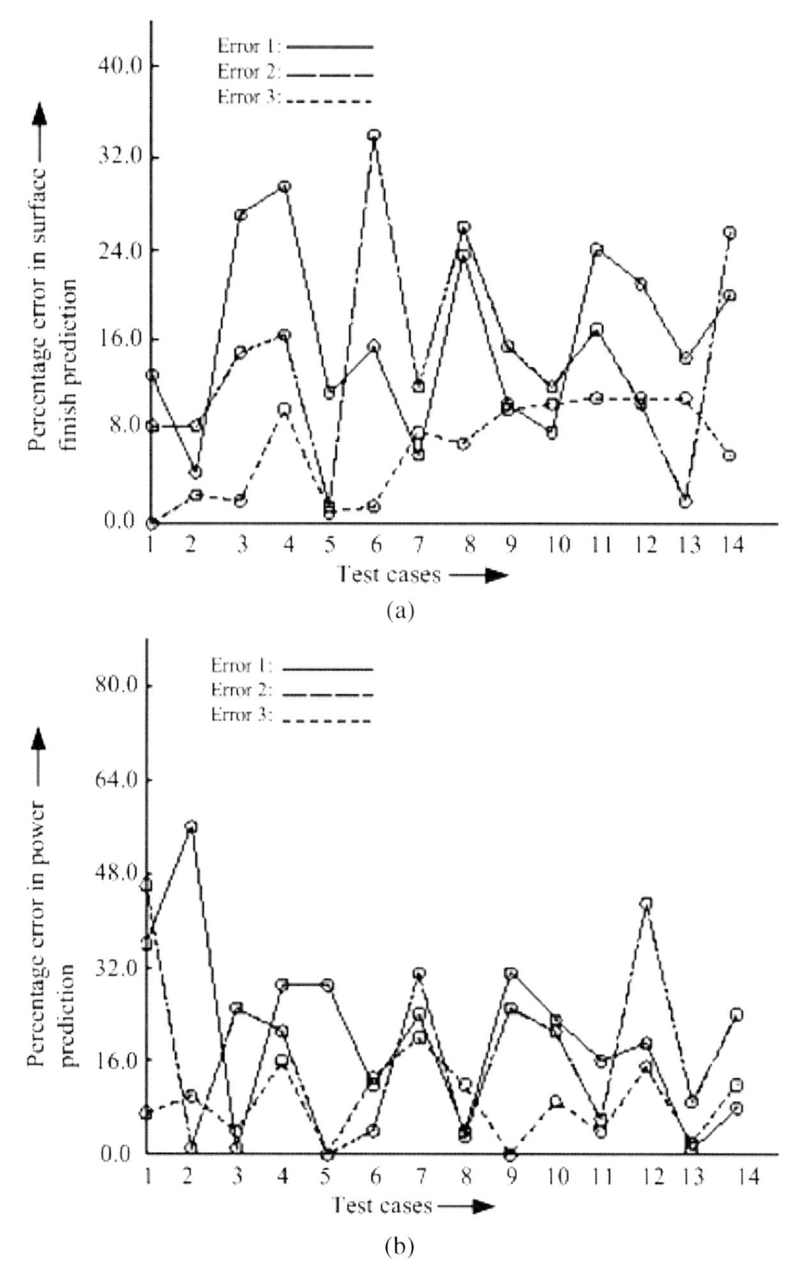

Fig. 4.27. Comparison of performances of FRBMs (with different types MFs) with those of experimental results (a) surface roughness (b) power requirement

In the above study, the GA was used to optimize the manually-defined KB of the FRBM. The manually-defined KB is designed based on the expert's knowledge of the process that may not be complete. Sometimes, it becomes difficult to gather knowledge of the process beforehand. To overcome this difficulty, the method for automatic design of fuzzy KB is adopted to model power requirement and surface roughness in plunge grinding process [10]. Table 4.1 describes the comparative results of root mean square (RMS) percentage deviations exhibited by fuzzy rule-based models (FRBMs) from those of real experimental values. It is found that the approach of automatic design of RB and tuning of MFs simultaneously using GA provides better result over the approach of tuned manually constructed RB and MFs simultaneously using GA. It happens because all the manually-designed fuzzy rules may not be good, whereas the GA has the capability of finding the good fuzzy rules through extensive search. Moreover, the main disadvantage of using later approach (the approach of tuned manually constructed RB and MFs) lies in the fact that the designer is required to have a thorough knowledge of the process to be controlled. Thus, a considerable amount of time is spent on manual construction of fuzzy RB. In the approach of automatic design of RB and tuning of MFs simultaneously using GA, no effort is made for designing the fuzzy rule-base manually and a good KB of the FRBM is designed automatically using a GA from a set of example (training) data.

Table 4.1. Comparison of RMS percentage deviations exhibited by fuzzy rule-based models from those of real experimental values

	Power requirement		
	Mathematical model	FRBM based on the approach of tuned rule base and MFs simultaneously using GA (approach 2)	FRBM based on the approach of automatic design of rule base and tuning of MFs simultaneously using GA (approach 1)
RMS percentage error	31.51	8.13	5.34
	Surface roughness		
RMS percentage error	16.44	10.22	6.32

4.5.2 Study of Drilling Performances with Minimum Quantity of Lubricant

The main objective of this study is to investigate the performances of FRBMs based on Mamdani-type and TSK-type of fuzzy logic rules, and different shapes (namely, 2nd order polynomial and trapezoidal types) of MFs for prediction and performance analysis of machining with minimum quantity of lubricant (MQL) in drilling of Aluminium (AA1050). In this study, predictions of surface roughness obtained in drilling and the corresponding cutting power and specific cutting force requirements for different amounts of lubricant rate will be carried out through a comparative analysis of the results of models with experimental results as well as those published in the literature. The approach of tuned RB and MFs simultaneously using GA and genetic linear regression method was adopted to construct the KB of Mamdani-type and TSK-type FRBM, respectively. The structure of rule-consequent function for TSK-type fuzzy rules is used as follows

$$y = c_1 V_c^{p_1} + c_2 F_r^{p_2} + c_3 L_r^{p_3} \tag{4.20}$$

where c_1, c_2 and c_3 are the function coefficients and p_1, p_2 and p_3 are the exponential parameters of rule consequent function. V_c, F_r and L_r are the input variables, cutting speed, feed rate and rate of lubricant, respectively.

A helical K10 drill (R415.5-0500-30) was manufactured according to DIN6537 by Sandvik(R). The drill has a point angle of 140°, 28 mm of flute length and is of 10% cobalt grade. The drills possess a diameter of 5 mm and are coated with TiAlN. A Kistler® piezoelectric dynamometer 9272 with a load amplifier was used to acquire the torque and the feed force. Data acquisitions were made through piezoelectric dynamometer by interfacing RS-232 to load amplifier and PC using the appropriate software, Dynoware Kistler(R). The surface roughness was evaluated (R_a according to ISO 4287/1) with a Hommeltester T1000 profilometer.

Here, 4 different models related to surface roughness and, other 4 different models for cutting power/specific cutting force requirement) are developed. The four different FRBMs are constructed using two different types of fuzzy logic rules (Mamdani-type and TSK-type) and two different shapes of MFs.

The comparative results of surface roughness, cutting power and specific cutting force with different lubricant flow rates for different cutting speed and feed rate are described in Figure 4.28, Figure 4.29 and Figure 4.30, respectively. In this study, nine different cases (of cutting speed and feed rate) are considered based on which the effects of lubricant flow rate on machining performances are analyzed. In these Figures, Model I indicates FRBM with Mamdani-type FLR and 2nd order polynomial MFs, Model II represents FRBM with Mamdani-type FLR and trapezoidal MFs, Model III shows FRBM with TSK-type FLR and 2nd order polynomial MFs and Model IV indicates FRBM with TSK-type FLR and trapezoidal MFs. In the following subsections the evolutions of the prediction performances of these models toward the effects of surface roughness, cutting power and specific cutting force with lubrication rate are discussed.

Surface roughness

In Figure 4.28, the predicted values of surface roughness by FRBMs are compared with the experimental values for 9 cases (Figure 4.28(i) to Figure 4.28(ix)). It has been observed that the performance of Model II is better than Model I for first 5 cases and case number 6. For other cases Model I outperforms over model II. But, the consistencies of deviations of results from that of the experimental values are not good for both the Model I and Model II in all the cases. In contrast, the results of Model III shows better than Model I and Model II for some cases (Figure 4.28 (iv), (v) and (vii)), but for other cases that are deteriorated compared to Model I and Model II. On the other hand it is noticed that the results of Model IV (FRBM with TSK-type FLR and trapezoidal MFs) yield less error (deviation from the

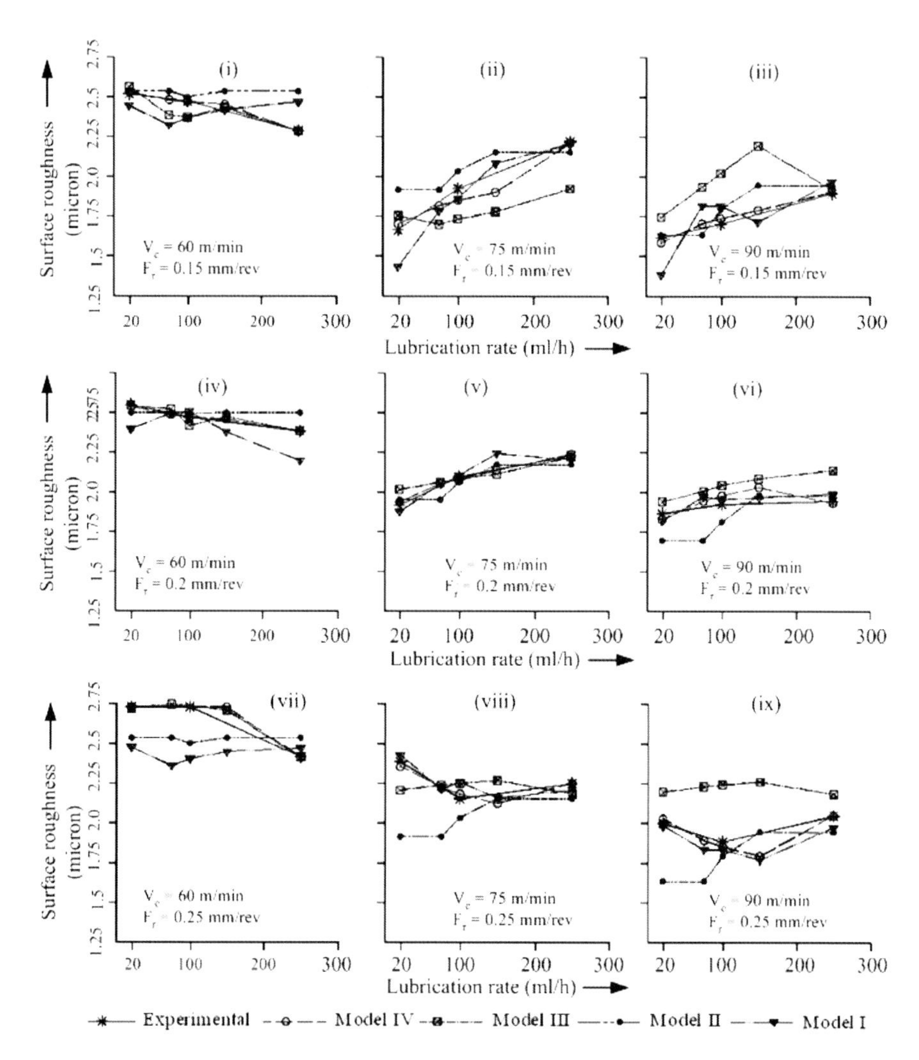

Fig. 4.28. Comparative results of surface roughness with different lubricant flow rates for different cutting speed and feed rate

experimental values) in majority than that of the other models for all (9) cases. Furthermore, it is noticed that the results of Model IV are also consistent for different values of lubrication rate. The maximum value of percentage error exhibited by Model IV is 2.2428, which is well accepted in industrial practice.

By analyzing the experimental values as well as results obtained by Model IV, it has been observed that the surface roughness is improved by increasing the flow rate for lower values of cutting speed (60 m/min) and constant feed rate. But, for higher values of cutting speed, the surface roughness deteriorated with increasing flow rate (75 and 90 m/min). In contrast, for a constant cutting speed, the rate of change of surface quality with flow rate is minimized as feed rate increase.

Cutting power

By analyzing the results of various models and experimental values as depicted in Figure 4.29, it has been observed that Model I as well as Model II provides poor results than other two models (Model III and Model IV). In contrast, it is found that both the models, Model III and Model IV obtain the best performance for predicting cutting power with the quantity of lubricant for a given cutting speed and feed rate. However, in cases V_c=90; f=0.15 and V_c=90; f=0.25, the Model IV shows better results than Model III.

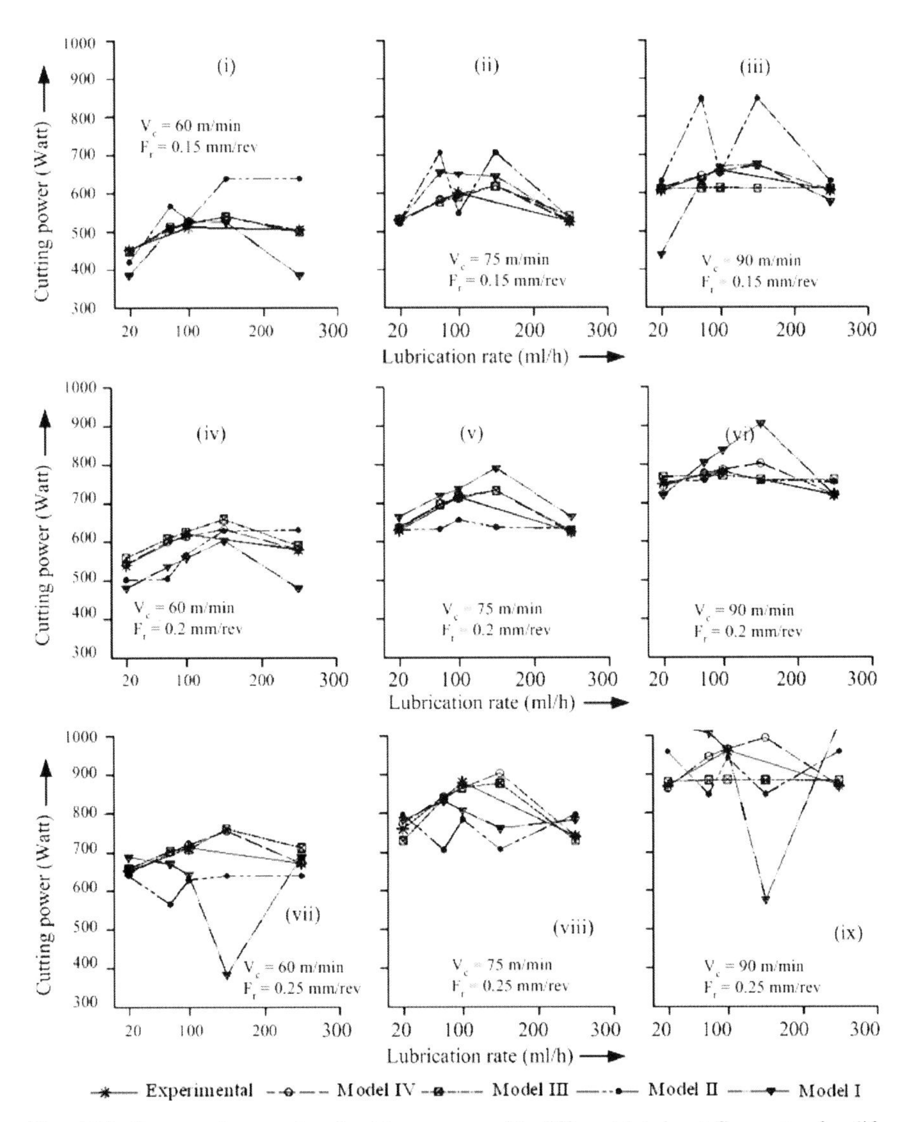

Fig. 4.29. Comparative results of cutting power with different lubricant flow rates for different cutting speed and feed rate

By analyzing the experimental values as well as results obtained by Model III and Model IV, it has been revealed that for a fixed value of cutting speed and feed rate, the cutting power increase to a certain value of lubrication flow rate. After that the value of cutting power decreases with increasing flow rate. From Figure 4.29, it is found that for constant cutting speed, the cutting power requirement increase with feed rate. It is also observed that the value of cutting power increases with cutting speed when feed rate is kept as a constant value.

Specific cutting force

As like cutting power, here also Model III and Model IV show the best result in predicting specific cutting force with the quantity of lubricant for a given cutting speed and feed rate (Figure 4.30). This is because; both the cutting power and specific force are depended on the same parameter, torque and a linear relationship is maintained among them. The variation of specific cutting force requirement with lubricant flow rate and other input parameters, cutting power and feed rate exhibit the same phenomena as found in cutting power.

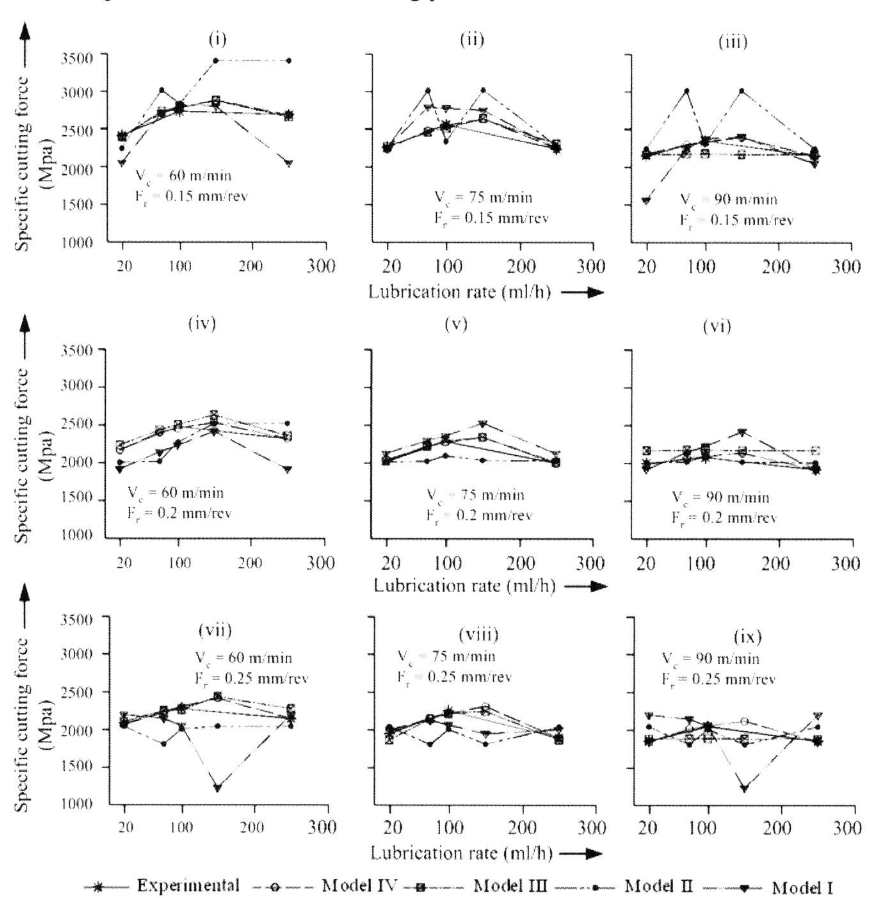

Fig. 4.30. Comparative results of specific cutting force with different lubricant flow rates for different cutting speed and feed rate

From above discussions, it may be pointed out that FRBMs with TSK type fuzzy logic rules provide best result in predicting surface roughness, cutting power and specific cutting force. Specifically for surface roughness, trapezoidal MFs is well suited, while trapezoidal as well as second order polynomial MFs give almost similar performances in predicting cutting power/specific cutting force requirements in drilling of Aluminium AA1050 with emulsion with oil Microtrend 231L lubricant. The above techniques may be adopted for developing FRBMs for other machining (drilling) performance parameters. Once the model is developed, it may be used online in drilling machine to control the MQL as per desired outputs.

References

[1] Groover, M.: Automation, Production System, and Computer Integrated Manufacturing. Prentice-Hall Int'l, Upper Saddle River (2001)

[2] Kosko, B.: Neural Network and Fuzzy Systems. Prentice-Hall, New Delhi (1994)

[3] Zadeh, L.A.: Fuzzy sets. Information and Control 8(3), 338–353 (1965)

[4] Mamdani, E.H., Assilian, S.: An experiment in linguistic synthesis with a fuzzy logic controller. International Journal of Man-Machine Studies 7(1), 1–13 (1975)

[5] Sugeno, M., Kang, G.T.: Structure identification of fuzzy model. Fuzzy Sets and Systems 28(1), 15–33 (1988)

[6] Tsukamoto, Y.: Fuzzy information theory. Daigaku Kyoiku Pub. (2004)

[7] Takagi, T., Sugeno, M.: Fuzzy identification of systems and its application to modeling and control. IEEE Transactions on Systems, Man, and Cybernetics 15(1), 116–132 (1985)

[8] Goldberg, D.E.: Genetic Algorithms in Search, Optimization and Machine Learning. Addison-Wesley, Reading (1989)

[9] Deb, K.: Multi-Objective Optimization using Evolutionary Algorithms. John Wiley & Sons Ltd, England (2001)

[10] Nandi, A.K., Pratihar, D.K.: Automatic Design of Fuzzy Logic Controller Using a Genetic Algorithm – to Predict Power Requirement and Surface finish in Grinding. Journal of Material Processing Technology 148(3), 288–300 (2004)

[11] Nandi, A.K.: TSK-Type FRBM using a combined LR and GA: surface roughness prediction in ultraprecision turning. Journal of Material Processing Technology 178(1-3), 200–210 (2006)

[12] Chandrasekaran, M., Muralidhar, M., Krishna, C.M., Dixit, U.S.: Application of soft computing techniques in machining performance prediction and optimization: a literature review. Int. J. Advance Manufacturing Technology 46, 445–464 (2010)

[13] Nandi, A.K., Pratihar, D.K.: Design of a Genetic-Fuzzy System to Predict Surface finish and Power Requirement in Grinding. Fuzzy Sets and Systems 148(3), 87–504 (2004)

[14] Nandi, A.K., Davim, J.P.: A Study of drilling performances with Minimum Quantity of Lubricant using Fuzzy Logic Rules. Mechatronics 19(2), 218–232 (2009)

5

Single and Multi-objective Optimization Methodologies in CNC Machining

Nikolaos Fountas[1], Agis Krimpenis[1], Nikolaos M. Vaxevanidis[1], and J. Paulo Davim[2]

[1] Department of Mechanical Engineering Technology Educators, School of Pedagogical and Technological Education (ASPETE), GR-14121, N. Heraklion Attikis, Greece
n.fountas@webmail.aspete.gr, a.krimpenis@webmail.aspete.gr, vaxev@aspete.gr
[2] Department of Mechanical Engineering, University of Aveiro, Campus Santiago, 3810-193, Aveiro, Portugal
pdavim@ua.pt

Aiming to optimize both productivity and quality in modern manufacturing industries, a wide range of optimization techniques and strategies has been presented by numerous researchers. Optimization modules such as Genetic Algorithms, Evolutionary Algorithms and Fuzzy systems are capable of exploiting manufacturing data with high efficiency and reliability, in order to provide optimal sets of solutions for machining processes. The main scope of this chapter is to present the fundamentals in formulating and developing optimization methodologies, which ultimately offer optimal cutting conditions for both prismatic and sculptured surface part machining and actually improve industrial practice.

5.1 Introduction

A large number of techniques have been implemented in the field of the manufacturing parameters optimization and others are considered candidate optimization tools. Optimization problems include product design, process engineering, quality control, process planning and different machining operations numerically, or conventionally, controlled. Moreover, optimization has also been investigated in other highly sophisticated and/or non-conventional manufacturing processes, such as Electro-discharge machining (EDM) or abrasive water jet machining (AWJM) [1, 2].

Initial approaches in the field of machining optimization involve classical optimization, Geometric Programming [3, 4], as well as various Graphical techniques [5]. Owing to the complexity of the machining problem – mixed real-valued and integer-valued parameters with unknown interactions – the classical optimization algorithms take too long to respond and have a high calculation cost, since they mainly involve deterministic and gradient-based operators. This practically makes them unsuitable for industrial usage. Thus, Artificial Intelligence algorithms appear to be a more suitable tool provided that they are well calibrated and carefully adjusted to the machining problem.

The main goal of this chapter is to provide an overview of the existing optimization methodologies in CNC machining, i.e., Neural Networks, Genetic and Evolutionary Algorithms, variations of optimization schemes and hybrid approaches. In order to better demonstrate the functionality of these systems, basic terms, definitions and standards about modeling of machining processes are analyzed, along with quality targets, objective functions and distinction between single- and multi-objective methodologies.

5.2 Modeling Machining Optimization

Depending on the number of optimization goals, optimization process can be qualified as single-objective, if a single quality objective is addressed, or multi-objective, if several quality objectives are undertaken. Since manufacturing problems have a vast solution space, more than one quality objectives are usually included in the optimization scheme. In case of multi-objective optimization, the problem should be carefully designed and the quality attributes suitably determined, according to their effect on the machining goals.

5.2.1 Definition of Optimization

Optimization of manufacturing processes is the effort to enhance performance of manufacturing environments by providing an elite set of solutions that perform best according to the target quality characteristics. Applying optimization methodology to a given machining task leads to finding maximum or minimum values of the corresponding fitness function (depending on whether it is maximization or maximization problem, respectively). This optimization philosophy can be applied to any scientific field were the aim is to enhance efficiency, productivity and quality. When optimization refers to a maximization problem with a quality goal that is measured by function f, the optimal solution is the one with the highest f value, or equivalently, the one with the lowest value of the opposite function, $-f$ [6].

In its mathematical description, an optimization process, if minimizing its quality goal, may be determined as follows:

For function f: $R^n \to R$, find $\hat{x} \in R^n$, such that $f(\hat{x}) \le f(x), \forall x \in R^n$. (1)

Similarly, for a maximization problem, the expression is the following:

For function f: $R^n \to R$, find $\hat{x} \in R^n$, such that $f(x) \le f(\hat{x}), \forall x \in R^n$. (2)

The domain R^n of f is the search space. Each parameter vector of this domain is a candidate solution in the search space, with $\hat{x}$ being the optimal solution. The value n represents the number of dimensions of the search domain and hence, the number of parameters involved in the optimization problem. The function f is the objective (or fitness) function that maps the search space to the function space. If the objective function has one output, then the respective function space is one-dimensional and thus provides a single fitness value for each set of parameters. This single fitness value specifies the optimality of the parameter set for the desired task. Usually, the function space can be directly mapped to the fitness space. However, the distinction between function space and fitness space is important in the case of multi-objective optimization problems, which include several objective functions drawing input from the same independent variable space [7, 8].

For a known differentiable function f, calculus may easily provide local and global optima of f. However, in machining problems this objective function f is not known. When addressing such problems, the objective function is treated as "black-box"; input and output parameter values are obtained. The result of a candidate solution evaluation is the solution's fitness. Then, the final goal is to point out parameter values that maximize or minimize this fitness function [7].

In order to suitably approach the optimization of a given problem, components of candidate solutions should subject to certain constraints. In manufacturing environments, such constraints may be the maximum available process values and limitations of involved machine tools and equipment, product quality demand, minimum manufacturing cost, etc. This statement implies that constraints can be either economical or technological depending on the nature of the optimization problem. Specifically, in machining processes optimization, constraints are both economical and technological. Examples of technological constraints are maximum available motor power of a CNC machine tool, its maximum capacity in terms of cutting force load, the generated heat during the machining process, the maximum available torque and the maximum range of feeds and speeds, etc; see for details [9].

5.2.2 *Objective Functions in Machining Problems*

An Objective Function measures a candidate solution's status or performance in an optimization algorithm and in that sense it is used for its evaluation. Objective Functions act as nodes between optimization modules and physical or digital systems. The Objective Function of each candidate solution is individually calculated, thus OF values vary in a solutions set. In order to sustain uniformity, Objective Value V can be mapped into Fitness Value using mapping matrix M, where f domain is usually greater than zero [10].

$$\mathrm{M} : \mathrm{V} \rightarrow f \tag{3}$$

When it comes to a minimization problem, individuals are considered better when low values of the Objective Function are obtained. Further on, the Fitness Function is used to transform the values obtained from the Objective Function to a relative Fitness expression [11]. Hence,

$$F(x) = g(f(x)) \tag{4}$$

where
f: Objective Function
g: Transformation of the objective function values to nonnegative
F: Resulting Fitness Function.

In CNC machining applications, objective functions vary as far as process characteristics and constraints are concerned. During different machining stages, objective functions tend to optimize attributes referring either to productivity or quality or even both. Posed process constraints are also chosen depending on the studied machining stage, namely roughing, semi-finishing or finishing stage.

In most cases, objective functions of machining processes are discontinuous and nondifferentiable. Stochastic optimization methodologies can be applied for quality objectives optimization, since they are unresponsive to discontinuities and they do not need derivative information to converge.

Quality objectives usually optimized by suitably developed objective functions are determined in each machining stage. Considering that in most case a machining process follows a two-stage scheme (roughing and finishing), the following quality objectives are specified:

- Roughing Operations:

During rough machining operations, the primary target is to rapidly remove material from the raw stock, until the roughed part geometry is close to its final shape. Thus, quality characteristics are mainly related to productivity and time. Targets, such as High Material Removal Rate, Minimum Roughing Time and Minimum Remaining Volume for the finishing process to take up on, are the most common optimization attributes.

- Finishing Operations:

During Finish machining Operation, the primary goal is to achieve final specifications of the part geometry in terms of surface quality (low surface roughness), dimensional accuracy and geometrical features within specified tolerance, regarding the 3D target model, the blueprints and the engineering drawings. As a matter of fact, quality targets like cutting forces, surface roughness and machining time are the quality objectives to be minimized.

5.2.3 *Mathematical Modeling of Process Parameters and Objectives*

Machining processes values and significant level of quality characteristics are usually different, according to the nature of the process. If a process strongly relies on productivity, e.g. high Material Removal Rate in roughing processes, then quality targets such as machining time play a vital role; much more than others such as surface quality (see Section 2.2). On the contrary, if a process relies on product quality, e.g. low Surface Roughness in finishing operations, then respective quality targets may be of greater importance. Moreover, these function values are different in order and magnitude. For instance, machining time expressed in seconds for a rough machining stage may be 10,000 – 100,000 sec, while remaining volume expressed in cubic millimeters may be of a much higher order, such as about 10,000,000 mm^3. It is imperative to stress out that normalization procedure should be carried out especially when it comes to contradictory quality characteristics put together in the same problem. If, for example, the values of machining time and remaining volume are summed up to formulate a single objective, then an inherent bias in favor of machining time appears, owing to its values' order. Obviously, results can be quite disorienting as far as the problem overall response is concerned.

Significance level of quality characteristics is determined according to user's demands through weighing coefficients. Simple mathematical operations for data normalization are applied to overcome this problem. Thus, linear combination of quality targets, and their respective weight coefficients according to their importance in the studied problem, can be used. This type of expressions offers no obvious bias, provided that values are normalized beforehand. In mathematical terms, it is described as follows:

Let X_1, X_2,......X_i be the given quality characteristics (as in multi-criterion optimization) and w_1, w_2,......w_j their respective weight coefficients. Practically, a weight coefficient's value – if it expressed as a percentage – represents the respective quality characteristic's significance in the problem. Significance, thus, is applied through w_i values in the Quality Characteristic Equation (5).

$$QC\ (X_1, X_2,......X_i, w_1, w_2,......w_j) = X_1.w_1 + X_2.\ w_2 + ...\ X_i.w_j \qquad (5)$$

Difference in magnitude of the quality attributes is yielded by their parameters from which the attributes are controlled. The most common machining parameters that are necessary for tool-path generation are listed here.

1. Machining Strategy / Toolpath Style: This option enables the user to determine what strategy will be used to machine a work piece. Each machining strategy offers additional parameters that define the way the tool moves and the tool-path style in more detail.
2. Stepover (Radial cutting depth): This parameter is the Radial Distance between two successive tool passes in X-Y plane (3-axis machining) or in the plane that is vertical to tool axis (5-axis machining). It can be determined either by an arithmetic value or as a ratio of cutting tool diameter (also called engagement).
3. Stepdown (Axial cutting depth): This parameter adjusts the maximum depth of cut in the vertical direction (or tool axis direction), between successive

machining planes. It can be determined either by setting the number of cutting planes, or by specifying the distance between machining planes.

4. Machining Tolerance: The value of the maximum allowable distance between the theoretical tool-path and the tool-path computed by the CAM system.
5. Machining Direction: This option adjusts the cutting direction whether to be of “Climbing” type (the front of the advancing tool cuts into the material first) or of “Conventional” type (the back of the advancing tool cuts into material first).

5.2.4 *Single-Objective and Multi-objective Optimization*

As mentioned previously, an optimization problem may be either single- or multi-objective, depending on the number of its quality targets and the type –scalar or vector– of the objective function. When dealing with machining optimization problems, quality characteristics are usually more than one (and often conflicting), due to high process complexity. The overall goal is to find solutions that are acceptable by the decision maker. Although in single-objective optimization problems, it is straightforward to find an optimal solution, multi-objective ones provide users with sets of solutions, out of which one should be chosen so as to apply in physical systems.

In multi-objective optimization, the objective function is a vector, thus belonging to a multi-dimensional space. For each solution x in the decision variable space X, there is a point in the objective space Z illustrated by $f(x) = Z = (Z_1, Z_2, \dots Z_n)^k$. Points in the objective space are considered to reflect optimal solutions when their coordinates lie in the nearest possible location to the coordinate system origin. The set of solutions having the smallest vectorial distance to the coordinate system origin is the optimal solution set; the so-called Pareto Front. When such a solution set is found, then the optimization algorithm has converged.

5.2.4.1 Globally and Locally Pareto Optimal Sets

As in the case of single-objective optimization, there are also local and global optimal solutions, or Global and Local Pareto-optimal sets. When the set S_P is the entire search space S_S (S_P=S_S) then the resulting non-dominated set S_P' is the Pareto-Optimal Set. The non-dominated set of the entire feasible search space S_S is the *globally Pareto-Optimal Set*. If for every solution x in a the set S_P no solutions y are existed in the neighborhood of x, $|y-x| \leq \varepsilon$, dominating any member of the set S_P, then S_P constitutes a locally *Pareto-Optimal Set*.

5.2.5 *Choosing Optimization Philosophy*

In general, manufacturing processes are dynamic procedures whose control variables (range of values) vary depending on the specific case. Manufacturing demands for products may be different, and so do manufacturing practices. Under this assumption, suitability of an optimization methodology should be tested and evaluated beforehand.

Differentiation in the structures of optimization algorithms establishes several optimization philosophies. Therefore, an optimization approach is chosen based on the quality objective (or objectives), the constraints, the process and the preferred workflow of problem handling. More specifically, an optimization module should be selected to be implemented by taking into account the following issues:

- The nature and the number of process parameters.
- The number of quality objectives
- Applying single- or multi-objective optimization philosophy
- The selection among stochastic or conventional optimization approaches
- The potential constraints on the solution space and on parameter values
- The computational cost
- The proper functionality per case study
- The iteration time (as short as possible).

5.3 Building Meta-Models for Machining Processes Using Artificial Neural Networks

5.3.1 Basics of Artificial Neural Networks (ANNs)

Although there is not a single, or universally accepted, definition of ANNs, most researchers agree that an ANN is a network of many simple processing "units" (also referred to as "elements", "nodes" or "neurons") each with a small amount of local memory and in some cases arranged in layers. Nodes are connected through communication "channels" (usually referred to as "connections", "connection weights" or "weights"), see Figure. 5.1. Connections bear encoded arithmetic data. Nodes process only local data and inputs brought to them through the connections.

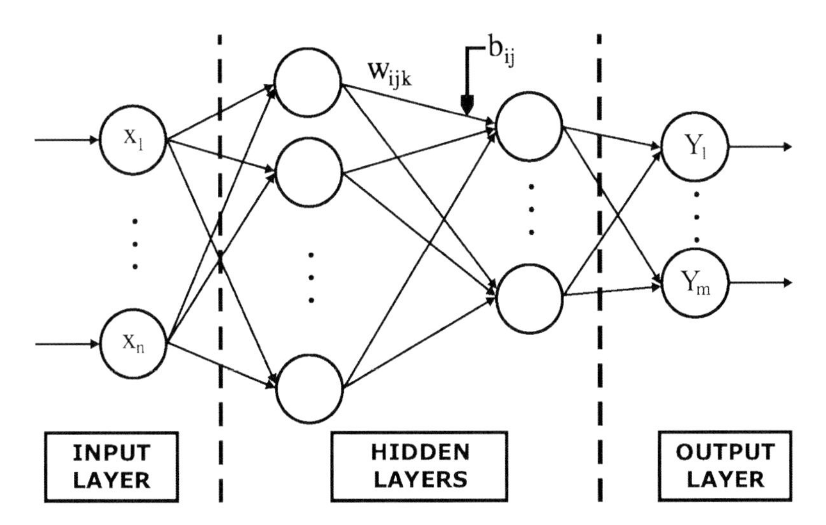

Fig. 5.1. Feed-forward multi-layer perceptron (MLP) with two hidden layers.

Scientifically, Artificial Neural Networks (ANNs) are systems consisting of a number of processing units or elements that calculate in parallel and whose functioning is determined by the network structure, the connection weights and the processing that is carried out on the elements or the nodes [4]. These systems tend to store empirical knowledge and to make it available for use similarly to the human brain. Knowledge is acquired through learning and it is stored in the connections between neurons [12].

Formulation of ANNs is the mathematical modeling of the biological neuron networks, so that they can perform complex, and possibly "intelligent", calculations similar to the ones performed by the human brain. Most ANN types need a "training" stage, during which connection weights are adapted according to a set of training data. Thus, ANNs "learn" or "are trained" by examples. When trained carefully, ANNs can "generalize" (e.g. predict output values) beyond the field of training data. Training algorithms commonly implemented are Gauss-Newton [13-15], Levenberg-Marquardt [16, 17], conjugate-gradients algorithms [18], etc. ANNs can operate on parallel computer systems, since calculations on nodes are highly independent to each other.

Researchers have developed a large number of ANN types [19]. In the field of machining applications, the most commonly used types are Multilayer Perceptrons (MLP) and Radial Basis Function networks (RBF), that belong to the Supervised learning - Feed forward ANN category, and Self-Organizing Maps (SOM) that belongs to the Unsupervised learning - Competitive ANN category.

Figure 1 depicts a feed-forward MLP that is typically applied to function approximation problems, such as machining time or surface roughness prediction. It consists of four layers; one input layer, one output layer and two hidden layers. Any node of a given layer is connected through connection weights (w_{ijk}) to all nodes of the previous and the next layer. Apart from weights, biases (b_{ij}) are fed to each node. A node's output is an activation function (nonlinear, such as tanh or $1/(1+\exp(-x))$) value, whose argument is the node's sum of weights, plus the node's bias. Training stage ultimately fixes w_{ijk} and b_{ij} values and it is concluded when all inputs and outputs are matched with a predetermined error.

ANNs are mainly implemented in two major problem categories: (a) Classification or sorting problems, and (b) Value Prediction or Estimation of unknown functions. In classification problems, ANNs are tolerant to inaccurate data and can cope with vast training sets at which strict and fast rules fail, such as in Expert Systems. Moreover, almost any vector function defined at compact spaces can be approximated with a given accuracy by a feed forward ANN, if adequate data and resources are available [20-23].

"Case" or "example" is a vector whose values are presented to all input nodes of an ANN. This vector can also include the objective or output values. An input vector value is also called "input variable" or "independent variable", while an output vector value is called "output", "objective" or "dependent variable". "Data sets" are matrices containing a number of "cases".

The main objective of ANNs is to generalize, thus to accurately perform using new data. A data set presented to an ANN at any time is called "sample" and is divided in three subsets: training subset, which is used for parameter fitting (learning), validation subset, which is used for network architecture tuning, and test subset, which is used for accessing the generalization ability of a trained network. In literature, the use of validation and test sets is often reversed [24, 25].

Finding an ANN that performs optimally in new cases, while it does not just memorizes the already known cases with which it was trained, means that its performance is measured by an error function (e.g. mean square error, total absolute error, etc) when unknown –independent– data is presented to the network [24]. The validation set consists of these new cases. However, ANN efficiency is measured by a third –test– set, since validation procedure may lead to ANN over-fitting (data memorizing).

5.3.2 Machining Data Sets for Training, Validation and Test of ANNs

For specific parts, it is possible to immediately correlate machining parameters to Quality Characteristics through ANN. Machining parameters involve tool geometry, tool-path strategy, feed-rates, etc., while machining time and surface quality are considered Quality Characteristics. Based on this concept, properly trained ANNs are utilized as prediction functions. Machining parameters are the independent variables and Quality Characteristics are the dependent variables.

Training back-propagation feed forward multi-layer perceptrons with Levenberg-Marquardt algorithm leads to precisely predicting functions, under the assumption that there are sufficient training, validation and test sets. Using Design of Experiments [26], sets of data are systematically gathered. If ANOVA is carried out on these sets and no interdependencies among machining parameters are found, all parameters can be considered independent variables.

All necessary data can be accumulated through CAM software or by direct experiments on a CNC machine tool. In the first case, an application which interacts with the software must be developed for this purpose, so as to provide it with machining parameter values and to extract machining time and remaining volume values for further exploitation.

Accumulated machining data can be unequally divided into the three required subsets, provided that the data selection for each is unbiased. Random choice of data is a simple and adequate procedure, in order to achieve proper set division, keeping in mind that the major part of existing data are needed as training subset and the remaining data formulate validation and test sets. Although data coming from actual experimentation on CNC machine tools could be mixed with data from CAM software simulation for training, validation and test sets, users are advised to avoid such practices; ANN training would be more complicated and probably it will be not able to generalize. The measured results of actual experiments on machine tools include "signal noise" owing to stochastic phenomena such as machine tool vibrations and stiffness, chattering, tool and material defects, etc which are usually not included in CAM software.

5.3.3 Quality Characteristics Predictions Using ANN Meta-models

In machining processes, the main goal of ANN usage would be the creation of prediction functions of quality targets such as the Machining Time, the Remaining Volume (after a roughing operation) and/or Surface Roughness (in finishing), as long interdependency among machining parameters and the aforementioned Quality Characteristics exists. It has been seen that such interdependency exists.

Output level of a trained ANN contains either one or all the predicted Quality Characteristics. It is noted here that Quality Characteristic's values must have been properly normalized prior to the ANN training process. Moreover, it is possible that an ANN contains a weighed sum of more than one Quality Characteristic in a single-node output level. Researchers use the approach of either a single ANN that predicts all Quality Characteristics at the same time, which means that a larger training is needed for training, or a number of separately trained ANNs that predict a single Quality Characteristic.

An inherent drawback of using ANNs as prediction functions is that training data are usually based on specific machine tools, tools, part materials and geometries. In other words, ANN performance appears case-dependent. To overcome this, researchers can use data sets stemming from experiments on groups of similar (or parametrically designed) parts or ones that include the philosophy of machining features. Thus, trained ANNs can predict Quality Characteristics on any part of a certain group.

There is a plethora of publications concerning the application of ANNs for the prediction and/or optimization of the resulted outputs (quality characteristics) in machining. The discussion of such an extensive literature is beyond the scope of this chapter. The readers interested in particular aspects of the topic may refer to [27-30].

5.4 Genetic and Evolutionary Algorithms

Genetic Algorithms (GAs) are optimization modules that imitate the theory of Natural Selection through evolution, as developed by the 19th century biologist, Charles Darwin. In computers, GAs operate on a population –a set of potential solutions– by applying the principle of "survival of the fittest" [31] to produce better approximations to a solution of a given optimization task. Moreover, GAs are characterized by their stochastic ability to seek for optimized solutions within a search space, called exploration. GAs were first proposed and investigated by J. Holland at the University of Michigan [32]. Populations evolved in iterations, called "generations". A new population, the offspring, is created in each generation by applying mathematical operators on the old population, such as crossover, mutation, inversion, etc.

Evolutionary Algorithms (EAs) are in fact generalized GAs. They include more sophisticated operators than the ones in GAs and possibly implement other Artificial Intelligence, deterministic or heuristic algorithms, in order to reduce evaluation cost or increase the overall algorithm performance in terms of solution quality and of high solution exploitation. In simple words, exploitation is a GA's or an EA's ability to rapidly reach global optima after finding local optima inside the solution space.

The criterion by which candidate solutions in EA populations are evaluated is the Objective Function, as described earlier in this chapter. If the Objective Function is scalar, then the method is called Single-objective optimization, regardless of the number of physical objectives included. In the same sense, if the Objective Function is a vector, EAs perform Multi-objective optimization and the solutions are most commonly ranked based on Pareto optimality. For further details on GAs and EAs the reader may refer to [9].

5.4.1 Basic Genetic Algorithm Structure

Basic Genetic Algorithms (GAs) have a simple architecture, but they have to be suitably extended for a given optimization problem to be solved. A number of genetic operators have been proposed to apply various optimization problems. The fundamental operations a simple genetic algorithm may perform under its basic architecture are the following:

1. GAs work with strings coded the candidate solution
2. GAs work on a population of strings
3. GAs need only fitness value, not derivative information.

A basic GA is composed of Roulette Wheel Selection (reproduction), one-point Crossover (recombination) and a simple mutation (see Section 4.1.1). Each of these basic mechanisms works on strings in a population only with simple bit operations. The procedure of a GA may be as follows:

```
begin
   Initialization
   repeat
         Roulette Wheel Selection
         Crossover
         Mutation
         Evaluation
   until Termination_condition = True
end.
```

5.4.1.1 Encoding

When an optimization problem is solved with GAs, the solution space should be encoded into a string space. In other words, encoding means to map variables from the solution space into a finite-length string space. Good encoding schemes are thus required, so as to efficiently solve an optimization problem. Several encoding methods have been proposed so far; see [33-35]. The most important issue during encoding is to cover all the solution space with the mapped string space without redundancy. Consequently, the phenotype of the string space should be equal to the problem solution space, in order to make the problem simpler. What is more, string space should be generated as a set of feasible candidate solutions in order to avoid unnecessary search from the algorithm.

Genetic operators (see Section 4.1.2), however, directly work on the genotype (often called "chromosome") of GAs. Therefore, the performance of the genetic search highly depends on symbolic operations for the genotype. GAs perform better when substrings have consistent benefit throughout the string space. This is based on the concept that an encoding method is deeply involved in crossover operators. The neighborhood of a candidate solution in the solution space should be similar to the neighborhood of the respective string in the string space. If the offspring generated by a crossover operator is not similar to its parents, the population would proceed towards a different evolutionary direction and genetic search would probably result in failure. On the contrary, if good offspring is generated, genetic search results in success. When relationships between encoding methods and genetic operators are taken into account, two principles of the encoding rule are applied [31]:

1. The smallest characters that represent a solution space should be selected.
2. An encoding method that generates consistent strings through genetic operators should be designed.

There are three types of variable encoding: (a) *real value*, (b) *binary* and (c) *Gray binary*. The binary types are usually preferred, since they are said to offer no bias during the evolution being strings of binary digits. In some cases, encoding with real values (between 0 and 1) tends to make the GA converge faster, but this way the GA solutions are susceptible to entrapment in local optima.

GAs perform best when elite substrings tend to appear more frequently in a population without having their genetic material altered by genetic operators. Although the search depends on genetic operators, genetic information inherited from parents to offspring is dependent on the encoding. Therefore, as it is already been mentioned, a good encoding method must be adopted, so that genetic operators can produce meaningful offspring.

5.4.1.2 Selection

A GA procedure starts with reproduction, recombination and mutation. Selection simulates the process of natural selection and GAs need a similar mechanism to make a population evolve toward a better direction of optimal solutions. The main procedure of a selection module is to reproduce a population by pre-selecting an individual with a selection probability proportional to its fitness value. In this kind of selection, an individual with a higher fitness can reproduce more offspring. Since selection is performed by stochastically, a number of identical (or the same exactly) individuals may be chosen by chance. As a result, some types of strings tend to occupy a population. This is called "*genetic drift*" [32] or "*random drift*" which often occurs in the case where the GA population size is relatively small and has strong elitist behavior. Several selection schemes have been proposed in order to prevent a population from genetic drift.

Selection schemes are classified into two main categories; the *proportional selection scheme* and the *competition selection scheme*. In the first category, selection is based on the fitness value of an individual compared to the total fitness value of the overall population. In the second category, selection is based on the fitness values of some other individuals. Some of the most commonly used selection schemes are presented next [31]:

1. Roulette wheel selection scheme (or Monte Carlo method): This scheme selects an individual with probability in proportion to its fitness values.
2. Elitist selection scheme: This scheme preserves the fittest individual through all generation t, that is, the fittest individual is certainly selected prior to others into the next generation.
3. Tournament selection scheme: the individual with the highest fitness value between m randomly pre-selected individuals is selected. Note that m is the number of competing members.
4. Ranking selection scheme: This scheme is based on the rank of the individual's fitness value. Additionally, the individuals are selected by the number of their reproduction into the next generation based on the ranking table predefined earlier.
5. Expected value selection scheme: This scheme is based on the expected value of the individual's fitness. According to a respective probability, the expected value of an individual is calculated. Then, the probability of the selected individual is decreased by 0.5. Thus, this selection scheme relatively prevents an individual from being selected more than twice in the population.

5.4.1.3 Genetic Operators

The genotype of an individual is changed by genetic operators, such as combination, inversion and duplication, while evolution is controlled under natural selection. GAs require genetic operators to make the evolution feasible to all successive populations. Recombination generates new individuals with crossover operators. Mutation generates new individuals with some perturbation operators. Each operator contributes to evolution by performing specialized actions.

5.4.1.3.1 Crossover

The Crossover operator generates new individuals as solution candidates in GAs [31]. GAs can search the solution space mainly by using one of the crossover operators. With the absence of crossover operators, GAs would be random search algorithms. The crossover operator exchanges each substring between two individuals and replaces old individuals with others of a new genotype. The recombination between two strings is performed according to the type of crossover operator. Depending on the number of break-points among individuals, the crossover mechanism recombines the strings of the genotype.

Some of the crossover operators are the following:

- ***One-point Crossover*** is one of the basic crossover operators. By choosing a break-point as a cross site, the one-point crossover recombines substrings between two individuals.

Example 100|11110 → 100|10010
101|10010 → 101|11110

- ***Multi-point Crossover*** is a crossover operator with more than one break-points on a string. Multi-point crossover recombines some substrings which cut by some break-points between individuals.

Example: 1|11|1|1 10010
0|00|0|0 01101

- ***Uniform Crossover.*** In the Uniform Crossover operator, a mask pattern is randomly generated including "0" and "1". Recombination between two individuals is then achieved by exchanging the characters in the genotype string according to the mask pattern.

Example: 11|11|1 11001
00|00|0 00110 (Mask Pattern: 00110)

- ***Cycle Crossover*** operator first selects a starting point (not a crossing site). Then, it makes a closed round of substrings between individuals. The following example describes the workflow of Cycle Crossover:

Example: Consider the Parents *P1* and *P2* as follows:
P1: 31245
P2: 24351

Firstly the starting point on the string is chosen and then the selection of locus 2 of the *P1* parent is done. A closed round of substring begins from this starting point. The characters 1 and 4 in the locus 2 of the *P1*, *P2* to the same locus of offspring are copied, respectively:

O1: *1***
O2: *4***

Next, the character of the locus existed the character 4 in the locus 2 of *P2* is copied as follows:

O1: *1*4*
O2: *4***

The process is repeated until the first chosen character of *P1* is reached again. Consequently, the closed round of substring is achieved as follows:

O1: *1*45
O2: *4*51

Finally, characters are filled from the former parents to the latter offspring. The complete offspring becomes:

O1: 21345

O2: 34251

- ***Partially Matched Crossover (PMX).*** The PMX Crossover Operator is performed by executing two procedures. Like the Cycle Crossover operator, PMX Crossover is a specialized mechanism which may be applied in order to overcome permutation problems [31]. The workflow of the PMX Crossover is described below in the example:

Example: Consider the parents *P1* and *P2* having genotypes as follows:

*P*1: 31|24|5

P2: 24|35|1

The exchange of substrings is performed between two break-points to both parents without the overlapping of characters

O1: **35**

O2: **24**

The character 2 in the locus 3 of *P1* is exchanged with the character 3 in the same locus of *P2*, that is, the positions of the characters 2 and 3 in *P1* and the characters 3 and 2 in *P*2. The characters on the substring rounded by two break-points are also operated by this exchange, respectively. Consequently, partially matching substrings can be obtained as follows:

O1: 2*354

O2: 3524*

The remained character is copied from *P1* to *O1* and from *P2* to *O2*. As a result, the complete offspring is generated as follows:

O1: 21354

O2: 35241

In this way, genetic algorithms can generate new individuals of feasible candidate solutions satisfying the constraint of permutation problems [36]. It can be clearly seen that an independent relation exists between coding and crossover. When it comes to the design of crossover operation, one should take into account the inheritance of genetic information from parents to offspring. The offspring should inherit adequate genetic information from parents.

5.4.1.3.2 Mutation

Mutation occurs as the replicating error in nature. In GAs, the mutation operator replaces a randomly selected character on the string with the other one [31]. Mutation is performed regardless of individual fitness values. A classic mutation operation is the one-point changing per individual. Several mutation types are occurred in nature such as inversion, translocation and duplication. These are also the mutating mechanisms applied to GAs to simulate these phenomena.

- ***Inversion.*** Inversion partially changes a character sequence from one direction to the opposite. At first, two points are chosen randomly and the string is cut at these points. Then the substring is linked in the reverse direction into the remained one.

Example:

Assume a string included 5 characters. Two points on the locus 2 and 4 are selected:

1|234|5

By linking the substring, inversion results in the following state:

14325

- ***Translocation (or Shift).*** This operator changes the character sequence with moving to the different position. In translocation, a substring is chosen randomly as occurs in Inversion. The substring is then translocated to a randomly chosen break point (or locus).

Example:

Assume a string included 5 characters. The segment from the locus 1 to 2 is chosen as a substring.

|12|345

The substring is inserted from the locus 4 to 5 while the remained substring of the string shifts to the leftmost point of the string as follows:

34512

- ***Duplication.*** This mechanism overlaps the same substring on a string. the substring is randomly chosen as it occurs in inversion. The substring is then overwritten to a randomly selected locus.

Example:

Assume a string included 5 characters. The segment from the locus 1 to 2 is chosen as a substring.

|12|345

The substring is overwritten over the locus 4 and 5 and the remained substring in not operated as follows:

12312

However the above generated individual does not satisfy the constraint for permutation problems. Therefore, duplication is not suitable to address this problem.

Mutation operators have great influence on the GA performance because they seriously affect populations. GAs are able to search a global solution space, since the mutation mechanism randomly changes the strings of individuals. If the probability of mutation is high, the mutation operator often breaks important substrings on individuals. Therefore, it is essential that a very low mutation probability should be used to overcome this drawback.

5.4.1.4 Advanced Reproduction Models

Reproduction operations are applied to individuals during their replacements by others which have better fitness values. There are currently two reproductive techniques in use. The first one is called the Generational reproduction model and the second the Steady-State reproduction model.

5.4.1.4.1 Generational Reproduction Model

The Generational Reproduction model is probably the most widely used in the applications of GAs. The generational reproduction mechanism replaces all the population at once with a new population. This is achieved by selecting an individual from the current population according to fitness value and adding this individual to the next population repeatedly. This approach of elitist reproduction has the drawback of low convergence speed.

5.4.1.4.2 Steady-State Reproduction Model

The Steady-State reproduction model replaces only a few individuals in a generation. Further on, crossover and mutation is applied on these individuals. The generated individuals are immediately available for generating and selecting next individuals, hence the convergence speed increases. Generally, an individual is replaced with one generated by genetic operators in each generation. Therefore, methods that select individuals to be deleted should be introduced. Alternative deletion methods were presented by [37]:

1. ***Delete Least Fit***. Deletion of the least fit individual from the population.
2. ***Exponential Ranking***: The worst individual has a probability p of being deleted. If it is not selected, then the next to the last also has a probability of p chance and so on.
3. ***Reverse Fitness***: Each individual has a probability of being deleted according to its fitness value.

5.4.2 Evolutionary Algorithms

5.4.2.1 Evolution Algorithm Structure

An Evolutionary Algorithm's optimization philosophy is more or less the same to a GA's. In fact, there is not a discrete distinction between the definition of EAs and GAs. One could say that GAs are a subset of the general EA family. EAs utilize a population of solutions/individuals that successively evolves in generations and constantly produces better solutions until the algorithm converges and the global optima is obtained. Operators help EA population find potentially better solutions by recombining, mutating and other operators that change the original solutions, while they substitute bad/old individuals with better/younger ones in the population.

EAs, however, can be different to GAs in a number of manners. One common differentiation is that operators can be functions of time, in other words probabilities changing with respect to the current generation number or have completely different schemes that are chosen according to a specific probability or to the current generation number.

A major trend that minimizes calculation cost is to divide EA population appropriately in subpopulations. Each subpopulation optimizes a certain objective function, thus the EA subpopulations are competitive to one another and they exchange optimal values at the end of each generation. This notion comes from Game Theory, at which players in a game behave competitively to each other and try to make maximum profit, until a certain point is reached at which no player makes more profit by changing his playing strategy (Nash Equilibrium point). Alternatively, all subpopulations can optimize the same objective function, but calculations are carried out on different computers or on different CPUs. This is called parallelism of computers (or parallel systems, see Section 5.4.2.2.1) and algorithms and can lead to serious cut-down of calculation cost. In either case, subpopulation are parts of a single population, thus there is need for intercommunication among them. This is performed introducing migration operators that control the way that individuals move to other subpopulations carrying all genetic information along with them.

In some optimization problems, such as machining or flow optimization, the evaluators are very costly and take up too many resources, in order to calculate objective function values. This means that the EA must "pay" high calculation cost for all solutions, even for the bad ones. However, if there is a way to classify solutions as "bad" or "good" prior to evaluation, then only the "good" would be evaluated and cost associated with "bad" solutions/individuals would almost be eliminated. During the last years, classification problems are commonly solved using Artificial Neural Networks (ANNs). Both supervised and unsupervised ANNs can perform decently in solution classification, but it has been shown that the unsupervised ones work best at such problems in very small response times. Including ANN operators in the EA optimizes its performance in terms of calculation time. Nevertheless, solutions/individuals classified as "good" should always be exactly evaluated by the EA. There is an inherent drawback in ANN implementation; there is always the possibility that a "good" solution is classified as "bad" and vice versa. This is overcome by the nature of EA, since "good" genetic material is never lost or ignored through the use of the other EA operators, such as mutation, crossover etc., that force it again back in the population either in the current or a following generation.

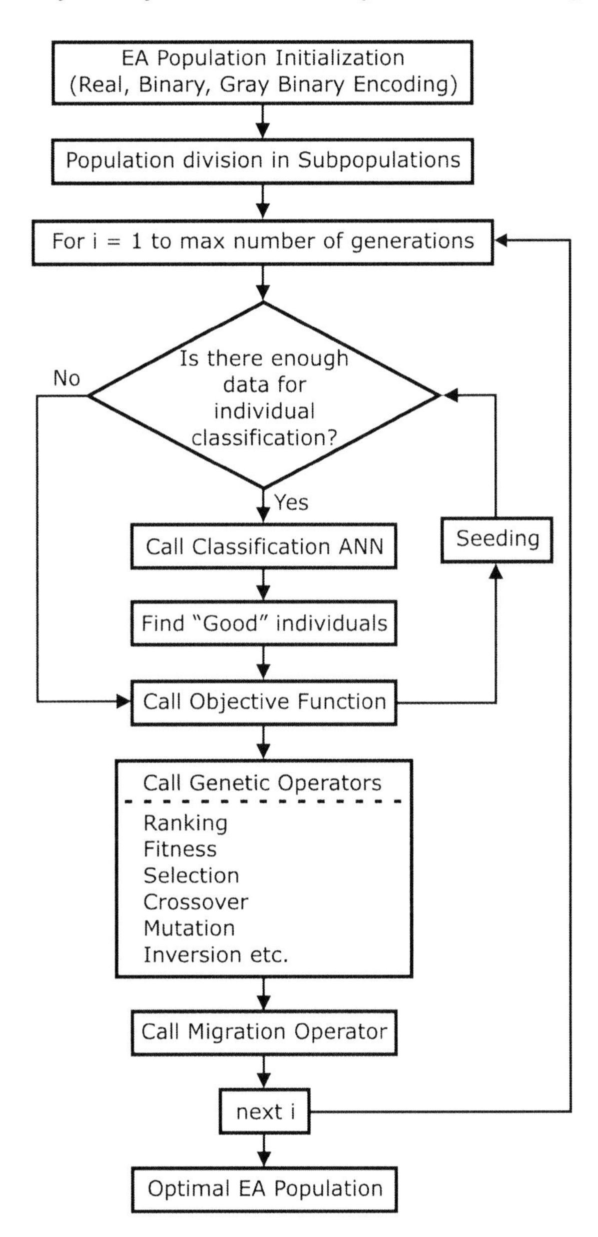

Fig. 5.2. Indicative flowchart of an EA.

Figure 5.2 depicts the flowchart of an EA that works on the philosophy described in this section as an example. As it is obvious, many variations and differentiations can be met in the relative literature. Apart from Game Theory and ANNs, heuristic methods are popular in the field of EAs. Heuristics are based on experience; they help in minimizing calculation cost and facilitate the algorithm in finding global optima rapidly, in other words augmenting both exploration and exploitation abilities of the algorithm. In the same sense, other knowledge-based systems can be implemented in EAs, such as Expert Systems.

5.4.2.2 Operators in Genetic Algorithms

In order to improve efficiency and speed of GAs, several advanced genetic operators have been introduced and applied so far. Mechanisms of genetic operators are based on biology, genetics and ecology. Moreover, advanced operators in GAs are applied to address difficulties related to premature convergence and minimal deceptive problems. The following sections represent some of the most known and implemented schemes that integrate the architecture of Genetic Algorithms.

5.4.2.2.1 Parallelism

Parallelism of GAs has been proposed by several researchers; see for example [38, 39]. Parallelism can be divided into two types. The first type works with a population divided into several sub-populations. The genetic operators in a sub-population prevent local minima from widely propagating to other sub-populations. The second type is to facilitate the rapid computation by working with parallel computer systems. Both types may be realized simultaneously. A Parallel Genetic Algorithm where individuals in a population are placed on a planar grid was proposed in [38]. Both Selection and Crossover are limited to operate on individuals in neighbourhoods on that grid. During the next generation, individuals from a specific location are selected. An individual of the old population is replaced with the selected one. Crossover is performed by mating individuals from the same neighborhood.

The main difference to the basic GA is the selection by replacement Operator. Each individual is replaced with a selected individual with a higher fitness value. The pseudocode of this parallel GA is:

```
begin
  Initialization
  repeat
        Selection by Replacement
        Crossover
        Mutation
        Evaluation
  until Termination_condition = True
end.
```

Construction of the grid presented above, which constitutes an advanced structural part of Parallel GAs, permits the local selection among individuals found in sub-populations; this imitates the natural phenomenon of mating individuals in local environments.

5.4.2.2.2 Migration

In Parallel GAs, genetic operations are performed locally and independently in the divided subpopulations. Each subpopulation tries to locate good local minima. New offspring is generated by genetic operators within each subpopulation. Therefore, each subpopulation evolves toward different direction, much like a hill-climbing algorithm. After some generations, the best solutions in a locus (subpopulation) are propagated to neighboring subpopulations. This is called *Migration*.

During Migration, the best individual in a generation is sent to its neighbors as a migrating individual. The migration frequency is an essential parameter for the efficient performance of this special operation. If the migration phenomenon appears with high probability, the group of sub-populations works equivalently to a single population. On the contrary, if no migration appears for a number of generations, sub-populations only perform local hill climbing.

The division of a population into sub-populations prevents premature convergence to local optima. GAs consisting of sub-populations are often run on parallel computer systems with multi-processors, since each subpopulation can easily be assigned to a different processor.

5.5 Variations of Evolutionary Algorithms

Except for the aforementioned modules of Genetic and Evolutionary algorithms, a wide range of variations for process optimization has been presented in the relative literature. Related evolution-based optimization algorithms are Particle Swarm Optimization, Simulated Annealing, optimization with Tribes, Tabu Search algorithms, etc. Some of the most known and extensively applied modules for optimization are demonstrated and described in the following sections.

5.5.1 Particle Swarm Optimization

Particle Swarm Optimization (PSO) is an optimization strategy that explores the search space in order to maximize a specific quality target. Kennedy and Eberhart [40] were the first to introduce PSO. Their concept was based on the swarming habits of certain kinds of animal species and the field of evolutionary computation.

PSO algorithm has the ability of the simultaneous maintenance of several candidate solutions in a given search space. In each algorithm iteration, a candidate solution is evaluated by the objective function being optimized, thus determining the fitness of that solution. The candidate solutions in PSO algorithm are treated as particles "flying" through the fitness range searching for the minimum or maximum of the respective objective function. The initial selection of candidate solutions within the search range is random. Every "'particle", which composes the candidate solution, sustains its position, its evaluated fitness and its velocity. The best fitness value that may be achieved is referred to as the individual's best fitness and the solution, from which the best fitness is obtained, is referred to as the individual best position (or individual's best candidate solution). Further on, PSO algorithm maintains the best solution in the group of particles until the optimization procedure terminates. This best solution is known as the global best position or global best candidate solution [41].

There are three major steps in the PSO algorithm development. These steps are repeated until termination/convergence criteria are met:

- Evaluation of particles' fitness values. This step is conducted by feeding the objective function with candidate solutions.
- Updating individuals, global best fitness values and candidate solutions. This is achieved via the comparison between the newly evaluated fitness values against the previous individuals and global best fitness values, replacing values when it is appropriate.
- Updating velocity and position of each particle. This step determines the optimization ability of PSO algorithm. The velocity of the particles in the swarm is updated using the following equation:

$$v_i(t+1) = wv_i(t) + c_1r_1[\hat{x}_i(t) - x_i(t)] + c_2r_2[g(t) - x_i(t)] \quad (6)$$

In Equation (6), i represents the index number of each particle. Moreover, $v_i(t)$ is the velocity of the *i-th* particle at time t, whereas $x_i(t)$ denotes the position of *i-th* particle at time t. Parameters w, c_1, c_2, ($0 \leq w \leq 1.2$, $0 \leq c_1 \leq 2$ and $0 \leq c_2 \leq 2$) are coefficients adjusted by the user. The values r_1 and r_2 ($0 \leq r_1 \leq 1$ and $0 \leq r_2 \leq 1$) are random values regenerated for each velocity update. $\hat{x}_i(t)$ represents the individual best candidate solution for particle i at time t, and $g(t)$ is the swarm's global best candidate solution at time t.

As for the rest three terms existed in the velocity Equation (6), it should be mentioned that they have different roles in the PSO algorithm. The term $wv_i(t)$ is the *inertia component.* The main role of this term is to keep the particle's movement in the same direction with the one that it was initially heading. The term w typically ranges between 0.8 and 1.2 and can either decelerate the particle's inertia or accelerate it along its original direction [42].

The second term $c_1r_1[\hat{x}_i(t) - x_i(t)]$ is the *cognitive component,* acting as the particle's memory, causing it to return to the regions of the search space in which it experienced high individual fitness values. The cognitive coefficient c_1 is usually close to 2, affecting the step size of the particle when approaches to its individual best candidate solution $\hat{x}_i$. The third term of Equation (6), i.e., $c_2r_2[g(t) - x_i(t)]$, is called the *social component.* The social component moves the particle to the best region that the algorithm has found so far. The social coefficient c_2 is close to 2, representing the step size of the particle when it moves toward the global best candidate solution $g(x)$ that the swarm has found up until that point.

The r_1 and r_2 values in the cognitive and social component respectively enable them to have a stochastic influence on the velocity update. The stochastic nature of the two components causes each particle in the swarm to move in a semi-random manner greatly influenced in the directions of the particle's individual best solution and hence, the swarm's global best solution.

The *velocity clamping* technique offers the ability of keeping the particles from moving too far beyond the search domain. This is achieved by limiting the maximum velocity of each particle [7]. If $[-x_{max}, x_{max}]$ is a specific search space, then velocity clamping tends to limit the velocity to the range $[-v_{max}, v_{max}]$, where $v_{max} = k * x_{max}$. The k value is a user-defined parameter and represents the velocity clamping vector taking values in the range $0.1 \leq k \leq 1.0$. It has been noticed that the search space in many optimization issues is not centered around 0 and, hence, the range $[-x_{max}, x_{max}]$ is not an adequate definition of the search domain. When it comes to such problems, one may define $v_{max} = k * (x_{max}-x_{min})/2$. After the calculation of the particles velocities, the positions are updated by applying the new velocities to the particles' previous positions. Finally,

$$x_i(t+1) = x_i(t) + v_i(t+1) \tag{7}$$

This procedure is repeated until some stopping criteria are met. Some common stopping conditions include a specific number of iterations of the PSO algorithm, a number of iterations since the last update of the global best candidate solution, or a predetermined fitness value of a quality target.

5.5.2 Simulated Annealing

Simulated Annealing (SA) methodology is a stochastic approach used in a wide range of optimization tasks especially in Engineering. Simulated Annealing was introduced in 1953, but actually first Kirkpatrick, Gelatt and Vecchi [43] used this approach in computer applications and related tasks. The respective methodology mimics the way thermodynamic systems go from one energy level to another [44]. In fact, the Simulated Annealing optimization algorithm is inspired by the metallurgical annealing process where a controlled heating and cooling procedure of a material through a temperature T is intended to produce a uniform distribution of crystals with the lowest possible internal energy [45, 46].

In computational terms, SA tries to approach a domain's global optima with predetermined constraints, traversing its different sub-optimal solutions within a neighbourhood, starting at the highest T. The parameter T enables the random search over the largest possible range. If the initial T and "cooling" rate are optimally selected, the number of sub-optimal solutions may be restricted, thus stabilizing the system and shrinking the neighborhood barriers. As neighborhood turns smaller (T parameter tends to 0), SA algorithm degenerates from stochastic to deterministic for the reason that only improvements are accepted. However, if parameter T is rapidly decreased, then the algorithm clusters to a local minimum. The algorithm is capable of escaping local minimum if this may be imperative. This also is the major advantage of this particular approach compared to other stochastic optimization algorithms.

Davidson and Harell [47], summarize the parameters that are needed in the implementation of SA algorithm. These parameters are listed below.

- Set of configurations or states of the system including an initial configuration randomly chosen.
- A generation rule for new configurations, which is usually obtained by determining the neighborhood of each configuration and choosing the next configuration randomly out of the current neighborhood.
- The target or the fitness function to be minimized over the search space (analogue of the energy).
- The "cooling" schedule of the target parameter, including initial values and rules for when and how to change it (analog to the temperature and its decreases).
- The termination criterion that is usually based on time, fitness function values and/or target values.

The basic Simulated Annealing Algorithm follows the 18 steps described in [48].

1. **Procedure** Simulated Annealing
2. **begin**
3. $t \leftarrow 0$
4. initialize T
5. select a current point $v_{current}$ randomly
6. evaluate $v_{current}$
7. **repeat**
8.   **repeat**
9.   select a new point v_{new} in the neighborhood of $v_{current}$
10.   **if** eval ($v_{current}$) < eval (v_{new})
11.   **then** $v_{current} \leftarrow v_{new}$
12.   **else if** random [0,1] < $e^{\frac{eval(v_{new})-eval(v_{current})}{T}}$
13.   **then** $v_{current} \leftarrow v_{new}$
14.   **until** (termination-condition)
15. $T \leftarrow g(T,t)$
16. $t \leftarrow t+1$
17.   **until** (halting-criterion)
18. **end**

5.5.3 Tabu Search

Tabu Search (TS) is a heuristic approach for solving optimization tasks, designed to prevent other methods or their components from being trapped in local optima [49, 50]. The Tabu Search approach offers the ability of solving combinatorial optimization problems whose applications vary on graph theory and matroid settings to general pure and mixed integer programming tasks. It is an adaptive procedure with the ability to utilize different methodologies, such as linear programming algorithms and specialized heuristics, which it directs to overcome the limitations of premature convergence to local optimality [49, 50].

The basic Tabu Search Algorithm follows the steps below:

1. **Procedure** Tabu Search
2. **begin**
3. select a current point $v_{current}$ randomly
4. evaluate $v_{current}$
5. $v_{tabu} \leftarrow v_{current}$
6. **repeat**
7. evaluate every point in the neighborhood of $v_{current}$
8. select a new point v_{new} in the neighborhood of $v_{current}$
9. **if** [eval ($v_{current}$) > eval (v_{new})] ^ $v_{new} \notin \vec{v}_{tabu}$
10. **then** $v_{current} \leftarrow v_{new}$
11. $\vec{v}_{tabu} \leftarrow \vec{v}_{tabu} \cup v_{current}$
12. $t \leftarrow t+1$
13. **until** (halting-criterion)
14. **end**

5.5.4 Ant-Colony Optimization

Ant-Colony Optimization (ACO) is a metaheuristic algorithm that tries to provide solutions to combinatorial optimization tasks. The function of ACO algorithm mimics the collective foraging behavior of ants. The first ACO algorithm was introduced in 1991 by Dorigo et al., [51], as a novel nature-inspired metaheuristic optimization approach. The vital component of ACO algorithm is the *pheromone module*. This particular module samples the search domain with a probabilistic manner [52].

The main characteristic of ACO algorithm is the explicit use of elements of previous solutions. In fact, ACO algorithm drives a constructive low-level solution and incorporates it in a population framework, while randomizing the construction in a "Monte Carlo" way. A "Monte Carlo" combination of different solution elements is suggested also by Genetic Algorithms, but in the case of ACO the probability distribution is explicitly defined by previously obtained solution components.

Considering the suggestions reported in [53], a combinatorial problem to be processed by implementing an ACO algorithm may be defined over a set $C = C_1, C_2, \ldots\ldots, C_n$ of basic elements. A subset S of elements represents a solution of the problem; $F \subseteq 2^C$ is the subset of feasible solutions, hence a solution S is feasible only if $S \in F$. A cost function f is determined over the solution domain, as f : $2^C \rightarrow R$, the objective being to find a minimum cost feasible solution S*, i.e. to find S*:S* $\in$ F and f(S*) $\leq f$(S), $\forall S \in F$. According to this model, a set of computational concurrent and asynchronous agents (an ant colony) moves through states of the problem corresponding to partial solutions of the solved problem. The colony moves by applying a stochastic local decision policy, relying on two parameters called *trails* and *attractiveness*. By moving each ant incrementally, a solution to the problem is obtained. When an ant completes a solution or during the construction phase, the ant evaluates the solution and modifies the trail value on the elements used in its solution. This pheromone information directs the search of future ants [51, 53].

Moreover, a typical ACO algorithm includes two more mechanisms: *trail evaporation* and, optionally, *daemon actions*. Trail evaporation decreases all trail values over time, in order to avoid unlimited accumulation of trails over some components. Daemon actions can be used to implement centralized actions that cannot be performed by single ants, such as the invocation of a local optimization procedure or the update of global information to be used for deciding whether to bias the search process from a non-local perspective [53].

As described in [52], a typical Ant Colony Optimization Algorithm follows the steps below:

1. **Procedure** Ant Colony Optimization
2. **Begin**
3. Input $P\ (S, f, Q)$
4. Initialize Pheromone Values (T)
5. $a_{bs} \leftarrow$NULL
6. **while** (termination criteria are not met) **do**
7. $k_{inter} \leftarrow \emptyset$
8. **for** $j = 1, \ldots, b_a$ **do**
9. $a \leftarrow$ Construct Solution (T)
10. **if** a (is a valid solution,) **then**
11. $a \leftarrow$ Local Search (a) {optimal}
12. **if** ($f(a) < f(a_{bs})$) or (a_{bs} = NULL) **then**
13. $a_{bs} \leftarrow a$
14. $k_{inter} \leftarrow k_{inter} \cup \{a\}$
15. **end if**
16. **end for**
17. Apply Pheromone Update (T, k_{inter}, a_{bs})
18. **end While**
19. **output:** The best-so-far solution, a_{bs}

5.5.5 Tribes

A tribe is a sub-swarm formed by particles which have the property that all particles inform all others belonging to the tribe (a symmetrical clique in graph theoretical language). The concept is therefore related to the "cultural vicinity" (information neighborhood) and not on "spatial vicinity" (parameter-space neighborhood). It should be noted that, due to this definition, the set of informers of a particle (its so-called *i-group*) contains the whole tribe but is not limited to it [54]. Note that Tribes mechanism can be also auto-parameterized.

The principles of Tribes are: 1) the swarm is divided in tribes; 2) at the beginning, the swarm is composed of only one particle; 3) according to tribes' behavior, particles are added or removed; and 4) according to the performances of the particles, their displacement strategies are adapted [54-56]. The so-called structural adaptation rules describe the time when a particle is created or removed and when a particle becomes the informer of another, whereas the moving strategies indicate how particles modify their positions. The Tribes algorithm is executed by the following steps [54].

- ***Swarm Initialization.*** If the initial iteration is $t = 1$ then a population of one particle and one tribe can be initialized, with random values generated regarding a uniform probability distribution to predefined upper and lower bounds.
- ***Evaluation of Each Particle in the Swarm.*** The Fitness (Objective Function) value of each particle is evaluated.
- ***Swarm Moves.*** Two moving strategies ("Simple Pivot" or "Noisy Pivot") can be applied to move the Swarm regarding the quality of the Particles ("Excellent", "Good" and "Bad").
- ***Setting the Adaptation Scheme.*** Determining the number of links in the population, after every $L/2$ iterations the structure of the swarm may be adapted by applying one of the adaptation rules.
- ***Stopping Criteria.*** The number of generations is specified for $t=t+1$. The procedure is continued with the evaluation of each particle in the swarm until a stopping criterion is met. Usually the stopping criterion may be a maximum number of iterations or a maximum number of evaluations of the respective objective function.

5.5.6 Hybrids of Evolutionary Algorithms

In the pursuit of EAs with optimized performance both in terms of solution quality and of calculation cost, researchers tend to blend various Artificial Intelligence algorithms together or to embody operators of one method into another. This leads to new evolutionary schemes, often referred to as "hybrids". Since there is not a specific definition of what exactly a hybrid is, even EAs described in Section 4.2 own properties of hybrids. A necessary step in the EA hybridization is to rearrange or formulate "non-EA" operators properly, so that they can work on populations of solutions/individuals instead of on single solutions, if a mixed optimization scheme is implemented for the optimization problem.

Owing to their high response speed, ANNs are commonly coupled with EAs either as surrogates for the Objective Function or as solution classification tools (see Section 4.2), especially when expensive evaluators should be used. In the field of machining processes, properly trained ANNs can predict values such as machining time, mean surface roughness, total remaining volume, etc. [57], which otherwise would be very time-consuming and costly to extract from experiments. This is because iterative algorithms involve a very high number of solutions, not all of them usable or exploitable.

Mixing and combining Evolutionary Algorithms have been investigated by researchers with approaches such as Simulated Genetic Annealing; see [58, 59], integrated SA, Tabu search and GA; see [60], coupled GA and PSO; see [61-63], genetic and estimation of distribution algorithms; see [64], etc. It is obvious that hybrids can be applied in many scientific fields and have already been implemented in optimization of CNC machining processes; see [2].

Another powerful hybrid is created using Game Theory and Hierarchical Games coupled with EAs. These have been proven very useful in optimization problems with vast numbers of heterogeneous process parameters [65]. Apart from controlling the subdivision of an EA population into subpopulations and the way these communicate with each other, it is possible to create an optimization scheme with various EAs behaving as leading and following players. "Leaders" establish the overall optimization strategy, while "followers" take up specific tasks or objectives to optimize. In this way, calculation load is distributed among the parts that the overall optimization algorithm consists of and calculations are carried out in parallel.

5.6 Conclusions

Machining processes, especially detailed CNC machining of high quality parts, involve a large number of process parameters. Classical optimization schemes often fail to produce global optima owing to the enormous calculation load. Thus, stochastic optimization methodologies offer great advantages in this field. This is mainly due to the probabilistic nature of their operators, which do not need any derivative information of –in most cases- unknown functions, while they have proven very efficient in searching the solution space. Artificial Neural Networks are a relatively simple tool in predicting machining process values after proper training or even classifying solutions. They can be utilized either as stand-alone methods for high speed calculations or as internal functions in other optimization algorithms.

Genetic and Evolutionary Algorithms are very powerful optimization tools that perform only in a fraction of the calculation time of classical methods. Provided that a machining problem is well defined, as far as independent and dependent variables are concerned, and appropriately constrained, they yield optimal results with low calculation cost. This chapter gives an overview of these algorithms, as well as their variations, such as Simulated Annealing, Tabu Search, Particle Swarm Optimization, Ant-Colony Optimization etc. Researchers have implemented these methods in many scientific fields, including the CNC machining field, providing users with optimal results in varying calculation times.

Owing to the large number and the different types of process parameters, machining optimization problems are highly sophisticated; especially when 5-axis CNC sculptured surface milling is involved. This calls for more complex optimization schemes, which are named "hybrids". Hybridization of stochastic algorithms is the procedure of blending together methods of Artificial Intelligence or lending operators from one algorithm to another. To that extend, Game Theory helps in declaring interactions among operators and gives them priorities, as well as defines the scope of each part of the algorithm. Overall, producing new hybrids, or improving the existing ones, is an open field and can push optimization algorithm domain to a whole new level.

References

[1] Vaxevanidis, N.M., Markopoulos, A., Petropoulos, G.: Artificial Intelligence in Manufacturing Research. In: Paulo Davim, J. (ed.) Artificial neural network modelling of surface quality characteristics in abrasive water jet machining of trip steel sheet, ch. 5, pp. 79–99. Nova Publishers (2010)

[2] Rao, R.V.: Advanced Modeling and Optimization of Manufacturing Processes. Springer, London (2011)

[3] Petropoulos, P.G.: Optimal selection of machining rate variable by geometric programming. International Journal of Production Research 11, 305–314 (1973)

[4] Sönmez, A.İ., Baykasoğlu, A., Dereli, T., Filiz, İ.H.: Dynamic optimization of multi-pass milling operations via geometric programming. International Journal of Machine Tools and Manufacture 39, 297–320 (1999)

[5] Kiliç, S.E., Cogun, C., Şen, D.T.: Short Note: A computer-aided graphical technique for the optimization of machining conditions. Computers in Industry 22, 319–326 (1993)

[6] Diwekar, U.: Introduction to Applied Optimization, 2nd edn. Springer (2008)

[7] Kennedy, J., Eberhart, R., Shi, Y.: Swarm Intelligence. Elsevier, Burlington (2001)

[8] Zitzler, E., Laumanns, M., Bleuler, S.: A tutorial on evolutionary multi-objective optimization. In: Metaheuristics for Multiobjective Optimisation, pp. 3–37. Springer (2004)

[9] Dixit, P.M., Dixit, U.S.: Modeling of Metal Forming and Machining Processes by Finite Element and Soft Computing Methods. Springer, London (2008)

[10] Melin, P., Castillo, O.: Hybrid Intelligent Systems for Pattern Recognition Using Soft Computing. Springer, Berlin (2005)

[11] De Jong, K.A., Spears, W.M.: A formal analysis of the role of multi-point crossover in genetic algorithms. Annals of Mathematics and Artificial Intelligence 5(1), 1–26 (1992)

[12] Haykin, S.: Neural networks, a comprehensive foundation. Prentice-Hall, Englewood Cliffs (1999)

[13] Bertsekas, D.P., Tsitsiklis, J.N.: Neuro-Dynamic Programming. Athena Scientific, Belmont (1996)

[14] Fletcher, R.: Practical Methods of Optimization. Wiley, NY (1987)

[15] Gill, P.E., Murray, W., Wright, M.H.: Practical Optimization. Academic Press, London (1981)

[16] Levenberg, K.: A method for the solution of certain problems in least squares. m Quarterly of Applied Mathematics 2, 164–168 (1944)

[17] Marquardt, D.: An algorithm for least-squares estimation of nonlinear parameters. SIAM Journal of Applied Mathematics 11, 431–441 (1963)

[18] Masters, T.: Advanced Algorithms for Neural Networks: A C++ Sourcebook. John Wiley and Sons, NY (1995)

[19] Jain, A.K., Mao, J., Mohiuddin, K.M.: Artificial neural networks: a tutorial. IEEE Computer 29(3), 31–44 (1996)

[20] Valiant, L.: Functionality in Neural Nets. In: Proceedings of the American Association for Artificial Intelligence, St. Paul, Minnesota, August 21-26, vol. 2, pp. 629–634 (1988)
[21] Siegelmann, H.T., Sontag, E.D.: Turing Computability with Neural Networks. Applied Mathematics Letters 4, 77–80 (1999)
[22] Orponen, P.: An overview of the computational power of recurrent neural networks. In: Proceedings of the 9th Finnish AI Conference - STeP 2000 (2000), http://www.math.jyu.fi/~orponen/papers/rnncomp.ps
[23] Sima, J., Orponen, P.: Computing with continuous-time Liapunov systems. In: Proceedings of the 33rd Annual ACM Symposium on Theory of Computing - STOC 2001, Heraklion, Crete, Greece, July 06 - 08, pp. 722–731 (2001)
[24] Bishop, C.M.: Neural Networks for Pattern Recognition. Oxford University Press, Oxford (1995)
[25] Ripley, B.D.: Pattern Recognition and Neural Networks. Cambridge University Press, Cambridge (1996)
[26] Montgomery, D.C.: Design and analysis of experiments, 5th edn. John Wiley and Sons, USA (2001)
[27] Chen, S.-L., Chang, C.-C., Chang, C.-H.: Application of a neural network for improving the quality of five-axis machining. Proceedings of the Institution of Mechanical Engineers, Part B: Journal of Engineering Manufacture 214(1), 47–59 (2000)
[28] Karpat, Y., Özel, T.: Multi-objective optimization for turning processes using neural network modeling and dynamic-neighborhood particle swarm optimization. International Journal of Advanced Manufacturing Technology 35(3-4), 234–247 (2007)
[29] Davim, J.P., Gaitonde, V.N., Karnik, S.R.: Investigations into the effect of cutting conditions on surface roughness in turning of free machining steel by ANN models. Journal of Materials Processing Technology 205, 16–23 (2008)
[30] Pontes, F.J., Ferreira, J.R., Silva, M.B., Paiva, A.P., Balestrassi, P.P.: Artificial neural networks for machining processes surface roughness modeling. International Journal of Advanced Manufacturing Technology 49(9-12), 879–902 (2010)
[31] Goldberg, D.E.: Genetic Algorithms in Search. Addison-Wesley, Reading (1989)
[32] Holland, J.: Adaptation in Natural and Artificial Systems, 2nd edn. The MIT Press, Massachusetts (1992)
[33] Fogel, D.B.: Phenotype, Genotype and Operators in Evolutionary Computation. The, IEEE International Conference on Evolutionary Computation, 193–198 (1995)
[34] Hinterding, R.: Mapping, Order-independent Genes and the Knapsack Problem. In: Proceedings of the 1st IEEE Conference on Evolutionary Computing, vol. 1, pp. 13–17 (1994)
[35] Tamaki, H., Kita, H., Shimizu, N., Maekawa, K., Nishikawa, Y.: A Comparison Study of Genetic Codings for the Travelling Salesman Problem. In: The 1st IEEE Conference on Evolutionary Computing, Florida, vol. 1, pp. 1–6 (1994)
[36] Bui, T.N., Moon, B.: A New Genetic Approach for the Travelling Salesman Problem. In: Proceeding of the 1st IEEE Conference on Evolutionary Computing, vol. 1, pp. 7–12 (1994)
[37] Syswerda, G.: A Study Reproduction in Generational and Steady-State Genetic Algorithms. In: Rawlins, G.J.E. (ed.) Foundations of Genetic Algorithms, pp. 94–101. Morgan Kaufmann Publishers, San Mateo (1991)

[38] Manderick, B., Spoessens, P.: Fine-grained parallel Genetic Algorithms. In: The 4th International Conference on Genetic Algorithms, Virginia, pp. 428–433 (1991)
[39] Wang, Z.G., Rahman, M., Wong, Y.S., Sun, J.: Optimization of multi-pass milling using parallel genetic algorithm and parallel genetic simulated annealing. International Journal of Machine Tools and Manufacture 45(15), 1726–1734 (2005)
[40] Kennedy, J., Eberhart, R.: Particle swarm optimization. In: Proceedings of the IEEE International Conference on Neural Networks, IV, pp. 1942–1948 (1995)
[41] Poli, R., Kennedy, J., Blackwell, T.: Particle swarm optimization - An overview. Swarm Intelligence 1(1), 33–57 (2007)
[42] Shi, Y., Eberhart, R.: A modified particle swarm optimizer. In: Proceedings of the IEEE International Conference on Evolutionary Computation, pp. 69–73 (1998)
[43] Kirkpatrick, S., Gelatt Jr., C., Vecchi, M.: Optimization by Simulated Annealing. Science 220, 671–680 (1983)
[44] Fleischer, M.: Simulated Annealing: Past, Present, and Future. In: Alexopoulos, C., Kang, K., Lilegdon, W., Goldsman, G. (eds.) Proceedings of the, Winter Simulation Conference, pp. 155–161. ACM (1995)
[45] Ryan, C.: Evolutionary Algorithms and Metaheuristics. In: Meyers, R.A. (ed.) Encyclopedia of Physical Science and Technology, 3rd edn., pp. 673–685. Elsevier (2001)
[46] Dréo, J., Pétrowski, A., Siarry, P., Taillard, E.: Metaheuristics for Hard Optimization: Methods and Case Studies. Springer, Berlin (2006)
[47] Davidson, R., Harel, D.: Drawing Graphs Nicely Using Simulated Annealing. ACM Transactions on Graphics 15(4), 301–331 (1996)
[48] Michalewicz, Z., Fogel, D.: How to Solve It: Modern Heuristics, 2nd edn. Springer, Berlin (2004)
[49] Glover, F.: Tabu Search, Part I. ORSA Journal on Computing 1(3), 190–206 (1989)
[50] Glover, F.: Tabu Search, Part II. ORSA Journal on Computing 2(1), 4–32 (1990)
[51] Dorigo, M., Stützle, T.: The ant colony optimization metaheuristic: Algorithms, applications and advances. In: Glover, F., Kochenberger, G. (eds.) Handbook of Metaheuristics, Kluwer Academic Publishers (2002)
[52] Dorigo, M., Blum, C.: Ant Colony Optimization Theory: A survey. Journal of Theoretical Computer Science 344, 243–278 (2005)
[53] Maniezzo, V.: Exact and approximate nondeterministic tree-search procedures for the quadratic assignment problem. INFORMS Journal of Computing 11(4), 358–369 (1999)
[54] Dos Santos Coelho, L., Alotto, P.: Tribes Optimization Algorithm Applied to the Loney's Solenoid. IEEE Transactions on Magnetics 45(3), 1526–1529 (2009)
[55] Chen, K., Li, T., Cao, T.: Tribe-PSO: A novel global optimization algorithm and its application in molecular docking. Chemometrics and Intelligent Laboratory Systems 82(1-2), 248–259 (2006)
[56] Cooren, Y., Clerc, M., Siarry, P.: Tribes-A parameter-free particle swarm optimization. In: The 7th EU/Meet, Adaptation, Self-Adaptation, Multi-Level Metaheuristics, Paris, France (2006)
[57] Krimpenis, A., Vosniakos, G.C.: Optimisation of roughing strategy for sculptured surface machining using genetic algorithms and neural networks. In: CD Proceedings 8th International Conference on Production Engineering, Design and Control, Alexandria, Egypt, December 27-29 (2004), paper ID: MCH-05
[58] Mahfoud, S.W., Goldberg, D.E.: Parallel recombinative simulated annealing: A genetic algorithm. Parallel Computing 21(1), 1–28 (1995)

[59] Wong, K.P., Wong, Y.W.: Genetic and genetic/simulated-annealing approaches to economic dispatch, Generation, Transmission and Distribution. IEEE Proceedings 141(5), 507–513 (1994)
[60] Mantawy, A.H., Abdel-Magid, Y.L., Selim, S.Z.: Integrating genetic algorithms, tabu search, and simulated annealing for the unit commitment problem. IEEE Transactions on Power Systems 14(3), 829–836 (1999)
[61] Esmin, A.A.A., Lambert-Torres, G., Alvarenga, G.B.: Hybrid Evolutionary Algorithm Based on PSO and GA Mutation. In: Proceedings of 6th International Conference on Hybrid Intelligent Systems, Brazil, 57I-62 (2006)
[62] Shi, X.H., Wan, L.M., Lee, H.P., Yang, X.W., Wang, L.M., Liang, Y.C.: An improved genetic algorithm with variable population-size and a PSO-GA based hybrid evolutionary algorithm. Machine Learning and Cybernetics 3, 1735–1740 (2003)
[63] Grimaccia, F., Mussetta, M., Zich, R.E.: Genetical Swarm Optimization: Self-Adaptive Hybrid Evolutionary Algorithm for Electromagnetics. IEEE Transactions on Antennas and Propagation 55(3), 781–785 (2007)
[64] Peña, J.M., Robles, V., Larrañaga, P., Herves, V., Rosales, F., Pérez, M.S.: GA-EDA: Hybrid Evolutionary Algorithm Using Genetic and Estimation of Distribution Algorithms. In: Orchard, B., Yang, C., Ali, M. (eds.) IEA/AIE 2004. LNCS (LNAI), vol. 3029, pp. 361–371. Springer, Heidelberg (2004)
[65] Krimpenis, A., Vosniakos, G.C.: Rough milling optimisation for parts with sculptured surfaces using Genetic Algorithms in a Stackelberg Game. Journal of Intelligent Manufacturing 20(4), 447–461 (2009)

6

Numerical Simulation and Prediction of Wrinkling Defects in Sheet Metal Forming

M.P. Henriques, T.J. Grilo, R.J. Alves de Sousa, and R.A.F. Valente

GRIDS Research Group
Centre for Automation and Mechanical Technology (TEMA)
Department of Mechanical Engineering
University of Aveiro
3810-193 Aveiro, Portugal
{marisaphenriques,a38296,rsousa,robertt}@ua.pt

The goal of the present work is to analyse distinct numerical simulation strategies, based on the Finite Element Method (FEM), aiming at the description of wrinkling initiation and propagation during sheet metal forming. From the FEM standpoint, the study focuses on two particular aspects: a) the influence of a given finite element formulation as well as the numerical integration choice on the correct prediction of wrinkling in walls and flange zones of cup drawing formed parts; and b) the influence of the chosen anisotropic constitutive model and corresponding parameters on the correct prediction and propagation of wrinkling deformation modes during forming operations. In this sense, this work infers about the influence of accounting for distinct planar anisotropy behaviours within numerical simulation procedures. Free and flange-forming examples will be taken into consideration, with isotropic and anisotropic material models. Additionally, the influence on wrinkling onset and propagation as coming from different numerical formulations will be accounted for shell and tridimensional continuum finite elements, along with implicit numerical solution procedures. Doing so, the present work intends to provide some insights into how numerical simulation parameters and modelling decisions can influence FEM results regarding wrinkling defects in sheet metal formed parts.

6.1 Introduction

Over the last years investigators have paid special attention to wrinkling defects in products coming from sheet metal-forming processes, particularly from the point of view of the numerical simulation and prediction of such problems in metallic parts. Nevertheless, when compared to other common defects in plastically formed components (such as springback and thinning), wrinkling onset and propagation during forming, in both single and multi-stage forming, is still an open problem from the point of view of robust numerical modelling.

From the fundamental point-of-view, wrinkling in metallic sheets or tubes subjected to plastic forming operations can be treated as localised buckling and, to some extent, can be dealt with in a similar way as classical buckling structural problems. This is, in fact, the basis of some analytical approaches, within the general theory of plastic bifurcation and loss of uniqueness, as can be seen in the classical works of [1] and [2].

From the bifurcation theory, at the onset of wrinkling two possible, physically-based, solutions can arise for equilibrium: the *unwrinkled state* and the *wrinkled state*, characterizing a non-unique solution at this point. The basic idea is that, for an unperturbed structure, wrinkling will start if the solution for the energy of the structure is not unique (i.e., a bifurcation point arises). After this bifurcation point, wrinkles may appear or the unwrinkled state may hold until the next bifurcation point. This analytical approach can be applied, as mentioned in [3], to the onset of wrinkling on *unconstrained* shell elements with double curvature, submitted to a biaxial plane stress state, and including plasticity (J2-flow theory). In this sense, it can deal to some extent with some types of wrinkling (wrinkling on unconstrained structures or surfaces). Doing so, analytical approaches can possibly provide useful estimates of elastic-plastic buckling for elementary shapes subjected to very simple boundary conditions. Nevertheless, the more complex the boundary conditions turn - as appearing, for instance, in flange-type wrinkling of blanks subjected to blank-holder forces during forming - the more difficult it is to provide a general analytical approach, although being still possible for very simple geometries. The previous references are just representative of decades of research works, and a comprehensive study can be found, for instance, in the work of [4].

Seeking for a higher generality in the analysis, numerical simulation (mainly by the *Finite Element Method* - FEM) takes place in the study of wrinkling effects in realistic sheet metal forming problems. For these cases, complexities in contact, loading, geometry and process parameters can completely impair the use of analytical tools. Focusing on numerical simulation procedures based on the FEM, two approaches are common: the so-called *bifurcation* and *non-bifurcation* types of analysis.

Non-bifurcation analysis employs models including initial imperfections, artificially introduced in the formulation as geometry, material or loading eccentricities and/or gradients [5] and [6]. This approach can lead to reasonable results - since real structures indeed have imperfections - but the fact that these imperfections are in some sense imposed into the formulation can lead to an inevitable sensitivity of the results to the imperfection level and type chosen. A kind of non-bifurcation procedure can be also retained, with good results, using explicit solution algorithms in the FEM, and thus benefiting from intrinsic round-up errors associated with this procedure [7]. These numerical errors can therefore act as the imperfections in the conventional non-bifurcation procedure. The main problem with explicit calculations in FEM is related to its arguable robustness in multi-stage sheet metal forming, particularly due to a poor evaluation of stress fields during forming. Additionally, high mesh sensitivity, typical on non-bifurcation procedures, is also observed in explicit analysis [8].

Bifurcation analysis, on the other hand, assumes no induced imperfections at the model level [9] and [10], thus benefiting from no need for the introduction of artificial empirical imperfection sets. Doing so, in order to catch wrinkling initiation, energy or geometry-based wrinkling indicators are needed, in order to point out the zones in the model that are prone to the appearance of wrinkling. This procedure has been used in conjunction with implicit solution techniques in FEM, in order to analyse the wrinkling behaviour of sheet metal in a more exact and rigorous way, either in free-form (side-wall) and constrained (flange) wrinkling, in both single-and multi-step forming [11]-[15].

The use of *wrinkling indicators* in bifurcation analysis with the FEM is also closely related to the attempts to establish a general *Wrinkling Limit Diagram* (WLD), in a sense similar to what happens to the well-known *Forming Limit Diagram* (FLD), in sheet metal forming. However, it turns out that seeking for a general WLD represents a much more demanding task than its FLD counterpart, as wrinkling limit diagrams, based solely on the final strain state of the formed part, may not be robust enough for general applications.

These difficulties are related to the fact that wrinkling onset and propagation depend, for instance, on the current local stress state during forming, the local curvature of the sheet as forming evolves and, finally, thickness values and variations during process, turning difficult to attain a general WLD. Attempts in determining wrinkling limit diagrams and curves, for predefined conditions, are nevertheless available in the literature, either for side-wall wrinkling [7,[16,[17] or flange wrinkling [18], but there is still place for research in the search of general-purpose robust WLD. An overview of the initial attempts in establishing limit diagram procedures can be found, for instance, in references [18] and [19].

Since wrinkling sometimes represents a transitory behaviour, adaptive meshing techniques based on wrinkling indicators are usually employed, as it would be computationally expensive to simply use a very refined mesh for a global wrinkling consideration. This local character of search for wrinkling zones in a mesh, together with the mentioned mesh sensitivity of non-bifurcation procedures [8], favours the use of bifurcation approaches with adaptive meshing during forming. However, successive mesh quality measurement and refinement, with proper state variable transfer procedures, can affect computational efficiency. Recently, alternative ways to improve mesh quality without refinement were used, such as the adoption of nodal enrichments [20] or mesh-free local sets [21] superposed to conventional meshes within the FEM [15]. Nevertheless, it was observed that the obtained solutions for wrinkling propagation were still dependent on the proper choice of functions to be added to the finite element mesh, in the latter method, or to the nodal enrichment level used, in the former approach. In this sense, none of these procedures can be truly considered as natural approaches.

From the experimental standpoint, some studies on wrinkling limits of commercially pure aluminium grades (subjected to different annealing treatments), when drawing through conical dies, were published [22[23]. From these studies, researchers concluded that the onset of wrinkling depends on the blank's geometry, the metal grading and the type of dies employed. Still relying on experimental data, it turns out to be difficult to avoid wrinkles in metal formed parts, particularly for some materials used in automotive industry, and therefore the use of blank-holder is important.

A study on variable blank-holder force, and its influence on wrinkling onset, was carried out, for instance, in reference [24]. Experimental results from two different shapes binder (a flat and cone-shaped ones) were compared with results coming from the finite element method and verified. Even though more researches were performed in the same study, only rigid shell modelling was used for the numerical simulation models of the deformable sheet. In this study a box specimen was the chosen shape, which, contrasting to cylindrical and conical cups, needed different amount of material to flow on different regions of the binder. To simulate the binder in the variable blank-holder force forming process, three models were built: flat-shaped binder with elastic solid elements, cone-shaped binder with also elastic solid elements and simplified segment binder with rigid shell elements. The cone-shaped binder was shown to provide a better control over the normal stress distribution, with a better formability of the blank in terms of the thickness and major strain distribution [24].

Morovvati *et al.* [25] investigated the plastic wrinkling of a circular two-layer blank to obtain the minimum blank-holder force required to avoid wrinkles, theoretically calculated by means of the energy method. The blank-holder force was dependent on the material properties and geometry, and the study concluded that a lower (a/b) ratio (where a is the punch plus die edge radius, while b is the blank radius) tends to increase the minimum blank-holder force required to avoid wrinkles. As a consequence, for a certain blank diameter, an increase of the punch diameter will decrease the blank-holder force. The study also concluded that the resultant yield stress of the circular two-layer blank was a function of the yield stress of the blank's components. Additionally, the effect of anisotropy was investigated for three different cases: a) the same material orientation for the two layers; b) a 45° difference in direction for the two materials' orientation; and c) a 90° difference in that direction. The results showed that, for the materials used, the minimum blank-holder force required has reduced by about 20% from case a) to case c).

An investigation carried out by Stoughton and Yoon [26] tried to find an efficient method for analysis of necking and fracture limits for sheet metals, combining a model for the necking limit with fracture limits in the principal stress space by applying a stress-based forming limit curve and the maximum shear stress criterion. The fracture model studied was applied on the opening process of a food can. Previous studies in this area demonstrated that stress-based forming limit curves calculated directly from the strain-based forming limit curve are substantially less sensitive to changes in strain path that normally happens in forming processes in industry. A new failure model was presented in that reference, taking into consideration the stress distribution through thickness direction and localised necking prior to fracture, being also capable of distinguishing forming processes where fracture could occur without necking. Also recently, a new method to test formability, and evaluating various modes of deformation, was investigated by Oh *et al.* [27]. The test consisted of three steps: drawing a blank-holder force *vs.* punch stroke diagram, measuring the strain level at the optimum condition on the stamped part and, finally, grading the test materials using a formability index. A new tool shape was designed and numerically simulated to optimise the dimensional details. The blank-holder force *vs.* punch stroke diagram had three failure loci to better evaluate the formability of the new tool and also to help find the optimum process condition and formability index. The numerical simulations were compared with experimental results and strain distribution in the forming limit diagram.

In reference [28], a user-defined material (UMAT) – accounting for an anisotropic material model based on non-associated flow rule and mixed isotropic-nonlinear kinematic hardening – was studied and implemented into the commercial finite element code ABAQUS. Two different forming processes were modelled: a cylindrical drawing and a channel drawing processes, in order to estimate the capability of the constitutive model in predicting earing, springback and sheet metal anisotropy effects. The achieved results demonstrated that applying the non-associated mixed-hardening material model both anisotropy and hardening descriptions significantly improved the prediction of earing in the cup drawing process and the prediction of springback in the side wall of drawn channel sections, even though a simple quadratic constitutive model and a single-backstress kinematic hardening model were used [28]. Although focusing on important geometric dimensional defects, the previous references lack in carrying out an in-depth study of wrinkling effects in the formed parts.

Although the onset of wrinkling takes place when the ratio of strain increments ($d\varepsilon_r/d\varepsilon_\theta$) or the ratio of strain ($\varepsilon_r/,\varepsilon_\theta$) reaches a critical value during forming, an attempt was made to try to find a theory that predicted wrinkling based on results obtained in the form of wrinkling limit diagram [29]. An aluminium alloy was studied, for four different annealing treatments, and it was found that the annealed sheet with higher *n*-values, *R*-value and UTS/σ_y ratios showed improved resistance against wrinkling, along with a clear curve separating the safe and wrinkling regions of service.

In reference [30], a study of different shapes for dies and blank-holders was performed in order to observe the distribution of the blank-holder force, the punch load at different drawing depths as well as the blank's thickness reduction during forming. The main goal of this investigation was to attempt to increase the deep drawing ratio along with decreasing the blank-holder forces involved. On the other hand, Port *et al.* [31] focused on surface defects of an industrial upper corner of a front door panel, as well as in an initially planar L-shaped part designed on purpose to reproduce at a small-scale surface defects after flanging. The simplified geometry was measured using a tri-dimensional measuring machine. The researchers achieved a good correlation between experiments and simulations, concerning the spatial position of the defects, and a buckling analysis during springback showed that the position of the defects effectively corresponded to a buckling mode.

In the present contribution, and within the numerical and experimental framework described, a sensitivity analysis of different numerical simulations, as well as the robustness of the respective results in characterising wrinkling defects, is carried out. The primary variables to be taken into account are related to numerical simulation models only and correspond to (i) distinct mesh densities; (ii) different finite element formulations; and (iii) basic and complex anisotropic constitutive models. It is seen that completely different quantitative and qualitative solutions (and, therefore, distinct wrinkling predictions) can be obtained for free and flange-forming examples with small perturbations or variations in the chosen input parameters, in a somewhat more severe way than the one that occurs in springback simulations. Also, and most noticeable, the correct prediction of wrinkling onset and propagation is more related to a given finite element formulation than to complex or more elaborate non-quadratic anisotropic constitutive models. All the analyses performed in this work were carried out using the finite element commercial package *Abaqus/Standard* [32], using fully implicit solution procedures, for both shell and solid elements. The anisotropic constitutive models adopted are not limited to those available in this commercial software, but also include more recent ones, implemented by the authors by means of user subroutines (UMAT).

6.2 Constitutive Isotropic and Anisotropic Models

Isotropic as well as anisotropic planar behaviours are considered in the following, and in both cases an isotropic hardening evolution was taken into account, relating the effective stress $\left(\bar{\sigma}\right)$ and equivalent plastic strain $\left(\bar{\varepsilon}^{p}\right)$.

Related to the numerical models for the characterisation of strain hardening effects, two isotropic hardening models were considered in the simulations. The first model is described by a Swift's law, in the form

$$\bar{\sigma} = K(\varepsilon_Y + \bar{\varepsilon}^p)^n, \tag{6.1}$$

where K is a material constant, ε_Y is the elastic strain at the yield state, n is the strain-hardening exponent and $\bar{\varepsilon}^p$ is given as

$$\bar{\varepsilon}^p = \bar{\varepsilon} - \bar{\varepsilon}^e, \tag{6.2}$$

where $\bar{\varepsilon}$ and $\bar{\varepsilon}^e$ represent the logarithmic total and elastic strain terms, respectively. On the other hand, the second model considered follows a Voce's law and is written in the form

$$\bar{\sigma} = \sigma_Y + R_{sat}[1 - \exp(-C_r \bar{\varepsilon}^p)] \tag{6.3}$$

where σ_Y is the uniaxial yield stress, C_r is a material constant and R_{sat} is given by

$$R_{sat} = \sigma_{sat} - \sigma^0. \tag{6.4}$$

In the present work, the anisotropic plastic behaviour was initially described by means of the Hill's quadratic yield criterion in its original version of 1948 [33], which represents a generalisation of the von Mises isotropic yield function, being expressed by a yield function $\phi = \phi(\sigma)$ in the form

$$\phi = \sqrt{F(\sigma_{22}-\sigma_{33})^2 + G(\sigma_{33}-\sigma_{11})^2 + H(\sigma_{11}-\sigma_{22})^2 + 2L\sigma_{23}^2 + 2M\sigma_{31}^2 + 2N\sigma_{12}^2}\,. \tag{6.5}$$

In this equation, the coefficients F, G, H, L, M and N are anisotropic constants coming from experimental test data. These coefficients can be defined as

$$\begin{gathered} F = \frac{(\sigma^0)^2}{2}\left(\frac{1}{\bar{\sigma}_{22}^2} + \frac{1}{\bar{\sigma}_{33}^2} - \frac{1}{\bar{\sigma}_{11}^2}\right), \quad G = \frac{(\sigma^0)^2}{2}\left(\frac{1}{\bar{\sigma}_{33}^2} + \frac{1}{\bar{\sigma}_{11}^2} - \frac{1}{\bar{\sigma}_{22}^2}\right) \\ H = \frac{(\sigma^0)^2}{2}\left(\frac{1}{\bar{\sigma}_{11}^2} + \frac{1}{\bar{\sigma}_{22}^2} - \frac{1}{\bar{\sigma}_{33}^2}\right), \quad L = \frac{3}{2}\left(\frac{\tau^0}{\bar{\sigma}_{23}^2}\right)^2 \\ M = \frac{3}{2}\left(\frac{\tau^0}{\bar{\sigma}_{13}^2}\right)^2, \quad N = \frac{3}{2}\left(\frac{\tau^0}{\bar{\sigma}_{12}^2}\right)^2 \end{gathered} \tag{6.6}$$

For the determination of these parameters, σ_Y is the normal yield stress stated before, while τ_Y is the corresponding yield stress on shear. The six anisotropic parameters involved in the last equation can be determined by three uniaxial tension tests, performed at 0º, 45º and 90º directions, respecting the rolling direction (RD). An alternative way of defining the anisotropic criterion is by means of the so-called Lankford's r-values (r_θ) for a specific (θ) direction, in the form

$$r_0 = \frac{H}{G}, \quad r_{45} = \frac{2N-(F+G)}{2(F+G)}, \quad r_{90} = \frac{H}{F}. \tag{6.7}$$

In order to perform the simulations in Abaqus/Standard software [32], it is also necessary to transform these three coefficients into the equivalent anisotropic yield stress ratios $(R_{11}, R_{22}, R_{33}, R_{12}, R_{13}, R_{23})$, required parameters for the input files of that FEM package. In planar anisotropy (plane stress conditions), r_0 and r_{90} coefficients are generally different, and so are the (R_{11}, R_{22}, R_{33}). Assuming σ_Y in the metal plasticity model to be equal to $\bar{\sigma}_{11}$, then it is nevertheless valid to assume that $(R_{11} = 1)$, and therefore

$$R_{22} = \sqrt{\frac{r_{90}(r_0+1)}{r_0(r_{90}+1)}}, \quad R_{33} = \sqrt{\frac{r_{90}(r_0+1)}{(r_0+r_{90})}}, \quad R_{12} = \sqrt{\frac{3(r_0+1)r_{90}}{(2r_{45}+1)(r_0+r_{90})}}\,. \tag{6.8}$$

Since the R_{13} and R_{23} coefficients refer to the thickness direction, and along this direction an isotropic behaviour is assumed (normal isotropy), it comes that $R_{13} = R_{23} = 1$.

The anisotropic yield criterion of Hill, on its version of 1948 as described before, is known to be well suited to the generality of steel alloys. Nevertheless, it provides poor results when characterising the behaviour of aluminium alloys sheets [34]. To fulfil this requirement in some of the following benchmark problems, non-quadratic anisotropic constitutive models were also implemented as user subroutines in Abaqus commercial software. For the sake of completeness, the implemented models will be summarised in the following.

The non-quadratic yield criteria described in the present work are suited for the description of anisotropic effects in aluminium alloys, and although a large number of yield criteria for this purpose exist in the literature (see, for a comprehensive survey on this topic, reference [34]), in the following only the criteria developed in the last decades by Barlat and co-workers (and particularly the 1991 [35] and 2004 [36] versions) will be described. The reason for the specific choice of those two yield criteria (*Yld91* and *Yld2004-18p*, respectively) is related to their ability to be numerically implemented in a three-dimensional framework, in opposition to other criteria that implicitly impose plane-stress conditions in the base formulation. Both three-dimensional criteria were afterwards implemented in Abaqus commercial finite element software, as UMAT subroutines [37].

The *Yld91* criterion is a generalisation of the isotropic criterion of Hershey [38] to anisotropic materials, with the anisotropic behaviour being considered in the formulation by replacing the principal values of the stress tensor by the principal values of an alternative tensor coming from linear transformations over the original stress fields. The anisotropy effects are subsequently described by the coefficients present in the linear transformation operator.

The fourth order linear operator $(\mathbf{L})$ therefore appears in the formulation as

$$\tilde{\mathbf{S}} = \tilde{\mathbf{L}}\mathbf{T}\sigma = \mathbf{L}\sigma, \tag{6.9}$$

where the operator $\tilde{\mathbf{L}}$ groups the anisotropy coefficients, where $\mathbf{T}$ is the operator that transforms the Cauchy stress tensor (σ) into the deviatoric stress tensor $(\mathbf{S} = \mathbf{T}\sigma)$. After this linear transformation, the components of the modified stress field $\tilde{\mathbf{S}}$ are introduced into the yield criterion and compared to the yield stress for uniaxial case, in the form

$$\phi = \left(\tilde{S}_1 - \tilde{S}_2\right)^{2k} + \left(\tilde{S}_2 - \tilde{S}_3\right)^{2k} + \left(\tilde{S}_3 - \tilde{S}_1\right)^{2k} = 2\sigma_Y^{2k}, \tag{6.10}$$

where k is a parameter that affects the yield surface shape [35], whereas the anisotropic effects may be reproduced by the knowledge of the coefficients affecting the operator $\mathbf{L}$, in the form

$$L=\frac{1}{3}\begin{bmatrix} c_2+c_3 & -c_3 & -c_2 & 0 & 0 & 0 \\ -c_3 & c_1+c_3 & -c_1 & 0 & 0 & 0 \\ -c_2 & -c_1 & c_1+c_2 & 0 & 0 & 0 \\ 0 & 0 & 0 & 3c_4 & 0 & 0 \\ 0 & 0 & 0 & 0 & 3c_5 & 0 \\ 0 & 0 & 0 & 0 & 0 & 3c_6 \end{bmatrix}, \tag{6.11}$$

as a function of the six anisotropic coefficients c_i $(i=1,\cdots,6)$. Therefore, and accounting for the exponent in the yield equation, the criterion is characterised by 7 coefficients, and the determination of the anisotropic coefficients can be carried out by conventional tension tests (and respective yield stress values) at 0º, 45º and 90º, and also from the yield stress value coming from a biaxial stress state, obtained, for instance, from a "bulge test". Despite the easy obtaining of these anisotropic coefficients, the ability to account for 3D stress states and the easy implementation of the criterion into FEM codes, the major drawback of the formulation would be the lack of reproducing distinct r_0 and r_{90} coefficients when uniaxial stresses in rolling and transverse directions are almost equal [34].

Seeking for a more general yield criterion, and after a succession of more evolved plane stress models directly applicable to thin aluminium sheets, Barlat and co-authors [36] have presented a more evolved formulation with a yield function proven to be convex and taking into account a large number of anisotropic coefficients (in this case, 18 parameters). According to this *Yld2004-18p*, suited for aluminium alloys and its anomalies (when compared to steel alloys), the yield function can be given in the form

$$\phi=\sum_{i=1,j=1}^{3,3}\left|\tilde{\mathbf{S}}_i^{(1)}-\tilde{\mathbf{S}}_j^{(2)}\right|^a=4\sigma_Y^a, \tag{6.12}$$

where index $i,j=1,\cdots,3$, whereas tensor fields $\left(\tilde{\mathbf{S}}_i^{(1)},\tilde{\mathbf{S}}_j^{(2)}\right)$ are defined by linear transformations of the type $\tilde{\mathbf{S}}^{(k)}=\tilde{\mathbf{L}}^{(k)}\mathbf{S}$, for operators $\tilde{\mathbf{L}}^{(k)}$ in the form

$$\tilde{\mathbf{L}}^{(k)}=\begin{bmatrix} 0 & -\tilde{L}_{12}^{(k)} & -\tilde{L}_{13}^{(k)} & 0 & 0 & 0 \\ -\tilde{L}_{21}^{(k)} & 0 & -\tilde{L}_{23}^{(k)} & 0 & 0 & 0 \\ -\tilde{L}_{31}^{(k)} & -\tilde{L}_{32}^{(k)} & 0 & 0 & 0 & 0 \\ 0 & 0 & 0 & \tilde{L}_{44}^{(k)} & 0 & 0 \\ 0 & 0 & 0 & 0 & \tilde{L}_{55}^{(k)} & 0 \\ 0 & 0 & 0 & 0 & 0 & \tilde{L}_{66}^{(k)} \end{bmatrix}. \tag{6.13}$$

Therefore, the two combined linear transformations allow for the characterisation of anisotropy based on a total of 18 parameters, which proves to lead to a very general 3D constitutive modelling. For the special case of plane stress analysis, the criterion degenerates in a simpler version involving 14 parameters. Furthermore, where these coefficients are equal altogether to 1.0, the criterion turns out to be equal to the isotropic criterion of Hershey [38], and for the particular case of $\tilde{\mathbf{L}}^{(1)} = \tilde{\mathbf{L}}^{(2)}$, then the *Yld91* anisotropic criterion is obtained.

The 18 coefficients involved in the *Yld2004-18p* yield criterion come from a series of experimental analyses seeking for the determination of the uniaxial yield stress values in tension (σ_Y), as well as the Lankford's coefficients (r_θ) for seven distinct directions, in the plane of the sheet, respective to the rolling direction (at 0º, 15º, 30º, 45º, 60º, 75º, 90º), the yield stress on biaxial loading solicitations (σ_b) and the anisotropy coefficient (r_c) for the compression loading experiment of a metallic disc.

The remaining factors are related to mechanical properties in the out-of-plane directions of the metallic sheet. Since standard experimental essays for out-of-plane properties are quite difficult to be obtained, a simplified version of the criterion, involving less material parameters, is also available (Yld2004-13p, see reference [36]), pointing to an yield function in the form

$$\begin{aligned}\phi = \left|\tilde{\mathbf{S}}_1^{(1)} - \tilde{\mathbf{S}}_2^{(1)}\right|^a + \left|\tilde{\mathbf{S}}_2^{(1)} - \tilde{\mathbf{S}}_3^{(1)}\right|^a + \left|\tilde{\mathbf{S}}_3^{(1)} - \tilde{\mathbf{S}}_1^{(1)}\right|^a - \left\{\left|\tilde{\mathbf{S}}_1^{(1)}\right|^a + \left|\tilde{\mathbf{S}}_2^{(1)}\right|^a + \left|\tilde{\mathbf{S}}_3^{(1)}\right|^a\right\} + \\ + \left|\tilde{\mathbf{S}}_1^{(2)}\right|^a + \left|\tilde{\mathbf{S}}_2^{(2)}\right|^a + \left|\tilde{\mathbf{S}}_3^{(2)}\right|^a = 2\sigma_Y^a\end{aligned} \tag{6.14}$$

where $\left(\tilde{\mathbf{S}}^{(1)}, \tilde{\mathbf{S}}^{(2)}\right)$ are now obtained by means of the transformation $\tilde{\mathbf{S}}^{(k)} = \tilde{\mathbf{L}}^{(k)}\mathbf{S}$, for the linear operators

$$\tilde{\mathbf{L}}^{(1)} = \begin{bmatrix} 0 & -1 & -\tilde{L}_{13}^{(1)} & 0 & 0 & 0 \\ -\tilde{L}_{21}^{(1)} & 0 & -\tilde{L}_{23}^{(1)} & 0 & 0 & 0 \\ -1 & -1 & 0 & 0 & 0 & 0 \\ 0 & 0 & 0 & \tilde{L}_{44}^{(1)} & 0 & 0 \\ 0 & 0 & 0 & 0 & \tilde{L}_{55}^{(1)} & 0 \\ 0 & 0 & 0 & 0 & 0 & \tilde{L}_{66}^{(1)} \end{bmatrix} \tag{6.15}$$

and

$$\tilde{\mathbf{L}}^{(2)} = \begin{bmatrix} 0 & -\tilde{L}_{12}^{(2)} & -\tilde{L}_{13}^{(2)} & 0 & 0 & 0 \\ -\tilde{L}_{21}^{(2)} & 0 & -\tilde{L}_{23}^{(2)} & 0 & 0 & 0 \\ -1 & -1 & 0 & 0 & 0 & 0 \\ 0 & 0 & 0 & \tilde{L}_{44}^{(2)} & 0 & 0 \\ 0 & 0 & 0 & 0 & \tilde{L}_{55}^{(2)} & 0 \\ 0 & 0 & 0 & 0 & 0 & \tilde{L}_{66}^{(2)} \end{bmatrix}. \tag{6.16}$$

It can be seen that for this simplified anisotropic criterion a lower number of anisotropic coefficients (13) are needed for a full 3D model, whereas for the reduction to plane stress problems this number turns out to be equal to 9. Although the approximation to experimental results is not completely perfect with these modified versions, the *Yld2004-13p* can be a valid alternative to the original *Yld2004-18p* yield criterion. More discussion and details on this can be found in references [36] and [37].

6.3 Benchmarks' Numerical Analyses

Numerous factors can influence wrinkling in plastically formed metallic parts, such as the material properties, friction, lubrication, the tool and blank geometries as well as the holding conditions imposed to the sheet to be formed. Although some researchers have experimentally investigated wrinkling effects and their relation to the process parameters, the present work will focus on the influence of decisions in the numerical modelling of the process and the consequent prediction of wrinkling initiation and propagation. The key idea is to infer about the robustness of numerically obtained solutions respective to the numeric input to FEM codes, by the user.

To achieve this goal, several numerical simulations were performed with the FEM code Abaqus/Standard ® [32], and attention was focused on the influence of mesh densities, finite element formulations (shell, solid and "solid-shell" element types) as well as constitutive material models (isotropic and anisotropic formulations) on the overall quality of the numerical results, against experimental references.

The following sections describe the most important features employed in the numerical simulations, for a set of examples involving unconstrained or free (conical cup) as well constrained or flange (cylindrical cup) forming.

6.3.1 Free Forming of a Conical Cup

The main idea of this test case is that when a conical die is used, there is no need to apply a blank-holder or clamping ring in the process of forming. Also, the conical form of the die, together with the absence of blank-holder or clamping devices, leads easily to the formation of unconstrained wrinkling modes in the plastically formed parts. Experimental results for this test case can be obtained, for instance, in the original paper of Narayanasamy and Sowerby [39], being further described in later references of the authors ([40[41]). On the other side, examples of numerical modelling and analysis for this specific problem can also be found in references [16[17], for instance.

Following reference [39], the experimental onset is represented in Figure 6.1, where can be seen the conical die as well as the cylindrical punch configurations as proposed by Narayanasamy and Sowerby. The metallic blank sheet to be formed has a circular drawing, and the plastic-forming process is defined as being completed as soon as the entire sheet is drawn into the die.

Following the indications and data from the previous experimental references, the diameter of the metallic sheet to be formed is taken equal to $\O = 104.94$ mm, with a thickness value equal to $t = 1.90$ mm. This specific combination of diameter and thickness values is prone to the formation of a predefined number of wrinkling zones along the perimeter of the circular blank, after forming, as experimentally verified in references [39[41]. From the point of view of geometric and FEM modelling, the tools (the punch and conical die) were considered as rigid bodies. The tools' dimensions follow those represented in Figure 6.1, that is: 49 mm for the punch diameter, 5 mm on the punch's radius and 260 mm of height, while the die has a 19º angle and an opening diameter of 54 mm. A detailed view of the tools involved can be inferred from Figure 6.2.

The punch stroke, responsible to form the final metallic part, was taken as equal to 60 mm. The metallic circular blank is therefore the only part to be considered as deformable, and being meshed by distinct configurations of finite elements, as seen in the following sections. Due to symmetry reasons, only one-quarter of the total blank will be discretized by finite elements.

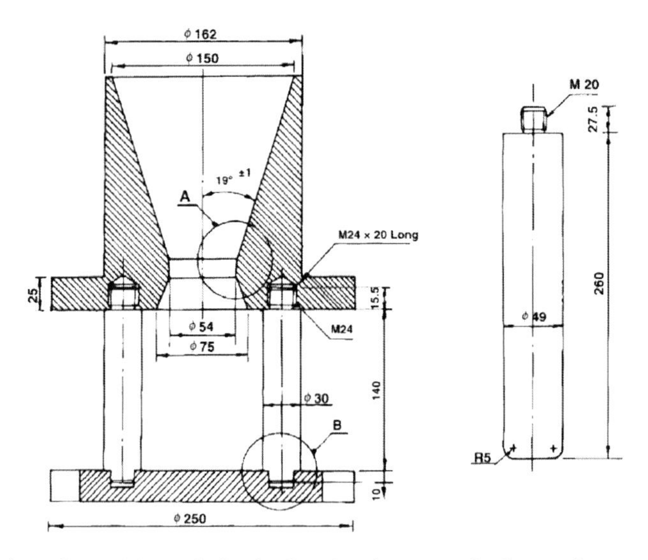

Fig. 6.1. Dimensions of the tools for the free-forming example, from references [39[41].

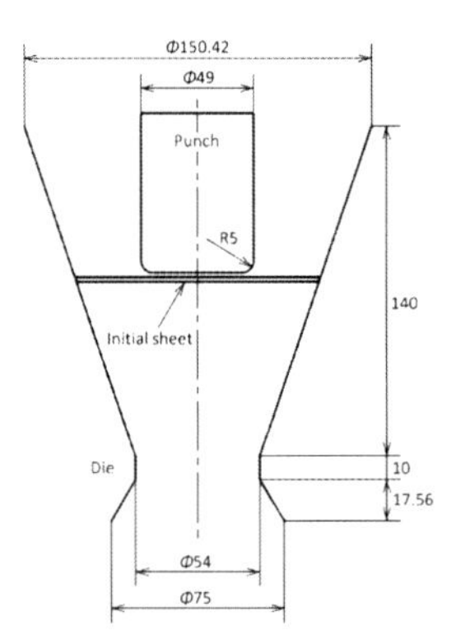

Fig. 6.2. Detailed description of the relevant dimensions for the tools in the conical cup drawing problem (dimensions in millimetres).

Regarding the constitutive modelling for the application of the anisotropic yield criterion of Hill (1948) [33], and following the references before, it was initially considered an aluminium alloy with anisotropy coefficients (r_θ) with values $r_0 = 0.17, r_{45} = 0.58, r_{90} = 0.46$, and, from equation (6.8), leading to values of $R_{12} = 1.09, R_{22} = 1.47, R_{33} = 0.92$. The elastic part of the constitutive behaviour is characterised by a Young's modulus and Poisson's ratio as E=69.0 GPa and υ=0.3, respectively. The effective plastic stress-strain relationship used in the numerical simulations is considered to be given by a Swift's law, with as main parameters $K = 127.83$ MPa, $\varepsilon_Y = 0.0003$ and $n = 0.03$.

On the other hand, for the case of modelling this problem by means of the *Yld91* and *Yld2004-18p* constitutive criteria, parameters available in the literature (experimentally obtained) for a 2090-T3 aluminium alloy were considered in the numerical simulations. The strain-based hardening model is considered to follow the parameters stated before for Swift's law, while in both cases a friction coefficient of $\mu = 0.15$ was assumed between all parts in contact. Furthermore, the forming process to be simulated involved only one work stage for achieving the final shape.

Since the tools are modelled as discrete rigid bodies, they were meshed by rigid elements in Abaqus/Standard FEM program. The FE mesh of the punch was composed of 5925 rigid finite elements of type R3D4 (4-node, 3-D bilinear quadrilateral, rigid element), while the die was discretized by 1770 elements of the same type. More information on this kind of FE formulation can be found in the program manual [32].

Initially, the blank to be plastically formed was assumed to be a solid shape for modelling purposes, since the adopted thickness values cannot be considered as extremely small compared to the overall blank dimensions and also in order to correctly describe the double-sided contact patterns involved in the process. In this sense, different three-dimensional finite elements were adopted, as available from the library of the FEM software: C3D8R (8-node, tri-linear 3-D solid element, reduced integration, that is, 1 integration point per element), C3D8 (8-node, tri-linear 3-D solid element, full integration, that is, 8 integration points per element) and C3D8I (8-node, tri-linear 3-D solid element, full integration and incompatible deformation modes) [32].

Nevertheless, in a second phase shell elements were also considered, in order it would be possible to infer about the influence of distinct finite element formulations in the obtained results. In this sense, a second group of simulations were considered, now including thin shell elements of type S4R (4-node, bilinear shell element, reduced integration, one integration point in the element reference plane and multiple integration points through thickness direction) as well as of type S4 (4-node, bilinear shell element, full integration within the element), for the sake of comparisons [32].

The blank zone to be meshed with finite elements included three mesh partitions in order a more refined mesh could be obtained in its central zone, as can be seen in Figure 6.3. The partition divides the blank into a square area in the middle of the blank, where the full contact with the punch will take place. A line from the corner of the square to the perimeter of the blank (45°) divides the rest of the partition in two parts. The adopted mesh density for both solid and shell elements can be shown in Table 6.1.

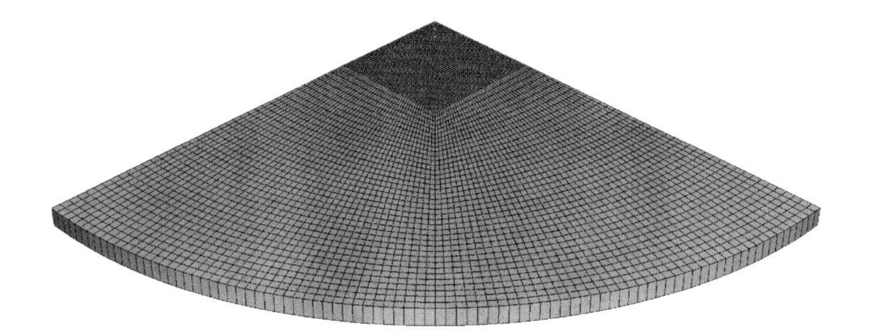

Fig. 6.3. Adopted mesh for the blank, with a refined centre zone.

Table 6.1. Mesh information for the deformable blank.

Element type	Number of elements along thickness (Gauss points for shell elements)	Total number of elements
C3D8R/C3D8/C3D8I	1 2 3	5451 12312 21312
S4R/S4	3, 5 , 7	7500

Besides the aluminium alloys mentioned before, the numerical simulations were also performed for a blank of stainless steel grade 301, now with a blank thickness of 1.60 mm and two different diameters: 110 mm and 130 mm. Doing so it would be possible to infer about the influence of different materials and geometries in the final results as predicted by numerical simulations, following the experimental analysis in [41]. For these setups and material, the adopted *r-values* are given as: $r_0 = 1.159$, $r_{45} = 1.147$, $r_{90} = 0.759$ ($R_{12} = 0.88$, $R_{22} = 0.90$, $R_{33} = 0.92$). The elastic parameters are given by the Young's modulus and Poisson's ratio, as E=212.0 GPa and υ=0.3 being, respectively [41]. The friction coefficient between the blank and tools was given as μ=0.10.

The simulations included shell (S4R) and solid elements (C3D8, C3D8R and C3D8I) and in addiction the "solid-shell" element SC8R, a eight-node, quadrilateral in-plane general-purpose continuum shell, reduced integration with hourglass control [32], which turns to be a finite element topologically identical to a solid one, but with a kinematics of a shell finite element. This last continuum shell element was also tested with 3, 5 and 7 Gauss points through thickness.

6.3.2 Flange Forming of a Cylindrical Cup

The flange-forming simulations (constrained wrinkling) performed in the present study considered two different materials (a mild steel DDQ and an aluminium alloy of the series 6111-T4). Three tools in each simulation were used: blank-holder, die and punch, whose dimensions can be seen in Fig. **6.4**, along with a circular blank, as in the previous example. Once again, and due to symmetry reasons, only one-quarter of the whole blank was considered. In Table 6.2 the constitutive properties considered for both materials, following reference [42] are shown.

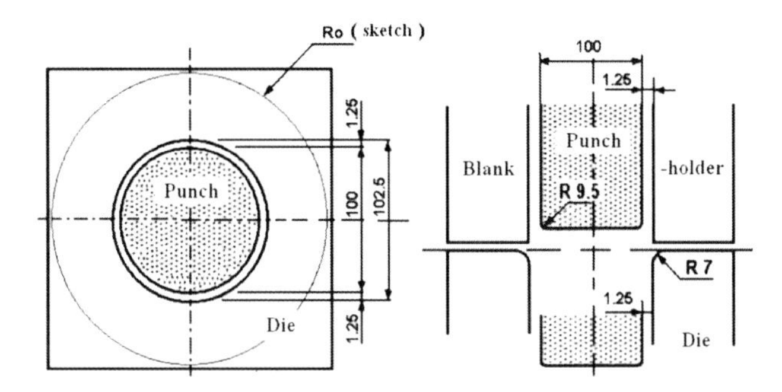

Fig. 6.4. Dimensions of the tools used in flange - forming (picture adapted from [42]).

Initially the tests start with the reference point of the die being fixed, with the punch being above 6 mm from the top surface of the blank, while the blank-holder is 5 mm from that surface. The simulations start with the blank-holder coming down 1 mm (so the gap between the blank and blank-holder is now 4 mm) and afterwards a vertical downward movement of 45 mm is imposed to the punch's reference point.

Table 6.2. Materials properties (mm / MPa).

Properties	Mild Steel DDQ	AA 6111-T4
Initial dimensions (mm)	Ø 180 x 1 (thickness)	Ø 180 x 1 (thickness)
Friction coefficient	0.0426	0.0096
Young modulus (MPa)	221370	70500
Poisson ratio	0.30	0.342
Yield strength (MPa)	151.70	187.53
Hardening model	Swift's law	Voce's law
Hardening parameters	C=544.27 n=0.2701	R_{sat}=420.86 C_r=8.448
Anisotropic parameters	F=0.25649 G=0.31646 H=0.68354 N=1.20949	F=0.71518 G=0.52798 H=0.47202 N=1.38115

Different anisotropic coefficients were calculated for the different materials used. For the mild steel (DDQ), replacing F, G, H and N (with L and M being considered as equal to 1.5) from Table 6.2 in equation **(6.7)**, obtaining the Lankford's r-values and using these values in equation **(6.8)** it is possible to calculate the coefficients R_{22}, R_{33} and R_{12} to be equal to 1.0314, 1.3211 and 1.1136, respectively (R_{13} and R_{23} are equal to 1 for the same reason as in free - forming).

The tools have been modelled as analytical bodies in this example, and therefore they have no mesh, while the blank is meshed with solid and shell elements as in the previous examples. In this case, the solid element used was the C3D8 (8-node trilinear 3-D brick, full integration), while the finite element S4 (4-node, bilinear shell element, full integration) was adopted when considering the blank to be a shell part [32].

6.4 Results and Discussions – Free - Forming of a Conical Cup

The first study to be carried out was a mesh dependency study, including three different mesh refinements, a coarse, base and refined mesh, each with one and two finite elements through the thickness direction. For all cases, the same finite element formulation (solid element, reduced integration C3D8R [32]) was considered. The material parameters, for the anisotropic case following Hill 1948 model, were specified in the last sections. The numerical results, for the evolution of the punch force along its displacement, can be seen in Fig. **6.5** (one element through thickness) and Fig. **6.6** (two finite elements through thickness). Due to the reduced formulation character of the solid element employed, the meshes in Fig. **6.5** have one single integration point along the thickness (constant stress field), while the meshes in Fig. **6.6** have two integration points along thickness direction (linear stress field). Meshes with more elements through the thickness direction were also analysed, but the obtained results were the same as those shown in Fig. **6.6**. For each graph, isotropic (von Mises yield criterion) and anisotropic (Hill 1948 yield criterion) results are represented.

What is interesting to note in this first set of simulation is that, although the mesh refinement in plane induces (as expected) different results, comparing Fig. **6.5** and Fig. **6.6** it can be seen that the description of wrinkling onset (and propagation) is strongly related to the number of integration points through the thickness direction, rather than to the in-plane refinement. It can be seen that a "poor" representation of the stress field along thickness (obtained with the single integration point in that direction for meshes in Fig. **6.5**) has led to the experimentally verified wrinkling pattern in the numerical solution, while the more refined modelling of Fig. **6.6**, under the same conditions (in-plane mesh, material model), leads to a numerical solution representative of a safe, wrinkling-free, metallic part.

It is also worth noting that the consideration (or not) of anisotropic effects does not improve the quality of results when compared to the simple von Mises isotropic constitutive model.

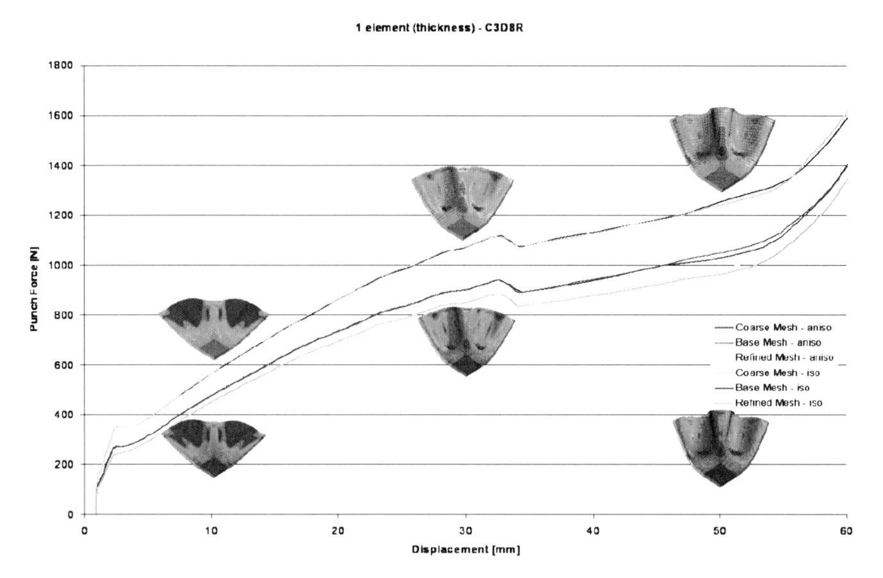

Fig. 6.5. The behaviour of the blank with different refinements, constitutive model and one element through thickness.

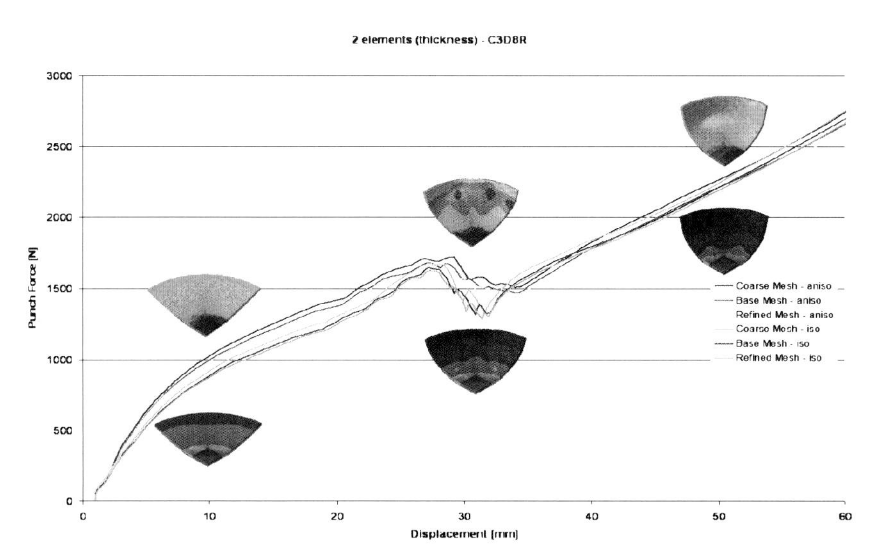

Fig. 6.6. The behaviour of the blank with different refinements, constitutive model and two elements through thickness (the same happens when using three elements along thickness).

Concerns might be raised at this time due to the fact that the Hill 1948 anisotropic yield criterion would not be the most appropriate constitutive modelling to correctly infer the plastic deformation of an aluminium sheet. Due to this fact, further analyses were carried out with the more advanced *Yld91* and *Yld2004-18p* anisotropic non-quadratic criteria detailed before.

To this end, and in order to have the appropriate anisotropic parameters available, it was assumed that the aluminium alloy was a representative of the series 2090-T3. From the literature, a representative hardening law for this alloy was selected [43[44] in the form

$$\sigma_Y = 646.0\left(0.025+\bar{\varepsilon}\right)^{0.227} \text{ (MPa)}, \tag{6.17}$$

and, following the same references, the anisotropic coefficients to be included in the Yld91 yield criterion are set as

$$\begin{gathered} c_1 = 1.0674,\ c_2 = 0.8559,\ c_3 = 1.1296, \\ c_4 = 1.000,\ c_5 = 1.0000,\ c_6 = 1.2970 \\ (a = 8). \end{gathered} \tag{6.18}$$

For the Yld2004-18p anisotropic constitutive model, and following reference [45], the coefficients for the 2090-T3 aluminium alloys are defined as

$$\begin{gathered} c_1 = -0.0698,\ c_2 = 0.9364,\ c_3 = 0.0791,\ c_4 = 1.0030,\ c_5 = 0.5247,\ c_6 = 1.3631, \\ c_7 = 1.0237,\ c_8 = 1.0690,\ c_9 = 0.9543,\ c_{10} = 0.9811,\ c_{11} = 0.4767,\ c_{12} = 0.5753, \\ c_{13} = 0.8668,\ c_{14} = 1.1450,\ c_{15} = -0.0792,\ c_{16} = 1.0516,\ c_{17} = 1.1471,\ c_{18} = 1.4046 \\ (a = 8). \end{gathered} \tag{6.19}$$

Focusing in this last yield criterion, and once again fixing the finite element formulation involved (reduced integration solid element C3D8R, in Abaqus commercial program), for different mesh densities in the plane of the blank, but keeping just one element (one integration point) along the thickness direction, the profile of the wrinkling appearance in the deformed blank is represented in Figure 6.7. In the picture, *RD* refers to "rolling direction", while *TD* refers to "transverse direction".

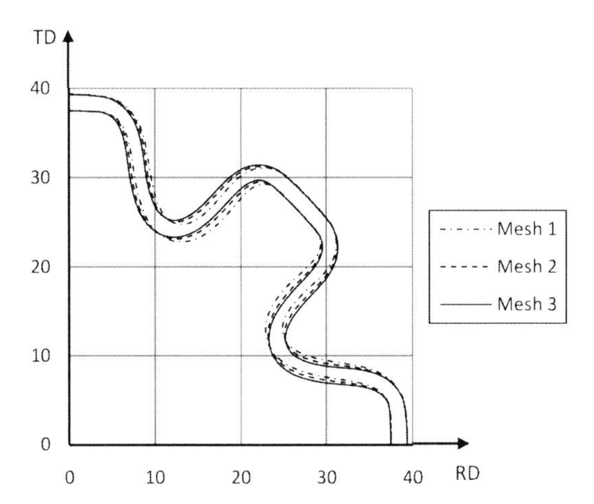

Fig. 6.7. Wrinkling profiles after forming for a one-quarter area of the initial circular blank. Results obtained with Yld2004-18p criterion, for 2739 (Mesh 1), 5368 (Mesh 2) and 8109 (Mesh 3) elements in the plane of the blank (one element, i.e., one integration point through thickness direction).

Taken the second refinement level (Mesh 2, in Figure 6.7), but now increasing the mesh refinement through thickness, the results for the deformed configuration after forming, and still accounting on the Yld2004-18p anisotropic criterion, are represented in Figure 6.8. In this graph, wrinkling profiles are shown for one, two and three elements along thickness direction, which directly corresponds to the same number of integration points.

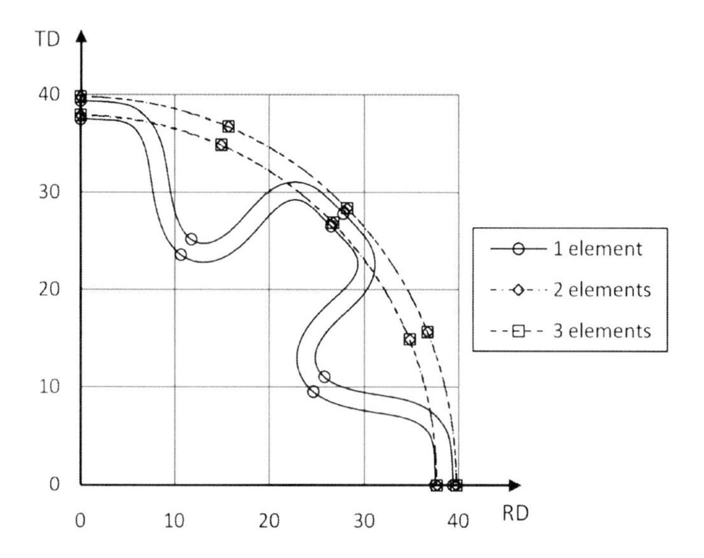

Fig. 6.8. Wrinkling profiles after forming for a one-quarter area of the initial circular blank. Results obtained with Yld2004-18p criterion, for one, two and three elements (integration points) through thickness direction.

From the last two graphs, it can be seen that even for a more sophisticated yield criterion, the more dominant effect in this example is the number of integration points through thickness direction. That is, the correct wrinkling profile after forming can only be numerically attained with a proper low order integration rule in the out-of-plane direction of the blank.

Since all the results shown before were valid for the same finite element formulation, it would be interesting to infer about the influence of distinct methodologies into the quality of the numerical results obtained. To this end, keeping the Yld2004-18p anisotropic criterion and once again the mesh density with one element along the thickness direction, in Figure 6.9 the deformed profile for the C3D8R (reduced integration) element, as well as the deformed configurations obtained in Abaqus by means of the formulations C3D8 (full integration) and C3D8I (full integration, enhanced strain modes), is again represented. Once again, the strong influence of the wrinkling appearance (or not) on the choice of the numerical integration rule adopted is visible.

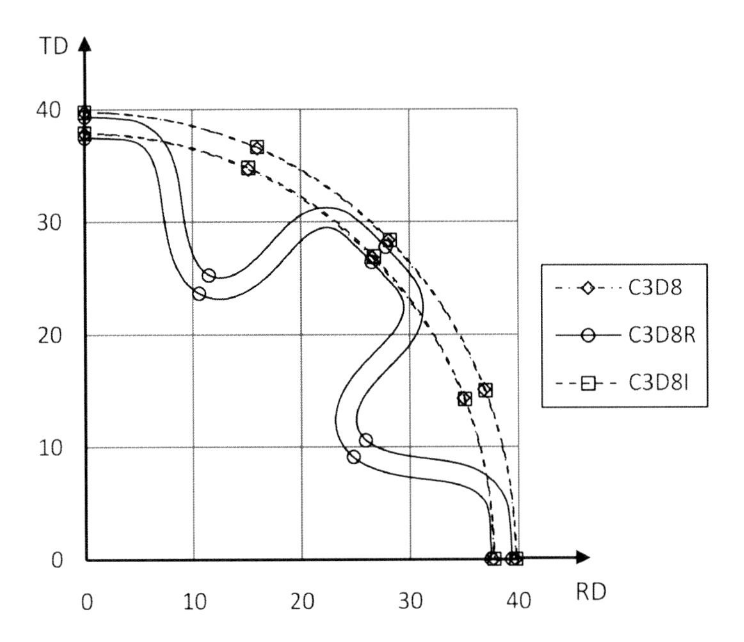

Fig. 6.9. Wrinkling profiles after forming for a one-quarter area of the initial circular blank. Results obtained with Yld2004-18p criterion, for reduced integrated (C3D8R) as well as fully integrated (C3D8 and C3D8I) formulations in Abaqus.

For the sake of confirmation of the conclusions being stated until now, a final analysis must be carried out for this example when analysed by means of the library of solid elements in Abaqus. In this case, it might be interesting to infer about the influence of Yld91, Yld2004-18p and an isotropic (von Mises) on the final configuration after forming, for the mesh system correctly inducing the wrinkling patterns. To this end, and starting from the Mesh 2 defined before (5368 elements in the plane of the blank), with one element along the thickness and adopting the C3D8R formulation (to ensure a single integration point in the out-of-plane direction), the obtained profiles after forming can be seen in Figure 6.10.

Once again, and as expected from the previous results, the numerical integration rule showed to be the dominant factor in the prediction of the correct wrinkling pattern in the final conical part, rather than the in-plane refinement or the constitutive model adopted. Increase on the order of integration along thickness, for the different yield criteria shown in Figure 6.10, will promote the complete disappearance of the wrinkling zones in the numerical solution.

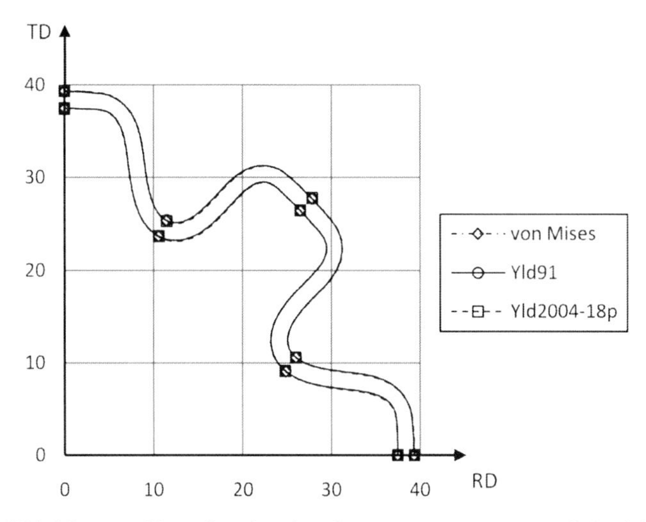

Fig. 6.10. Wrinkling profiles after forming for a one-quarter area of the initial circular blank. Results obtained with distinct yield criteria (isotropic and anisotropic), for a reduced integrated formulation in Abaqus.

Considering once again the Hill 1948 yield criterion, for simplicity reasons, but now focusing on alternative shell formulations in Abaqus, will lead to the analysis that follows. Although the relative dimensions of the blank would not directly point to a thin-shell problem, it might be useful anyway to infer about the performance of distinct shell element formulations and the corresponding obtained results. This can be seen in Figures 6.11, 6.12 and 6.13, where for the same in-plane mesh density, distinct number of integration points was adopted through thickness. Since dealing with shell elements, only one element is assumed along thickness, and it is possible to automatically define the integration order along that direction by an increase in the number of its integration points. Doing so, Figure 6.11 shows the deformed configuration and the evolution of the punch force during forming for meshes with three integration points, for both isotropic and anisotropic (Hill 1948) criteria, while Figures 6.12 and 6.13 do the same for 5 and 7 integration points in thickness direction. In each graph the variation of results as coming from a full in-plane integration rule (as in S4 shell element) against a reduced in-plane integration rule (as present in S4R shell element) [32] is also shown.

It can be seen from the pictures that none of the shell models was able to correctly predict the wrinkling pattern distribution, with some of the models even inducing a single non-physical wrinkling mode in the final obtained part (see Figures 6.11 and 6.12 for reduced integration S4R shell element).

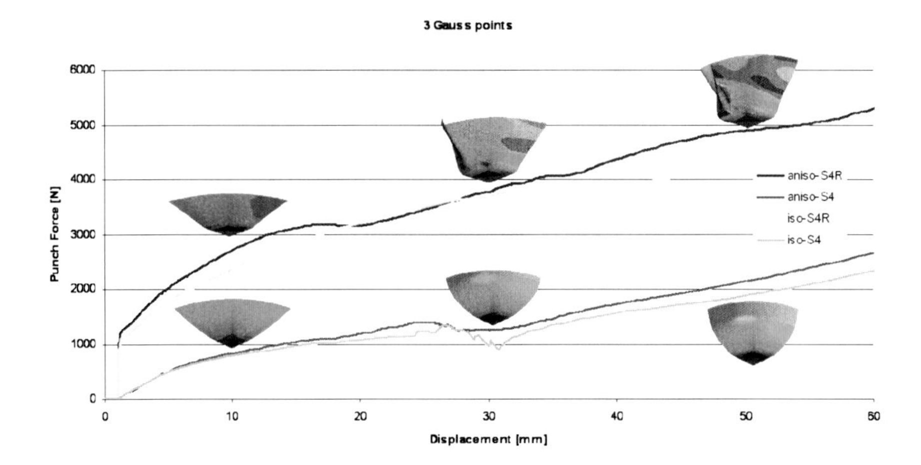

Fig. 6.11. Reaction force evolution and sheet metal deformation for elements S4 and S4R with three Gauss points and distinct constitutive behaviours (isotropic and anisotropic).

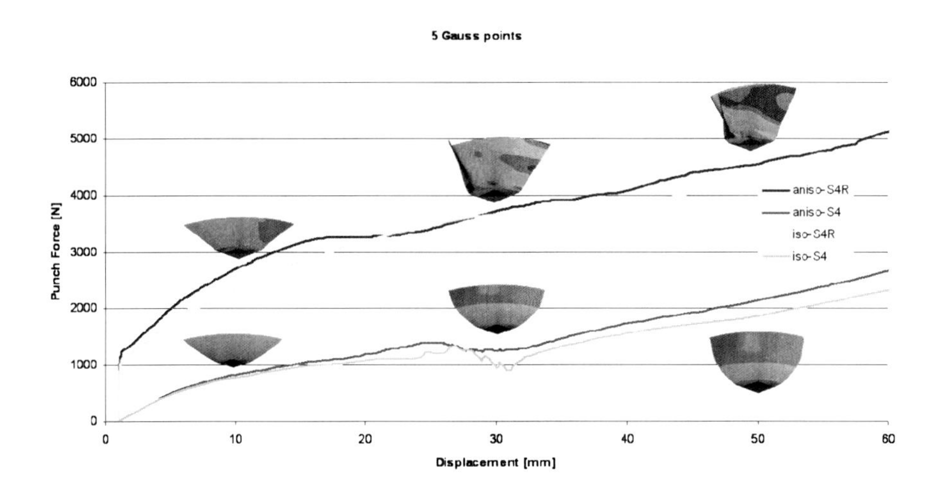

Fig. 6.12. Deformed shape of the blank for different elements formulation (S4 and S4R) with five Gauss points through thickness and distinct constitutive behaviours (isotropic and anisotropic).

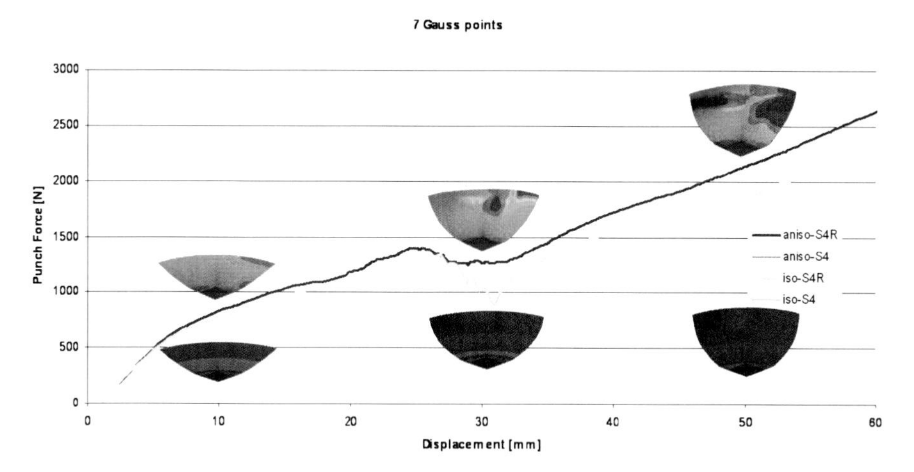

Fig. 6.13. Reaction force evolution and blank deformation for element S4R and S4R with seven Gauss points and both constitutive behaviour (isotropic and anisotropic).

For the sake of completeness, Figure 6.14 shows the deformed configuration of the conical cup inside the die after the whole displacement of the punch, while in Figure 6.15 the respective dimensions after forming are shown.

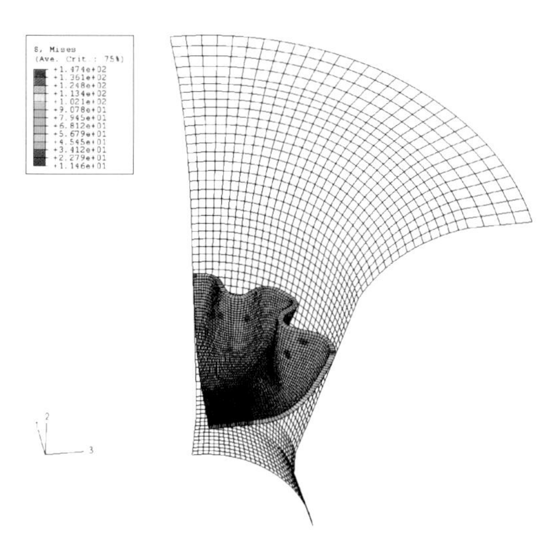

Fig. 6.14. Deformed configuration for a solid finite element formulation (C3D8R), one element through thickness direction and an anisotropic constitutive behaviour.

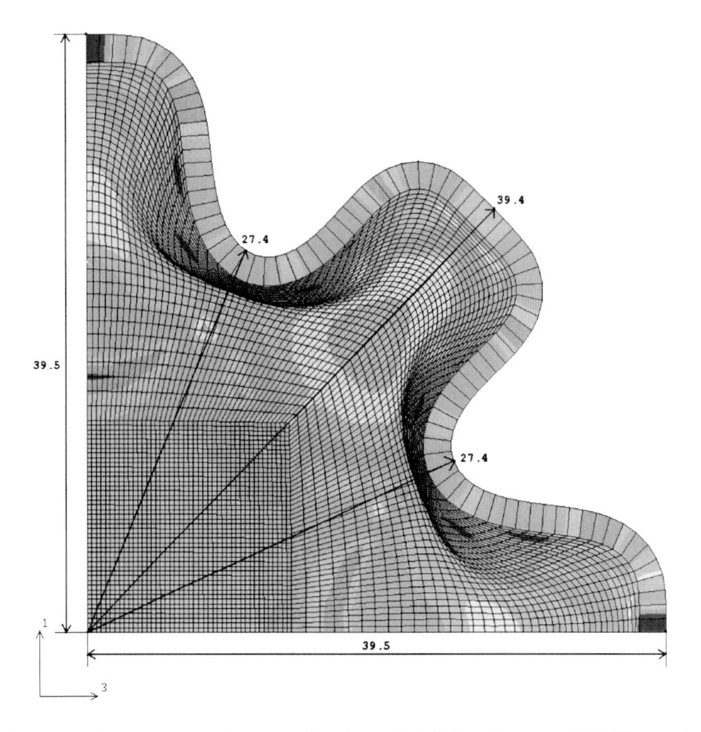

Fig. 6.15. Wrinkling patterns (in mm) for the solid finite element C3D8R mesh, with one element through thickness and anisotropic behaviour.

Following reference [41], in the following the forming of a circular blank of a distinct material (stainless steel grade 301) is considered, and anisotropic effects are taken into account by means of the Hill 1948 yield criterion. As specified in the last section, two distinct diameters are now analysed $(\phi_1 = 110\,\text{mm},\ \phi_2 = 130\,\text{mm})$, for a thickness value $t = 1.60$ mm. The anisotropic coefficients, as mentioned before, are equal to $r_0 = 1.159$, $r_{45} = 1.147$, $r_{90} = 0.759$ ($R_{12} = 0.88$, $R_{22} = 0.90$, $R_{33} = 0.92$).

In Figure 6.16 the evolution of the punch force against its displacement during forming, for both diameters and yield criteria, as a result of modelling the circular blank with a reduced integrated solid formulation (C3D8R, in Abaqus), is shown. As expected, that a geometric variation (i.e., the diameter) would induce distinct wrinkling patterns, while (as can be seen in the previous examples) the constitutive yield model does not induce noticeable differences in the results. The deformed final configurations can be seen in Figure 6.17, for the same punch displacement. Regarding the stress levels predicted in the final parts, the maximum equivalent stress level for the blank with a diameter of 110 mm is about 467 MPa, while for the blank with a diameter of 130 mm it reaches 377 MPa, with the first geometry attaining a higher number of wrinkles.

Once again, no noticeable wrinkling patterns were attained for those meshes including fully integrated formulations. The same happened for shell formulations employing three or more integration points along the thickness direction.

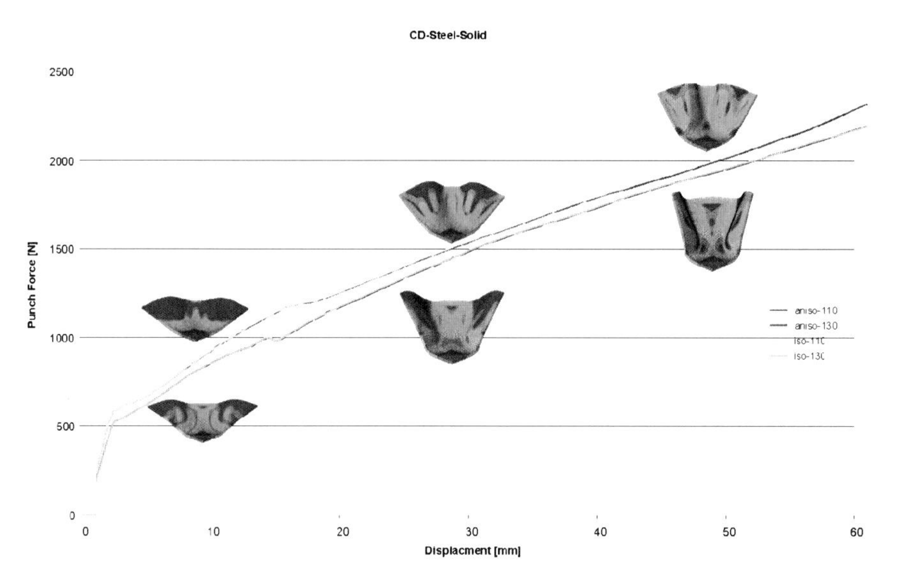

Fig. 6.16. Evolution of the punch force during forming, for both initial geometries (diameters of 110 and 130 mm), and isotropic and anisotropic (C3DR formulation, one element through thickness direction).

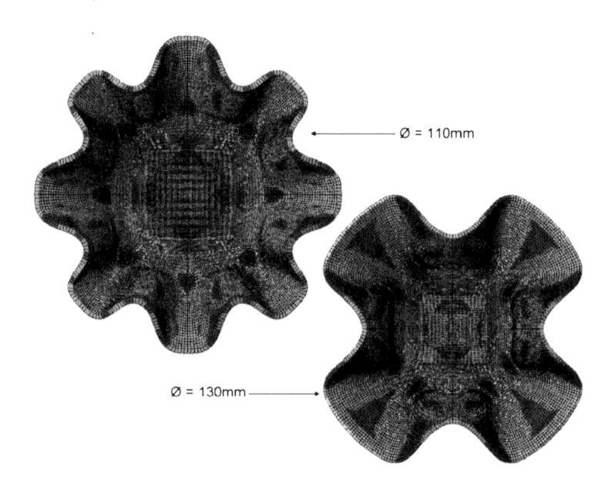

Fig. 6.17. The deformed shapes (360°) for the two diameters of the steel blank.

In this particular case, and when employing the "solid-shell" formulation available in Abaqus, some differences can be seen between the results coming from the use of isotropic or anisotropic models, although in neither case the wrinkling tendency is seen. Figure 6.18, for instance, shows the results for the circular blank with an initial diameter of 110 mm, while in Figure 6.19 the corresponding results for the larger diameter of 130 mm can be seen. In both cases, distinct number of integration points through the thickness direction was used, with no distinguishable differences between the results.

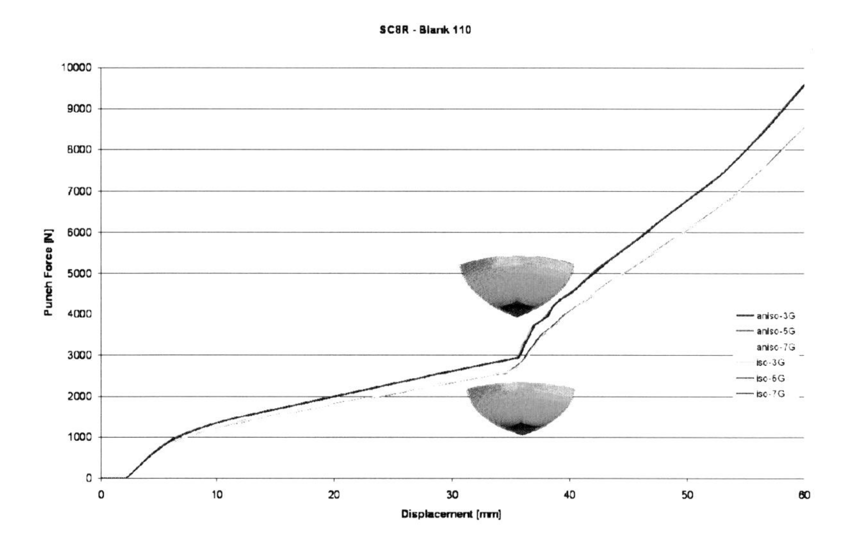

Fig. 6.18. Deformed shape of the 110 mm steel sheet with the element SC8R for 3, 5 and 7 Gauss points through thickness and isotropic and anisotropic constitutive behaviours.

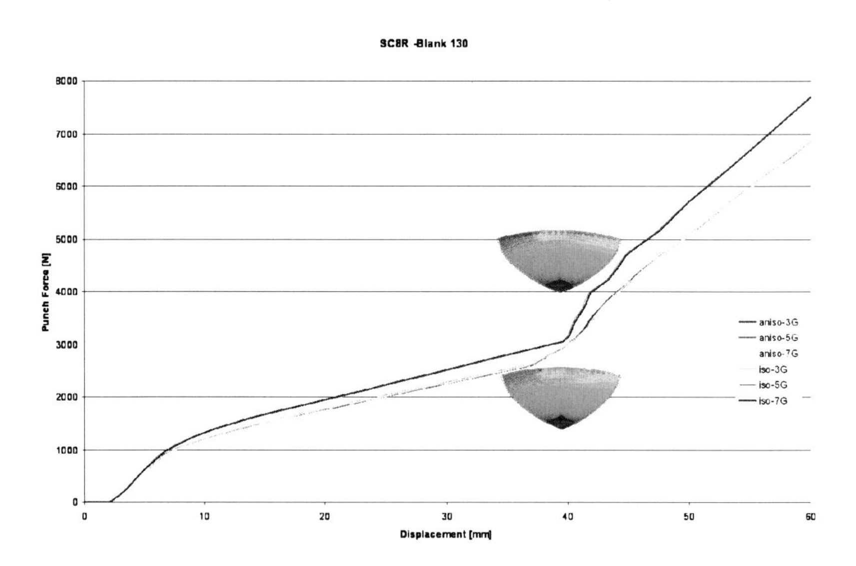

Fig. 6.19. Deformed shape of the 130 mm steel sheet with the element SC8R for 3, 5 and 7 Gauss points through thickness and isotropic and anisotropic constitutive behaviours.

6.5 Results and Discussions – Flange-Forming of a Cylindrical Cup

For this example, focusing on a constrained forming problem leading to flange type wrinkles, and schematically previously shown in Fig. **6.4**, different partitions were considered altogether in the meshing of the blank model, as can be seen in Fig. **6.20** and Fig. **6.21**.

In the following, two mesh refinements were adopted, including both 10 and 15 elements in each edge of each mesh area in Fig. **6.20** (for the case of solid elements in Abaqus) and a more elaborated mesh refinement procedure for shell elements, as schematically represented in Fig. **6.21**.

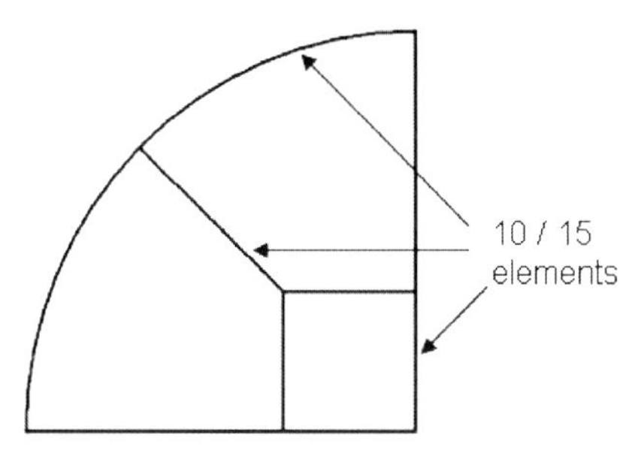

Fig. 6.20. Mesh partitions for solid elements, for the flange-forming example.

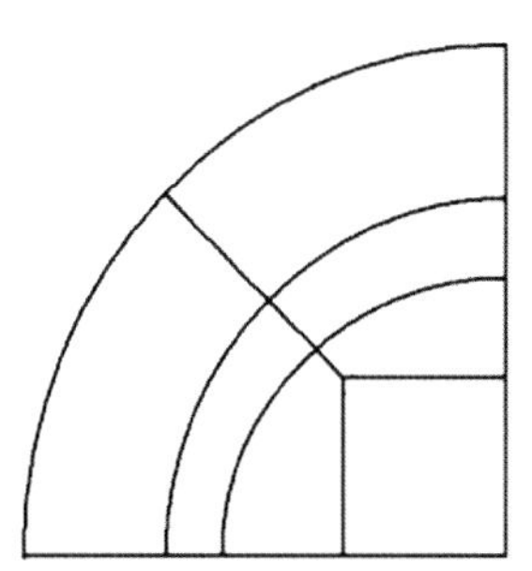

Fig. 6.21. Mesh partitions for shell elements, for the flange-forming example.

For the simulations considering solid finite element formulations, one element layer was taken into account, and in Fig. **6.22** and 6.23 the evolution of the punch's force throughout the analysis until the punch stroke is completed, for respectively, the first and second mesh systems considered can be seen.

The graphics lines reproduce the results obtained for both aluminium (AA 6111-T4) and mild steel (DDQ) alloys, also accounting for isotropic and anisotropic behaviours. It can be seen in these graphs that the force needed in the punch is larger for the aluminium alloy (6111-T4) than for the mild steel (DDQ). It is also shown that for the aluminium alloy with isotropic behaviour, the punch's force is bigger when compared with the anisotropic constitutive model. The opposite happens with the mild steel (DDQ) where the punch's force is lower for isotropic behaviour.

Here, and in opposition to the group of results in the last section, the influence between the isotropic (von Mises) and anisotropic (Hill 1948) criteria, for the anisotropic coefficients defined previously is more noticeable. The results were obtained with the fully integrated solid element of Abaqus library (C3D8), while the reduced formulation showed to suffer from severe hourglass effects, mainly in the regions where double-sided contact situations were dominant.

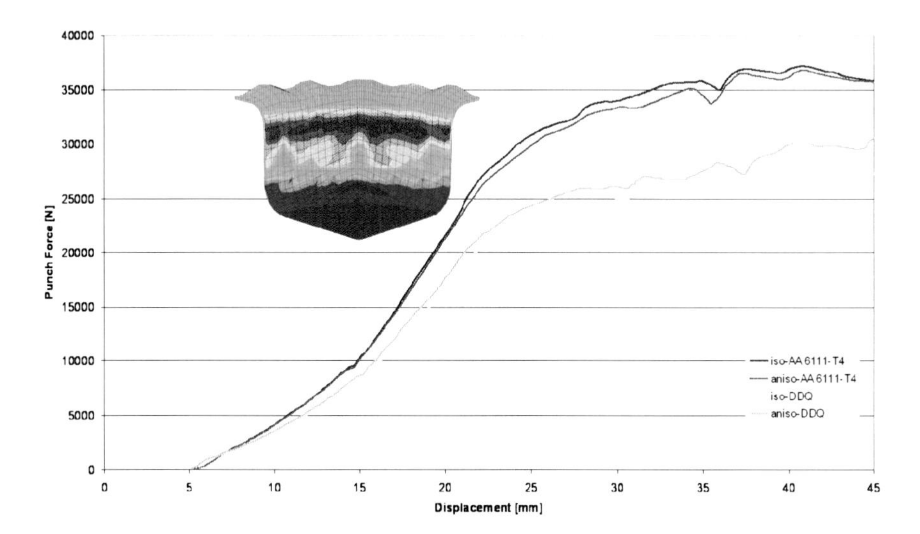

Fig. 6.22. Punch reaction during forming, for 10 solid elements in each edge (total the 700 elements).

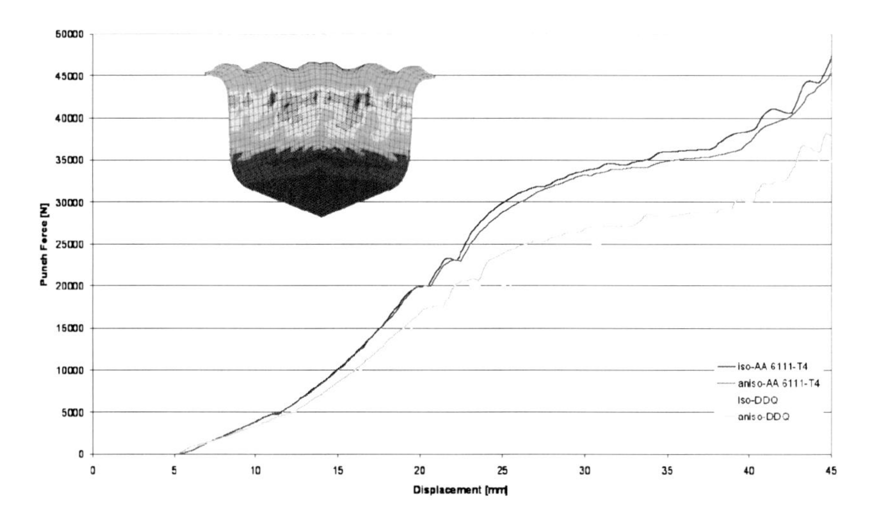

Fig. 6.23. Punch reaction during forming, for 15 solid elements in each edge (total the 1235 elements).

Fig. **6.24** and Fig. **6.25** show the deformed configuration for the same mesh, but accounting for different constitutive models, and it can be seen that, despite the differences in the evolution of the punch force against its displacement, the final aspects of the predicted plastically formed parts are quite similar for both isotropic and anisotropic models.

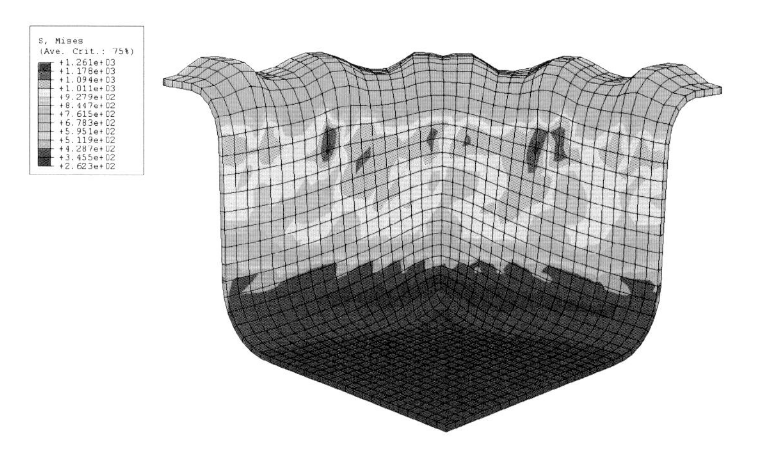

Fig. 6.24. Deformed shape for solid finite element formulation and an isotropic constitutive model.

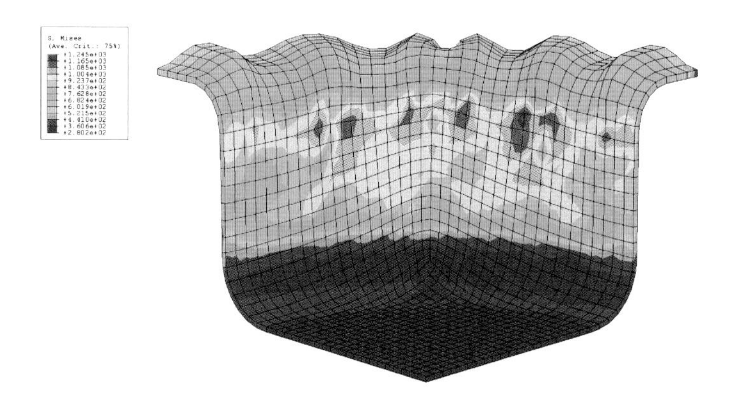

Fig. 6.25. Deformed shape for solid finite element formulation and an anisotropic constitutive model.

Focusing on the particular case of the aluminium alloy (6111-T4), and now taking into account numerical simulations considering shell elements (S4) in Abaqus, with full in-plane integration but five integration points along the thickness directions, the evolution of the punch force during forming (for an isotropic behaviour) can be seen in Fig. **6.26**, for the two mesh densities represented before in Fig. **6.21**.

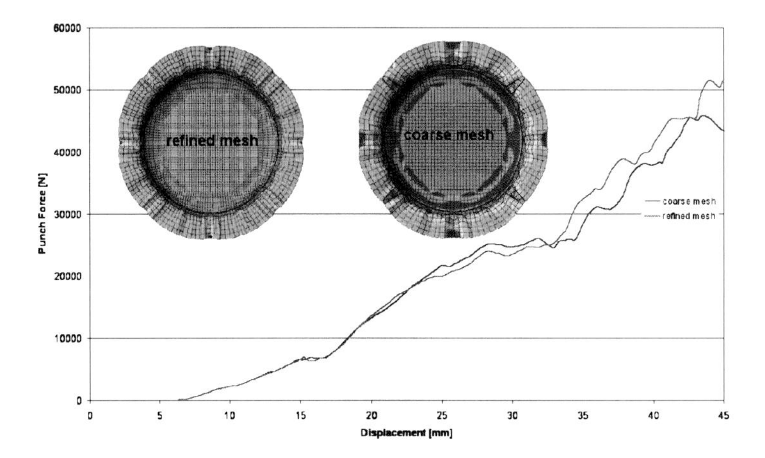

Fig. 6.26. Punch reaction during forming, for distinct mesh refinements and shell elements.

Distinct wrinkling patterns are formed, for this case, as can be seen in Fig. **6.27** and Fig. **6.28**, the first one (coarse mesh) being non-uniform and overlapped, while the second mesh system (refined mesh) gives rise to a coherent set of wrinkling patterns, qualitatively in accordance with published numerical and experimental results.

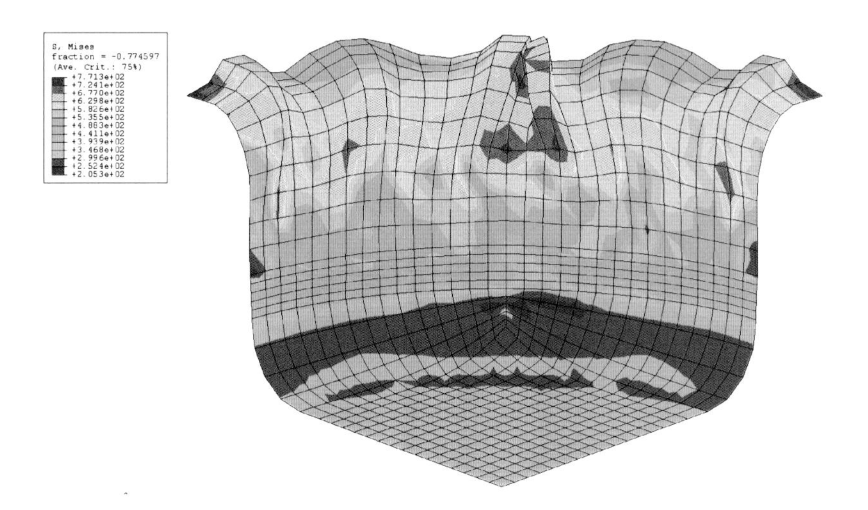

Fig. 6.27. Deformed shape for shell elements and a coarse mesh.

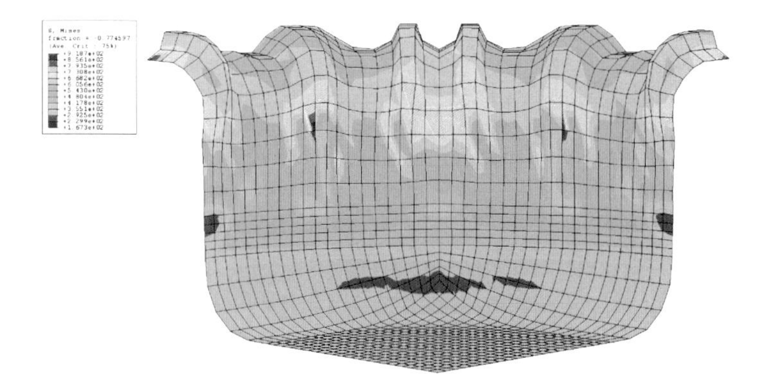

Fig. 6.28. Deformed shape for shell elements and a refined mesh.

It seems that for the constrained forming into a cylindrical cup, the results are quite dependent on the element type, and not only its numerical integration type (a situation not seen in the free - forming examples) mostly due to the more severe contact conditions involved in the flange area.

6.6 Conclusions

This work aimed to provide a preliminary insight into the influence of finite element formulation, finite element discretization through the thickness direction and constitutive material modelling in the onset and propagation of wrinkling patterns in sheet metal formed parts, as reproduced by numerical simulations based on the Finite Element Method.

It was seen that a correct prediction of wrinkling defects is very sensitive to the initial decisions on the modelling phase of analysis, and a conclusion coming from this work can be stated in the sense that - more than the correct constitutive modelling to be adopted - distinct finite element formulations and discretization levels show high influence on the quality of results obtained.

Nevertheless, and concentrating on the aspects related to the mesh systems and formulations to be adopted in a given numerical simulation, it is not yet clear what are the specific main driving effects when considering the correct prediction of wrinkling effects. Contrary to the correct prediction of springback effects in sheet metal formed products, where the dominant aspect to be taken into account is known to be the numerical integration procedure and number of integration points along the thickness direction, it is shown in the present work that for wrinkling effects a complex conjunction of (i) in-plane mesh refinement, (ii) out-of-plane mesh refinement (or, alternatively, increase of integration points through the thickness direction) and, finally, (iii) the finite element formulation itself (shell or solid elements) have a strong influence on the obtained simulation results, rather than the constitutive model adopted. Also, and most importantly, these conclusions seem to be extremely dependent on the examples chosen.

Based on that, proposals of future work are related to the research on alternative solid-shell finite element formulations in complex wrinkling prediction, where the main advantages of solid and shell formulations alone are gathered in the same formulation. Doing so, the sensitivity of the results to the mesh refinement levels would also be inferred. In particular, and trying to avoid the sensitivity to distinct numerical integration schemes, it would be useful for the development of a wrinkling criterion based on the use of enhanced-assumed strain finite solid-shell elements, following previous works of the authors in this field [46 - [48].

Acknowledgments. The authors would like to acknowledge the financial support coming from the *Fundação para a Ciência e a Tecnologia* (Portuguese Science and Technology Foundation), through the grants PTDC/EME-TME/66435/2006 and PTDC/EME-PME/113835/2009.

References

[1] Hill, R.: A general theory of uniqueness and stability in elastic-plastic solids. Journal of the Mechanics and Physics of Solids 6, 236–249 (1958)

[2] Hutchinson, J.W.: Plastic buckling. Advances in Applied Mechanics 14, 67–144 (1974)

[3] Hutchinson, J.W., Neale, K.W.: Wrinkling of curved thin sheet metal. Plastic Instability, pp. 71-78. Presses des Ponts et Chaussées, Paris (1985)

[4] Petryk, H.: Plastic instability: criteria and computational approaches. Archives of Computational Methods in Engineering 4, 111–151 (1997)

[5] Cao, J., Boyce, M.C.: Wrinkling behaviour of rectangular plates under lateral constraint. International Journal of Solids and Structures 34, 153–176 (1997)

[6] Cao, J.: Prediction of plastic wrinkling using the energy method. Journal of Applied Mechanics - Transactions of ASME 66, 646–652 (1999)

[7] Magalhães Correia, J.P., Ferron, G.: Wrinkling of anisotropic metal sheets under deep-drawing: analytical and numerical study. Journal of Materials Processing Technology, 155–156, 1604–1610 (2004)

[8] Kawka, M., Olejnik, L., Rosochowski, A., Sunaga, H., Maknouchi, A.: Simulation of wrinkling in sheet metal forming. Journal of Materials Processing Technology 109, 283–289 (2001)

[9] Wang, X., Lee, L.H.N.: Post-bifurcation behaviour of wrinkles in square metal sheet under Yoshida test. International Journal of Plasticity 9, 1–19 (1993)

[10] Wang, C.T., Kinzel, Z., Altan, T.: Wrinkling criterion for an anisotropic shell with compound curvatures in sheet forming. International Journal of Mechanical Sciences 36, 945–960 (1994)

[11] Nordlund, P.: Adaptivity and wrinkle indication in sheet metal forming. Computer Methods in Applied Mechanics and Engineering 161, 114–127 (1998)

[12] Wang, X., Cao, J.: On the prediction of side-wall wrinkling in sheet metal forming processes. International Journal of Mechanical Sciences 42, 2369–2394 (2000)

[13] Kim, J.B., Yang, D.Y., Yoon, J.W., Barlat, F.: The effect of plastic anisotropy on compressive instability in sheet metal forming. International Journal of Plasticity 16, 649–676 (2000)

[14] Kim, J.B., Yoon, J.W., Yang, D.Y.: Investigation into the wrinkling behaviour of thin sheets in the cylindrical cup deep drawing process using bifurcation theory. International Journal for Numerical Methods in Engineering 56, 1673–1705 (2003)

[15] Lu, H., Cheng, H.S., Cao, J., Liu, W.K.: Adaptive enrichment meshfree simulation and experiment on buckling and post-buckling analysis in sheet metal forming. Computer Methods in Applied Mechanics and Engineering 194, 2569–2590 (2005)
[16] Magalhães Correia, J.P., Ferron, G.: Wrinkling predictions in the deep-drawing process of anisotropic metal sheets. Journal of Material Processing Technology 128, 199–211 (2002)
[17] Magalhães Correia, J.P., Ferron, G., Moreira, L.P.: Analytical and numerical investigation of wrinkling for deep-drawing anisotropic metal sheets. International Journal of Mechanical Sciences 45, 1167–1180 (2003)
[18] Kim, Y., Son, Y.: Study on wrinkling limit diagram of anisotropic sheet metal. Journal of Materials Processing Technology 97, 88–94 (2000)
[19] Obermeyer, E.J., Majlessi, S.A.: A review of recent advances in the application of blank-holder force towards improving the forming limits of sheet metal parts. Journal of Materials Processing Technology 75, 222–234 (1998)
[20] Belytschko, T., Moes, N., Usui, S., Parimi, C.: Arbitrary discontinuities in finite elements. International Journal for Numerical Methods in Engineering 50, 993–1013 (2001)
[21] Belytschko, T., Lu, Y.Y., Gu, L.: Element-free Galerkin methods. International Journal for Numerical Methods in Engineering 37, 229–256 (1994)
[22] Narayanasamy, R., Loganathan, C.: Some studies on wrinkling limit of commercially pure aluminium sheet metals of different grades when drawn through conical and tractrix dies. International Journal of Mechanics and Materials in Design 3, 129–144 (2006)
[23] Loganathan, C., Narayanasamy, R.: Effect of die profile on the wrinkling behaviour of three different commercially pure aluminium grades when drawn through conical and tractrix dies. Journal of Engineering & Materials Sciences 13, 45–54 (2006)
[24] Wu-rong, W., Guan-long, C., Zhong-qin, L.: The effect of binder layouts on the sheet metal formability in the stamping with Variable Blank Holder Force. Journal of Materials Processing Technology 210, 1378–1385 (2010)
[25] Morovvati, M.R., Mollaei-Dariani, B., Asadian-Ardakani, M.H.: A theoretical, numerical, and experimental investigation of plastic wrinkling of circular two-layer sheet metal in the deep drawing. Journal of Materials Processing Technology 210, 1738–1747 (2010)
[26] Stoughton, T.B., Yoon, J.W.: A new approach for failure criterion for sheet metals. International Journal of Plasticity 27, 440–459 (2011)
[27] Oh, K.S., Oh, K.H., Jang, J.H., Kim, D.J., Han, K.S.: Design and analysis of new test method for evaluation of sheet metal formability. Journal of Materials Processing Technology 211, 695–707 (2011)
[28] Taherizadeh, A., Green, D.E., Ghaei, A., Yoon, J.W.: A non-associated constitutive model with mixed iso-kinematic hardening for finite element simulation of sheet metal forming. International Journal of Plasticity 26, 288–309 (2010)
[29] Ravindran, R., Manonmani, K., Narayanasmay, R.: An analysis of wrinkling limit diagrams of aluminium alloy 5005 annealed at different temperatures. International Journal of Material Forming 3, 103–115 (2010)
[30] Savaş, V., Seçgin, Ö.: An experimental investigation of forming load and side-wall thickness obtained by a new deep drawing die. International Journal of Material Forming 3, 209–213 (2010)

[31] Port, A.L., Thuillier, S., Manach, P.Y.: Occurrence and numerical prediction of surface defects during flanging of metallic sheets. International Journal of Material Forming 3, 215–223 (2010)
[32] Hibbitt, Karlsson, Sorensen: ABAQUS/Standard v.6.5 User's manual. Habbitt Karlsson & Sorensen, Inc., USA (1998)
[33] Hill, R.: A theory of the yielding and plastic flow of anisotropic metals. Mathematical and Physical Sciences 193, 281–297 (1948)
[34] Habraken, A.M.: Modelling the plastic anisotropy of metals. Archives of Computational Methods in Engineering 11, 3–96 (2004)
[35] Barlat, F., Lege, D.J., Brem, J.C.: A six-component yield function for anisotropic materials. International Journal of Plasticity 7, 693–712 (1991)
[36] Barlat, F., Aretz, H., Yoon, J.W., Karabin, M.E., Brem, J.C., Dick, R.E.: Linear transformation-based anisotropic yield functions. International Journal of Plasticity 21, 1009–1039 (2005)
[37] Grilo, T.J.: Study of anisotropic constitutive models for metallic sheets. MSc Dissertation, University of Aveiro, Portugal (2011) (in Portuguese)
[38] Hershey, A.V.: The plasticity of an isotropic aggregate of anisotropic face centered cubic crystals. Journal of Applied Mechanics 21, 241–249 (1976)
[39] Narayanasamy, R., Sowerby, R.: Wrinkling behaviour of cold-rolled sheet metals when drawing through a tractrix die. Journal of Materials Processing Technology 49, 199–211 (1995)
[40] Loganathan, C., Narayanasamy, R.: Wrinkling of commercially pure aluminium sheet metals of different grades when drawn through conical and tractrix dies. Materials Science and Engineering A 419, 331–343 (2006)
[41] Narayanasamy, R., Loganathan, C.: The influence of friction on the prediction of wrinkling of prestrained blanks when drawing through conical die. Materials and Design 28, 904–912 (2007)
[42] Alves, J.L.C.M.: Numerical Simulation of the Sheet Metal Forming Process of Metallic Sheets. PhD Thesis, University of Minho, Portugal (2003) (in Portuguese)
[43] Yoon, J.W., Barlat, F., Chung, K., Pourboghrat, F., Yang, D.Y.: Earing predictions based on asymmetric nonquadratic yield function. International Journal of Plasticity 16, 1075–1104 (2000)
[44] Yoon, J.W., Barlat, F., Dick, R.E., Chung, K., Kang, T.J.: Plane stress yield function for aluminum alloy sheets - part II: FE formulation and its implementation. International Journal of Plasticity 20, 495–522 (2004)
[45] Yoon, J.W., Barlat, F., Dick, R.E., Karabin, M.E.: Prediction of six or eight ears in a drawn cup based on a new anisotropic yield function. International Journal of Plasticity 22, 174–193 (2006)
[46] Valente, R.A.F., Alves de Sousa, R.J., Natal Jorge, R.M.: An enhanced strain 3D element for a large deformation elastoplastic thin-shell applications. Computational Mechanics 34(1), 38–52 (2004)
[47] Parente, M.P.L., Valente, R.A.F., Natal Jorge, R.M., Cardoso, R.P.R., Alves de Sousa, R.J.: Sheet metal forming simulation using EAS solid-shell elements. Finite Elements in Analysis and Design 42, 1137–1149 (2006)
[48] Alves de Sousa, R.J., Yoon, J.W., Cardoso, R.P.R., Valente, R.A.F., Grácio, J.J.: On the use of a reduced enhanced solid-shell (RESS) element for sheet forming simulations. International Journal of Plasticity 23, 490–515 (2007)

Manufacturing Seamless Reservoirs by Tube Forming: Finite Element Modelling and Experimentation

Luis M. Alves[1], Pedro Santana[2], Nuno Fernandes[2], and Paulo A.F. Martins[1]

[1] IDMEC, Instituto Superior Técnico, Universidade Técnica de Lisboa, Av. Rovisco Pais s/n, 1049-001 Lisboa, Portugal
luisalves@ist.utl.pt, pmartins@ist.utl.pt
[2] OMNIDEA, Aerospace Technology and Energy Systems, Tv. António Gedeão, 9, 3510-017 Viseu, Portugal
nuno.fernandes@omnidea.net, pedro.santana@omnidea.net

This chapter introduces an innovative manufacturing process that is capable of shaping industrial tubes into seamless, reservoirs made from different materials and available in multiple shapes. The process offers the potential for the manufacturing of medium to large batches of, for instance, high pressure vessels, for a wide variety of industrial and commercial applications. The elimination of weld seams allows for much less non-destructive test inspection requirements with the consequent reduction in production time and avoidance of potential failures. Special emphasis is given to practical aspects related to the tools and techniques used to fabricate spherical and cylindrical reservoirs (with hemispherical and semi-ellipsoidal domes). The presentation is supported by experimental data and numerical modelling, the latter based on independently determined mechanical properties aimed at understanding the deformation mechanics, identifying the formability limits and demonstrating the overall performance and feasibility of the proposed manufacturing process.

7.1 Introduction

Large-size reservoirs, like silos and tanks, are usually fabricated from curved steel panels joined by circumferential and meridional welds. The presence of welds adds defects, residual stresses and geometrical imperfections due to joint mismatching between panels that may lead to reduction in the overall performance, namely the buckling strength [1].

Medium-size reservoirs (diameters up to 1 m) are fabricated by joining panels or, alternatively, by multiple-stage fabrication processes. For instance, the central cylindrical section of medium-size cylindrical reservoirs can be fabricated by rolling a sheet into a cylindrical surface and then joining the two ends by meridional welding [2, 3], while medium-size spherical reservoirs can be fabricated in two-half shells by deep drawing or spinning and then joined by circumferential welding [4].

However, conventional fabrication processes running on panel joining or two-stage manufacturing technologies are only suitable for producing single or small numbers of reservoirs because they involve long production lead times and are usually not appropriate to fabricate reservoirs in other materials than steel. This prevents conventional fabrication processes from meeting the challenges imposed by the increased demand of small reservoirs, for a wide variety of applications such as anaesthetic and analgesic medical systems, supplemental and emergency oxygen needs for patients, scuba diving tanks, high altitude 'oxygen aid' vessels and compressed gas reservoirs for transportation systems, high pressure gas storage systems for aeronautical and space applications and compressed air tanks for paintball and other leisure equipment, among others.

Despite recent efforts in fabricating small-size reservoirs from stainless steel and aluminium by casting, conventional or hydromechanical deep drawing [5] as well as explosive forming [6] the need for other undeveloped manufacturing technologies, able to produce medium to large batches of small-size reservoirs in a wide range of materials exists, because casting is limited to simple shapes (e.g. cylinder liners) and its operating costs demand very high production rates, explosive forming suffers from industrialisation problems and conventional or hydromechanical deep drawing, although being a flexible solution, is required subsequently for joining operations for the produced half-shells into a reservoir by means of welding (tungsten inert gas or, for space applications, more frequently electron beam welding).

This chapter focuses on the above-mentioned problems and presents an innovative manufacturing process for producing seamless, low-cost, axisymmetric metallic reservoirs by tube forming (Fig. 7.1).

Fig. 7.1. Spherical and cylindrical reservoirs made from aluminium AA7050 and AA6063 fabricated by the proposed manufacturing process.

The proposed manufacturing process avoids joint mismatching and geometrical imperfections from half-shells welding, allowing utilisation of materials other than steel, enabling mass production but still maintaining manufacturing flexibility to allow possible customisation requirements from smaller batch production sizes, such as the ones usually required by specialised markets such as the space market. In other words, the process is flexible enough to fulfil the requirements of the growing trend for the development of manufacturing processes which can demonstrate very short life- cycles and production lead times as well as medium development time requirements.

The chapter is organized into five parts; the first part describes the proposed manufacturing process, the second part presents material characterisation, the third part introduces the tools and techniques employed in the numerical and experimental modelling of the process, the fourth part characterises the mechanics of deformation and the influence of the major operating parameters on the workability window of the process and the final part is centred on demonstrating the feasibility of the process for fabricating small-size cylinder reservoirs made from aluminium and stainless steel.

The overall approach draws from accumulated knowledge in tube end forming [7] and is based on flexible tool design, experimental trials performed under laboratory-controlled conditions, independently determined mechanical properties of the raw materials and process modelling using the in-house finite element computer program i-form [8].

The chapter is expected to effectively contribute to transferable knowledge on an innovative manufacturing process to produce small-size, low-cost, seamless reservoirs.

7.2 Innovative Manufacturing Process

7.2.1 Tooling Concept

The new manufacturing process for producing small-size, seamless, metallic reservoirs is schematically shown in Fig. 7.2a. As seen, the forming operation is carried out by axial pressing the open ends of a tube against two profile shaped dies, while providing internal support by means of a low melting point recyclable mandrel, until the desired shape of the reservoir is achieved.

The basic components of the process, which is protected by an international patent request [9], are; (i) the upper and lower profile shaped dies, (ii) the container, (iii) the mandrel and (iv) the tubular preform.

The upper and lower dies are the active tool components and the sharp - edge of the upper die is protected against collapse due to circumferential tensile stresses by means of the container which acts as a shrink fit tool part. The dies are dedicated to

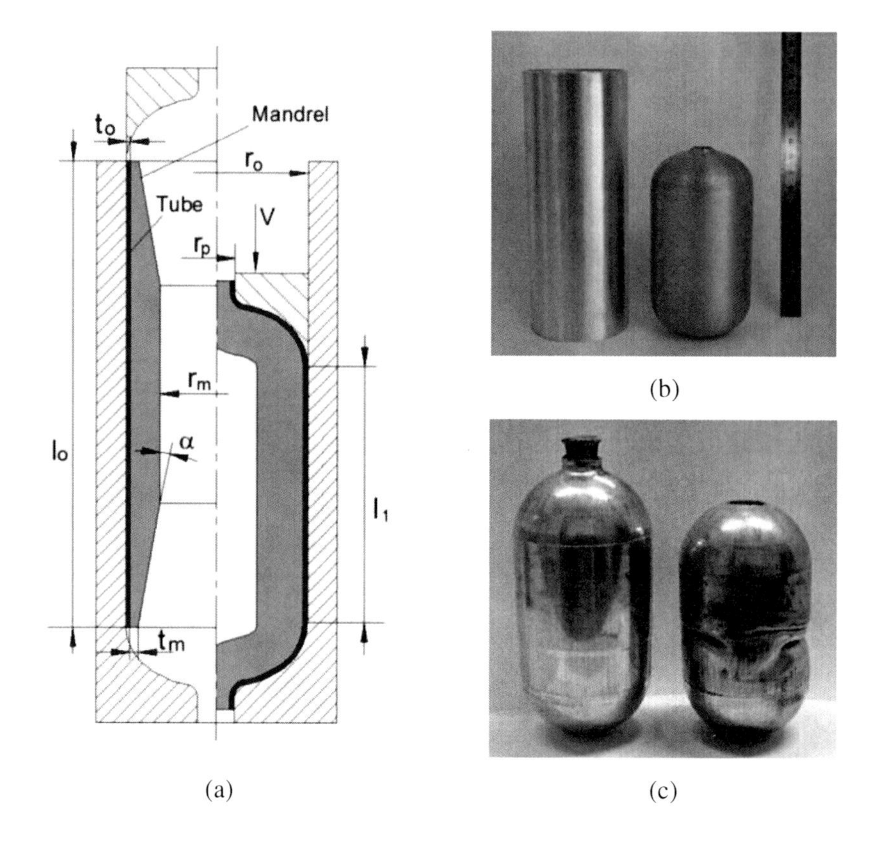

Fig. 7.2. Forming a tubular preform into a seamless cylindrical reservoir with profile shaped ends. The photograph in (b) shows the preform and the reservoir with semi-ellipsoidal ends and the photograph in (c) shows successful and non-successful modes of deformation that were obtained when forming the reservoir with and without internal mandrels.

a specific outside radius of the tube r_0 and its profile defines the geometry of the reservoir. The container constrains material from flowing outwardly in order to avoid the occurrence of buckling and helps minimizing the errors due to misalignment between the tubular preforms and the individual dies. The mandrel provides internal support to the tubular preform during plastic deformation in order to avoid collapse by wrinkling and local instability at the equatorial region. Figure 7.3 shows an exploded view drawing and a picture of the tool with its major active components.

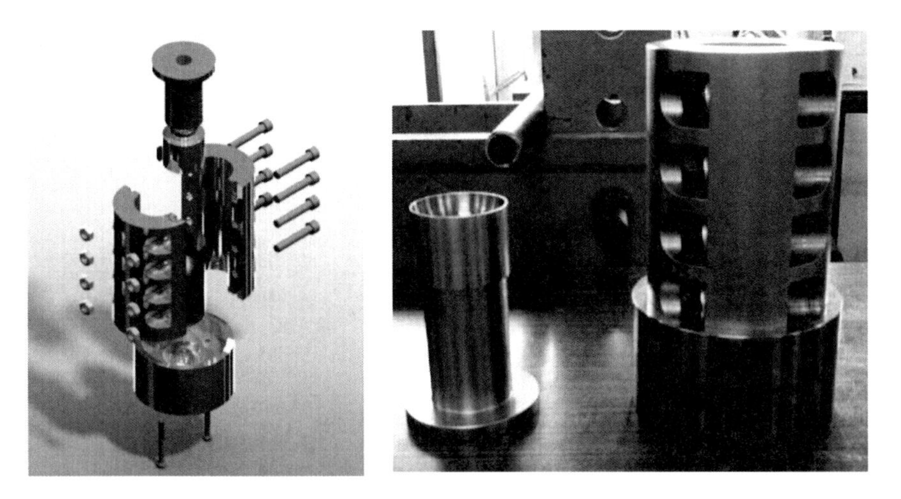

Fig. 7.3. Tool for fabricating seamless reservoirs with profile shaped ends from tubular preforms.

7.2.2 Preforms and Mandrels

In case of bending thin-walled tubes over tight radii, internal support with mandrels or loose fillers is commonly utilized to prevent the collapse by flattening or wrinkling. Mandrels are preferentially chosen for large batch sizes (mass production) and can be rigid, flexible or articulated. The last two types are employed when support of the tube is needed far into the bending region. Loose fillers such as sand and low-melting alloys are mainly chosen for small-batch production.

End forming of thin-walled tubes may also require internal support. Simple operations such as one-side nosing and reduction can take advantage of conventional mandrels to prevent collapse by local instability and wrinkling. However, other operations might not benefit from the advantages of having internal support if, for example, the flow of material creates difficulties (or even, renders it impossible) to withdraw the mandrels. This often limits the mass production of sound thin-walled tubular formed parts to components having geometrical features within a compact range.

In case of small-batch applications, difficulties of extracting the mandrels from the tubular formed parts can, whenever the geometry is favourable, be overcome by the use of loose fillers, expanding mandrels or state-of-the-art shape memory mandrels [10] that fully recover their original (undeformed) shape when heated above the transition temperature. However, if the objective is, for instance, to manufacture spherical or cylindrical seamless reservoirs from tubular preforms by means of the manufacturing process described in the previous section, there is a need for a new type of internal support that combines the effectiveness of conventional mandrels with the easy extraction of loose fillers.

A possible solution is to employ 'sacrificial polymer mandrels' that consists of a polymer tube shrunk into the inner diameter of a metallic tubular preform (Fig. 7.4a). The resulting bi-material assembly ensures that the sacrificial polymer mandrel is tightly held in position and the inner surface of the metallic thin-walled tubular preform is under the action of tensile stresses. Since plastic deformation of the sacrificial mandrel during forming is connected to the plastic deformation of the metallic preform, a firm internal support is guaranteed and, therefore, wrinkling and local instability at the equatorial region can be avoided. However, these mandrels are difficult to extract because their removal by melting is time consuming and likely to produce toxic gases. The alternative of using solvents for removing the mandrels at the end of the forming process should not be considered because it is not environmentally acceptable.

An alternative environmental friendly solution that was recently developed by the authors consists of employing mandrels made from low melting point alloys comprising bismuth, lead, tin and cadmium, among other materials [11]. These mandrels are capable of continuously adapting its shape to that of the formed tube and are easily removed (and recyclable) by heating slightly below or above 100ºC, while leaving the reservoir intact, at the end of the process (Fig. 7.4b).

Fig. 7.4. Tubular preforms with internal mandrels made from (a) polyvinyl chloride (PVC), (b) low melting point alloy MCP70 and an (c) aluminium alloy.

Typical commercial alloys utilized in mandrels made from low melting point alloys are MCP70 and MCP137 with melting temperatures equal to 70ºC and 137ºC, respectively. The mandrels are cast and their edges deburred with slotted angles (around 15º, Fig.7.4b) in order to avoid premature flow of the low melting point alloy into the polar openings during forming. This would greatly increase the compression force required at the end of the process and give rise to undesirable material flow at the poles.

In case of forming reservoirs made from stainless steel AISI 316 it is sometimes needed to employ internal mandrels made from aluminium alloys (Fig. 7.4c).

7.2.3 Lubrication

Forming seamless metallic reservoirs by means of the proposed manufacturing process is consistent with the three basic mechanisms governing material flow behaviour in tube processing; (i) bending, (ii) compression along the circumferential direction and (iii) friction. Bending takes place where the tubular preform contacts the dies while circumferential compression and friction develop gradually as the preform deforms against the profile shaped dies.

In case of friction, previous research work on tube forming put into evidence that operating parameters giving rise to successful modes of deformation can easily lead to unsuccessful modes of deformation if lubrication is inexistent or simply inappropriate [7]. In case of the process development described in this chapter, lubrication with zinc stearate proved efficient in a wide range of operative conditions.

7.3 Mechanical Testing of Materials

7.3.1 Flow Curve

The raw materials utilized in the investigation consisted of commercial tubes of aluminium AA6063-T0 and AA7050-T0, stainless steel AISI 316 and ingots of commercial low melting point alloysMCP70 and MCP137. The aluminium and stainless steel tubes were split into preforms that were formed into seamless reservoirs, whereas the low melting point alloys were utilized to cast the mandrels using a permanent mould by taking advantage of the good recycling properties of the material and mould.

The stress-strain curves of the materials were determined by means of conventional and stack compression tests [12] that were carried out at room temperature. The stack compression tests made use of multi-layer cylinder specimens that were assembled by pilling up circular discs cut from the commercial tubes. The preparation of the discs was critical for ensuring that all the layers were concentric and had identical cross-sectional area in order to ensure homogenous deformation under frictionless conditions along the contact interface with the platens. The test specimens were machined from the supplied material stock.

Figure 7.5 presents the stress-strain curves for the aluminium and low melting point alloys. Two different stress responses with increasing strain are observed; (i) the aluminium alloys present strain hardening while (ii) the low melting point alloys present evidence of strain softening for values of strain above 0.2.

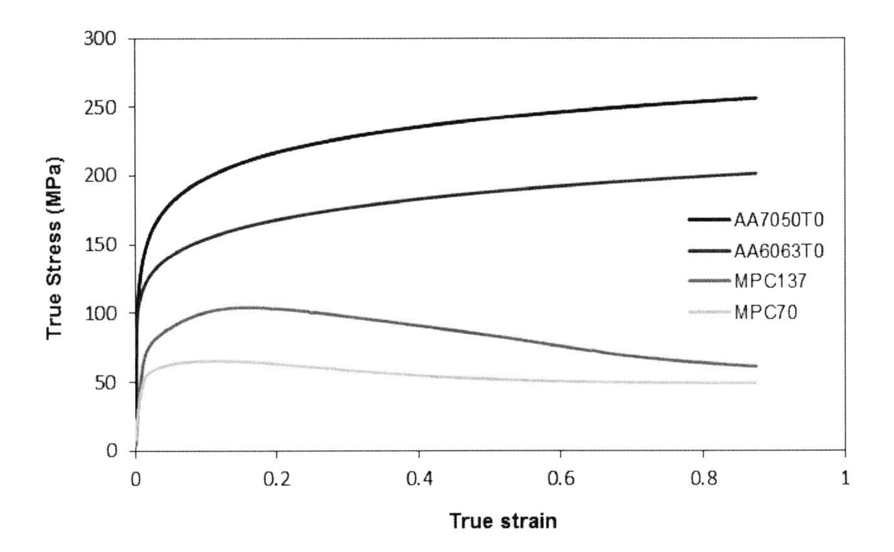

Fig. 7.5. True stress-strain curves obtained from conventional and stack compression tests of aluminium alloys AA6063-T0, AA7050-T0 and low melting point alloys MCP70 and MCP137.

7.3.2 Critical Instability Load

Axial compression of tubes gives rise to either symmetrical or unsymmetrical instability mode of deformation. The latter, usually referred to as buckling, happens when a tube is long and has relatively thick walls, while the former, usually referred to as local buckling, takes place when a tube, either short or long, has thin walls.

The understanding of the physics behind local buckling of thin-walled tubes was first given by Alexander [13] on the basis of inward and outward movements of the tube, later improved by Allan [14] and more recently enhanced by Rosa et al.[15] who illustrated the contact between successive instability waves as well as the progressive changes taking place at the contact region between the tube and the dies. The critical instability load that originates local buckling on a thin-walled tube can be determined analytically, numerically or experimentally [15].

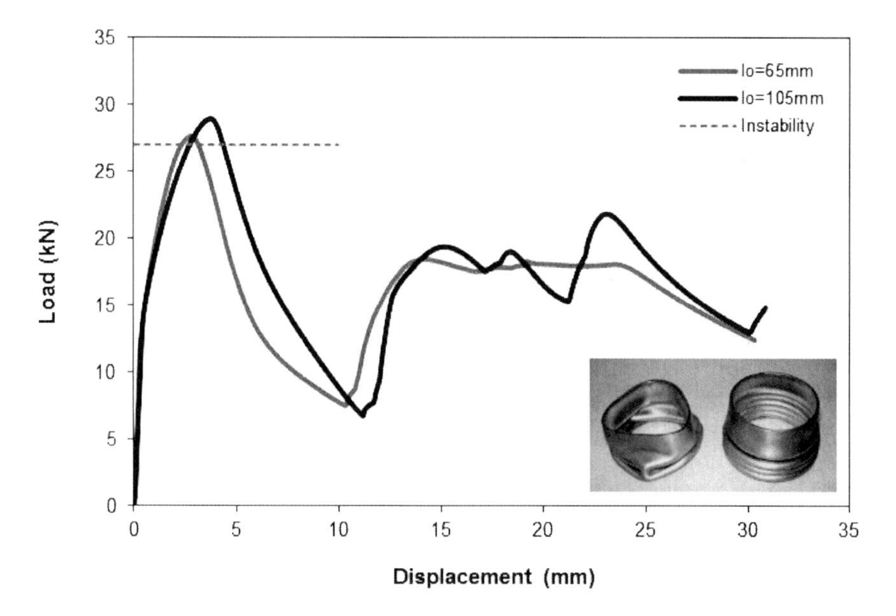

Fig. 7.6. Experimental evolution of the load–displacement curve for the axial compression of thin-walled AA6063T0 short and long thin-walled tubes between flat dies.

The experimental value of the critical instability load for the occurrence of local buckling in thin-walled tubes subjected to axial loading can be determined by compressing tubular specimens with different initial lengths between flat dies.

Figure 7.6 shows the critical instability load as a function of the displacement of the upper flat die. As can be seen, the load increases sharply from zero and local buckling occurs upon reaching a critical experimental value equal to 27 kN for aluminium tubes AA6063T0 with 60 mm diameter and 2 mm thickness. The picture placed inside Fig. 7.6 shows that a diamond shaped instability prevails over conventional axisymmetric instability when the ratio of initial tube length to diameter is small (say, close to 1).

7.4 Theoretical and Experimental Background

7.4.1 Finite Element Flow Formulation

Because the experiments in forming tubes into small size, seamless, reservoirs were performed at room temperature under a quasi-static constant displacement rate (100 mm/min) of the upper-table of the universal testing machine, no inertial effects on forming mechanisms were likely to occur and, therefore, no dynamic effects in the deformation mechanics were needed to be taken into account. These operative conditions allowed numerical modelling of the manufacturing process to be performed with the finite element flow formulation and enabled the authors to utilize the in-house computer program i-form that has been extensively validated against experimental measurements of metal-forming processes since the end of the 1980s [8].

The finite element flow formulation giving support to i-form is built upon the following weak variational form expressed entirely in terms of the arbitrary variation in the velocity,

$$\delta\Pi = \int_V \bar{\sigma}\,\delta\dot{\bar{\varepsilon}}\,dV + K\int_V \dot{\varepsilon}_V\,\delta\dot{\varepsilon}_V\,dV - \int_{S_F} t_i\,\delta u_i\,dS = 0 \tag{7.1}$$

where V is the control volume limited by the surfaces S_U and S_T where velocity and traction are prescribed, respectively, and K is a large positive constant penalizing the volumetric strain rate component $\dot{\varepsilon}_v$ in order to enforce incompressibility.

The utilisation of the flow formulation based on the penalty function method offers the advantage of preserving the number of independent variables, because the average stress σ_m can be computed after the solution is reached through,

$$\sigma_m = K\,\dot{\varepsilon}_v \tag{7.2}$$

The effective stress and the effective strain rate are defined, respectively, by,

$$\bar{\sigma} = \sqrt{\frac{3}{2}\sigma'_{ij}\sigma'_{ij}} \tag{7.3}$$

$$\dot{\bar{\varepsilon}} = \sqrt{\frac{2}{3}\dot{\varepsilon}'_{ij}\dot{\varepsilon}'_{ij}} \tag{7.4}$$

where σ'_{ij} is the deviatoric stress tensor and $\dot{\varepsilon}'_{ij}$ is the deviatoric strain-rate tensor.

The spatial discretization of the weak variational form by means of M finite elements with constant pressure interpolation, linked through N nodal points, results in the following set of nonlinear equations [8, 16],

$$\sum_{m=1}^{M}\left\{\int_{V^m}\frac{\bar{\sigma}}{\dot{\bar{\varepsilon}}}\delta\mathbf{v}^T\mathbf{K}\,\mathbf{v}\,dV^m + K^m\int_{V^m}\delta\mathbf{v}^T\mathbf{C}^T\mathbf{B}\mathbf{v}\mathbf{C}^T\mathbf{B}\,dV^m - \int_{S_T^m}\delta\mathbf{v}^T\mathbf{N}\mathbf{T}dS^m\right\} = 0 \tag{7.5}$$

which can be written in the following compact form,

$$\sum_{m=1}^{M}\{[\bar{\sigma}\,\mathbf{P} + K^m\mathbf{Q}]\{\mathbf{v}\} = \{\mathbf{F}\}\} \tag{7.6}$$

where,

$$\mathbf{P} = \int_{V^m}\frac{1}{\dot{\bar{\varepsilon}}_{n-1}}\mathbf{K}\,dV^m \tag{7.7}$$

$$\mathbf{K} = \mathbf{B}^T\mathbf{D}\,\mathbf{B} \tag{7.8}$$

$$\mathbf{Q} = \int_{V^m} \mathbf{C}^T \mathbf{B} \mathbf{C}^T \mathbf{B} \, dV^m \tag{7.9}$$

$$\mathbf{F} = \int_{S_T^m} \mathbf{N} \mathbf{T} dS^m \tag{7.10}$$

The symbol **N** denotes the matrix containing the shape functions of the element, **B** is the velocity-strain rate matrix, **C** is the matrix form of the Kronecker symbol and **D** is the matrix relating the deviatoric stresses with the strain rates according to the rate - form of the Levy-Mises constitutive equations.

The nonlinear set of Eq. 7.6 derived from the flow formulation based on the penalty function approach can be efficiently solved by a numerical technique resulting from the combination between the direct iteration and the Newton–Raphson methods.

The direct iteration method, which considers the Levy–Mises constitutive equations to be linear (and therefore constant) during each iteration, is to be preferentially utilized for generating the initial guess of the velocity field required by the Newton-Raphson method. The Newton-Raphson method is an iterative procedure based on a Taylor linear expansion of the residual force vector **R(v)** of the nonlinear set of Eq. 7.6,

$$\mathbf{R}^n = \sum_{m=1}^{M} \left\{ \left[\bar{\sigma} \mathbf{P} + K^m \mathbf{Q} \right]^n \{\mathbf{v}\}^n - \{\mathbf{F}\}^n \right\} \tag{7.11}$$

near the velocity estimate at the previous iteration,

$$\mathbf{R}\left(\mathbf{v}^n\right) \cong \mathbf{R}^n = \mathbf{R}^{n-1} + \left[\frac{\partial \mathbf{R}}{\partial \mathbf{v}} \right]_{n-1} \Delta \mathbf{v}^n = \mathbf{0} \tag{7.12}$$

where $\Delta \mathbf{v}$ is the first-order correction of the velocity field, the symbol n denotes the current iteration number,

$$\{\mathbf{v}\}^n = \{\mathbf{v}\}^{n-1} + \alpha \{\Delta \mathbf{v}\}^n \qquad \alpha \in \left]0, 1\right] \tag{7.13}$$

and α is a parameter that controls the magnitude of the velocity correction term $\Delta \mathbf{v}$. This procedure is only conditionally convergent, but converges quadratically in the vicinity of the exact solution.

The aforementioned numerical techniques are designed in order to minimise the residual force vector **R(v)** to within a specified tolerance and control and assessment is performed by means of appropriate convergence criteria.

The numerical evaluation of the volume integrals included in Eq. (7.6) is performed by means of a standard discretization procedure. Due to the rotational symmetry and as no anisotropy effects were taken into account, the finite element models set up to replicate the experimental test cases were accomplished by discretizing only the cross-section of the tubular preform and mandrel by means of axisymmetric quadrilateral elements (Fig. 7.7).

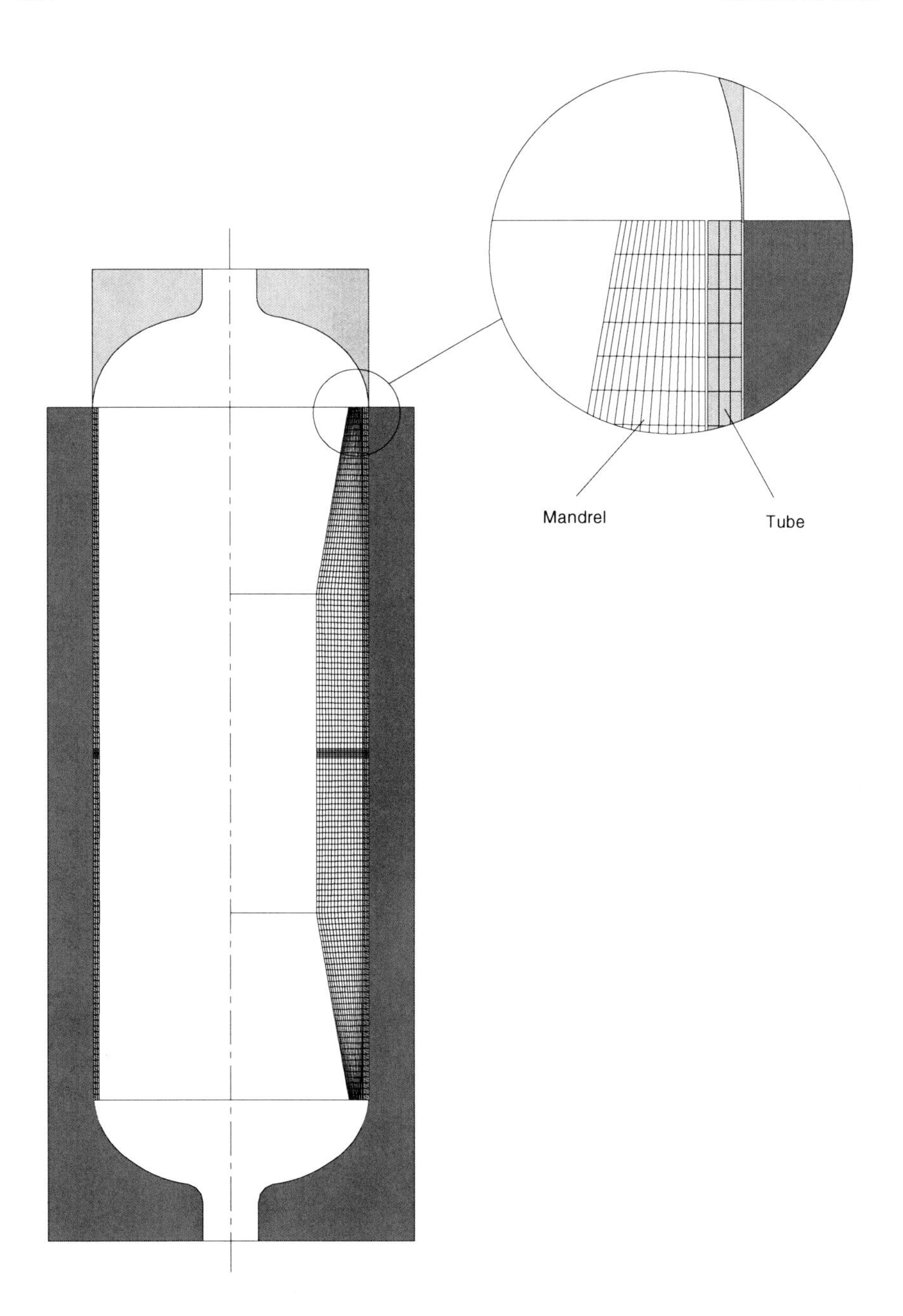

Fig. 7.7. Finite element model of the manufacturing process. Discretization of the preform and mandrel by means of quadrilateral elements.

7.4.2 Friction and Contact

Friction at the contact interface S_f between tubular specimens and tooling is assumed to be a traction boundary condition and the additional power consumption term is modelled through the utilization of the law of constant friction $\tau_f = mk$, where k is the shear yield stress in pure shear.

Implementation of friction as a traction boundary condition is performed by extending the weak variational form in Eq. 7.5 as follows [16],

$$\sum_{m=1}^{M}\left\{\begin{array}{l}\displaystyle\int_{V^m}\frac{\bar{\sigma}}{\dot{\bar{\varepsilon}}}\mathbf{K}\,\mathbf{v}\,dV^m + K^m\int_{V^m}\mathbf{C}^T\mathbf{B}\,\mathbf{v}\,\mathbf{C}^T\mathbf{B}\,dV^m - \int_{S_T^m}\mathbf{N}\,\mathbf{T}\,dS^m \\ \displaystyle\qquad + \int_{S_{FR}^m} mk\frac{2}{\pi}\mathbf{N}\tan^{-1}\left[\frac{\mathbf{N}\mathbf{v}_r}{v_0}\right]dS^m\end{array}\right\} = 0 \tag{7.14}$$

where v_r denotes the relative sliding velocity between the tube.

The approximation of the frictional stress $\tau_f = mk$ by an arctangent function of the relative sliding velocity eliminates the sudden change of direction of the frictional stress at the neutral point,

$$\tau_f = mk\left\{\frac{2}{\pi}\arctan\left(\frac{|u_r|}{u_0}\right)\right\}\frac{u_r}{|u_r|} \tag{7.15}$$

where v_0 is an arbitrary value within the range from 10^{-3} to 10^{-4} in order to avoid numerical difficulties.

The contact algorithm implemented in the finite element computer program solves the interaction between the tubular specimens and tooling by means of an explicit direct method. The algorithm requires the discretization of the tool surface into contact–friction linear elements and is based on two fundamental procedures; (i) identification of the nodal points located on the boundary of the mesh and (ii) determination of the minimum increment of time Δt_{min} for a free nodal point located on the boundary of the tubular preform to go in contact with the surface of the tool. The minimum increment of time Δt_{min} can be computed in accordance with the procedure described elsewhere [8].

The contact interface between tubular preforms and recyclable mandrels was modelled by means of a nonlinear procedure based on a penalty approach. The approach is built upon the normal gap velocity g_n^k for a nodal point k, contacting an element side ab of the adjacent element, (Fig. 7.8a),

$$g_n^k = v_n^k - \beta v_n^a - (1-\beta)v_n^b \tag{7.16}$$

where subscript n indicates normal direction and β and $1\text{-}\beta$ are the fractions of the element side l_{ab} defining the velocity projection of nodal point k on the element side ab. The penalty contact approach adds the following extra term to Eq. (7.1),

$$\delta \Pi_c = \gamma \sum_{k=1}^{N_k} g_n^k \, \delta g_n^k \tag{7.17}$$

where N_k is the total number of contacting points and γ is a large positive constant enforcing the normal gap velocity $g_n^k \geq 0$ in order to avoid penetration.

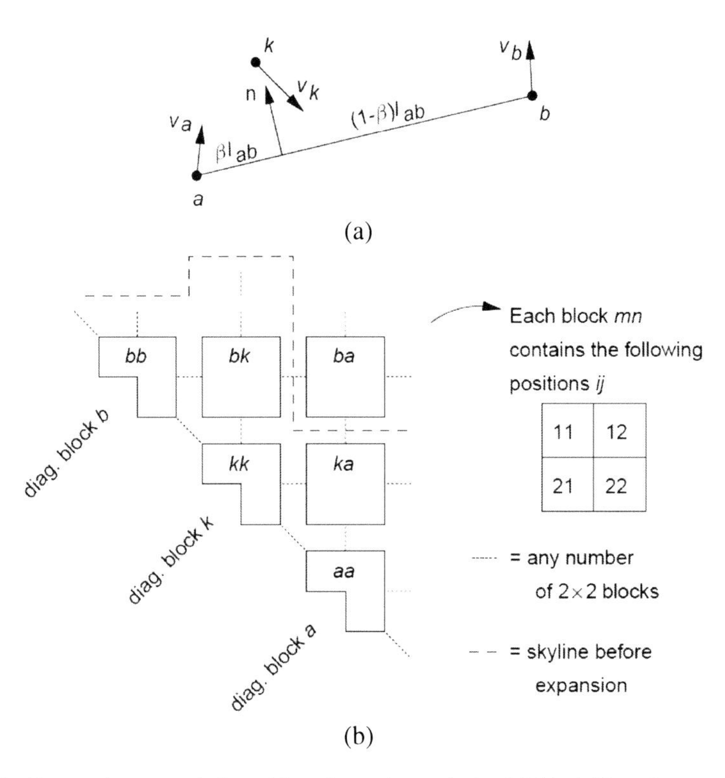

Fig. 7.8. Contact between deformable tube and mandrels. (a) Modelling the contact between nodal point k of the tubular preform (or mandrel) and element side ab of the mandrel (or tubular preform) and (b) schematic illustration of the modifications that are performed on the global stiffness matrix of the finite element model due to the contact between nodal point k and element side ab.

The extra term in Eq. (7.17) gives rise to additional contact stiffness terms $\mathbf{K}_c$ in the original stiffness matrix $\bar{\sigma}\,\mathbf{P} + K^m\mathbf{Q}$ resulting from the minimization of Eq. (7.1),

$$\begin{aligned} &\mathbf{K}_c^{ijmn} = \gamma \alpha_m \alpha_n n_i n_j \\ &(i,j) = 1,2 \quad (m,n) = k,a,b \quad \alpha_k = 1, \; \alpha_a = -\beta, \; \alpha_b = -(1-\beta) \end{aligned} \tag{7.18}$$

The positions *ijmn* of the contact stiffness terms in the overall stiffness matrix are schematically illustrated in Fig. 7.8b for typical skyline storage. It is worth noting that skyline storage usually needs to be expanded during numerical simulation in order to include new contacting pairs. The penalty contact method has the advantage of being purely geometrically based and therefore no additional degrees of freedom have to be considered as in case of alternative approaches based on Lagrange multipliers.

The numerical simulation of the manufacturing process was accomplished through a succession of displacement increments each of one modelling approximately 0.1% the initial height of the test specimens. No remeshing operations were performed and the overall CPU time for a typical analysis containing around 2500 elements was below 5 min on a standard laptop computer.

7.4.3 Experimental Development

The experimental work plan was designed in order to meet two different objectives; (i) to understand the influence of the major operating parameters on the workability window of the process and (ii) to demonstrate the feasibility of the new manufacturing process for producing seamless reservoirs with profile shaped ends.

The first objective was accomplished by means of an investigation centred on the modes of deformation and strain loading while the second objective was accomplished by fabricating small size reservoirs with hemispherical and semi-ellipsoidal ends and ends made from aluminium and stainless steel by means of the proposed manufacturing process.

The experiments were performed at room temperature in a universal testing machine under a constant cross-head displacement rate of 100 mm/min (1.7 mm/s) and the overall work plan is listed in Table 7.1.

Table 7.1. The experimental work plan

Material	Geometry	Mandrel	End Shape	t_0 (mm)	l_1 (mm)	r_0 (mm)
AA6063T0	spherical	-	-	1.1-2.5	-	8-50
AA6063T0	spherical	MCP70	-	0.9-2.2	-	20-50
AA7050 & 7075T0	spherical	MCP70	-	1.0-1.5		25-35
AA6063T0	cylindrical	MCP137 and MCP70	spherical	1-2	0-150	25-35
AA6063T0	cylindrical	MCP137	ellipsoidal	0.8-1.5	0-150	30
AISI 316	cylindrical	AA6063, MCP137 and MCP70	spherical	0.5-3	0-175	30

7.5 Mechanics of the Process

7.5.1 Modes of Deformation

Forming a tubular preform into a spherical reservoir by means of the proposed manufacturing process is the result of three basic mechanisms that compete with each other: plastic work, local bucking and wrinkling. Plastic work is caused by compression along the circumferential direction which gradually deforms the tube against the dies. Local buckling and wrinkling are associated with compressive instability in the axial and circumferential directions and limit the overall formability of the process by giving rise to non-admissible modes of deformation.

Figure 7.9 shows that forming a tubular preform with a deformable mandrel made from a low melting point alloy successfully avoids collapse by local buckling and wrinkling in order to produce a sound spherical reservoir. The finite element predicted geometry of the spherical shell and mandrel at the end of the forming process (Fig. 7.9b) compares very well with the experimental part shown in Fig. 7.9a. In fact, even the small depression at the poles that is caused by the relative sliding between the mandrel and the tubular preform is effectively predicted by finite element modelling.

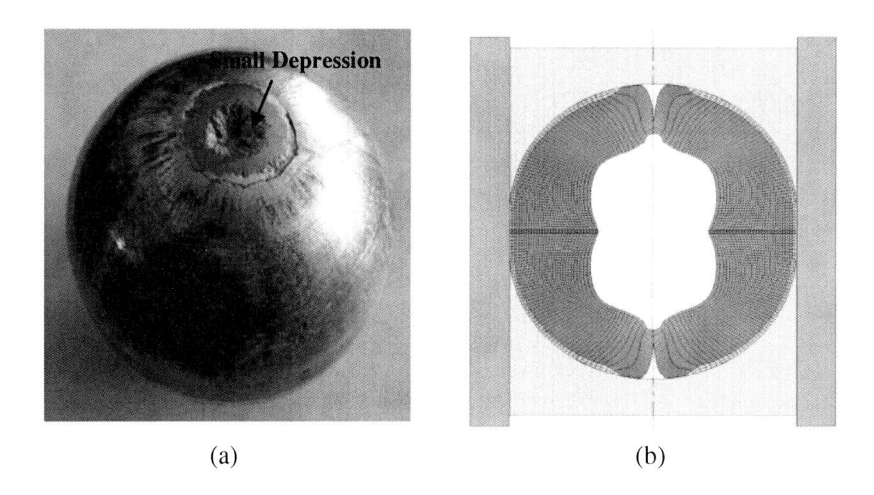

Fig. 7.9. Spherical reservoir fabricated by means of the proposed manufacturing process. (a) Spherical reservoir and internal mandrel after being formed and (b) finite element predicted geometry at the end of the process.

Subsequent removal of the mandrel by melting, while leaving the shell intact, installation of the upper valve and the lower end cap, polishing and painting, results in the spherical reservoir depicted in Fig. 7.1.

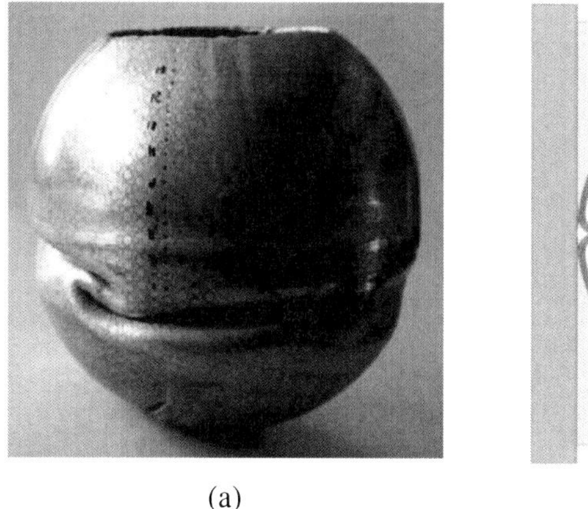

(a)

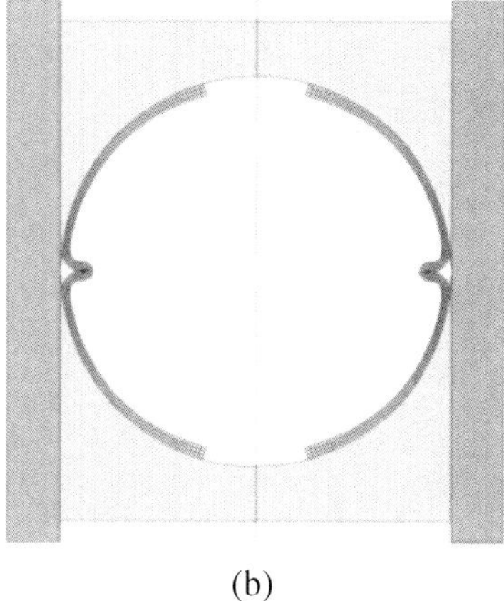

(b)

Fig. 7.10. (a) Experimental and (b) finite element predicted collapse by local bucking due to compressive instability in the axial direction.

Figure 7.10 shows the specimen and the predicted finite element geometry resulting from an attempt to shape a tubular preform into a spherical reservoir by means of the proposed manufacturing process without using an internal mandrel. As seen, formability is limited by local buckling due to compressive instability in the axial direction.

7.5.2 Formability

The technique utilised for obtaining the experimental strain loading paths in the principal strain space involved electrochemical etching of a grid of circles with 1 mm initial radius on the surface of the preforms before forming and measuring the major and minor axes of the ellipses that result from shaping the tubes into spherical shells. The experimental values of the in-plane strains were determined from (Fig. 7.11),

$$\varepsilon_1 = \ln\left(\frac{a}{2R}\right) \qquad \varepsilon_2 = \ln\left(\frac{b}{2R}\right) \tag{7.19}$$

where the symbol R represents the original radius of the circle and the symbols a and b denote the major and minor axes of the ellipse.

The in-plane components of strain resulting from the application of the above mentioned procedure in different locations taken along the meridional direction of the reservoir and plotted in the principal strain space allow us to determine the strain loading path resulting from the gradual deformation of the tubular preform against the dies (Fig. 7.12).

The principal strain space is of major importance in the analysis of forming processes because it allows foreseeing if a loading path resulting from a manufacturing process is likely to produce admissible or inadmissible modes of deformation. In case of the proposed manufacturing process, measurements and finite element

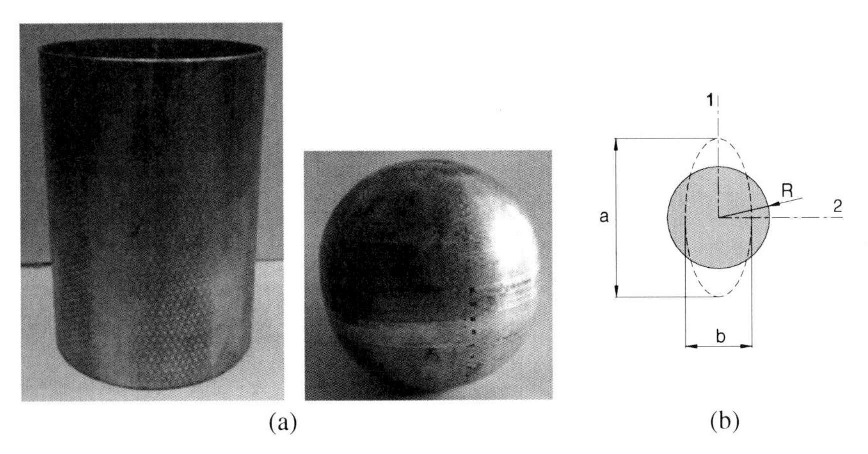

Fig. 7.11. (a) Grid of circles that were utilized for obtaining the local values of strain and (b) schematic deformation of a circle into an ellipse during the forming process.

predicted values of strain allow us to conclude that the strain loading path is similar to that of pure compression. So lying thus, close to the onset of wrinkling (refer to the photograph included in Fig. 7.12 that shows a spherical reservoir with wrinkles at the upper end).

Under these conditions, the internal mandrel plays a key role in the proposed manufacturing process because it is capable of avoiding collapse by local buckling due to compressive instability in the axial direction and impeding the strain loading path to approach the onset of wrinkling. These conclusions apply to other shapes of reservoirs than spherical as will be seen in the following section of this chapter.

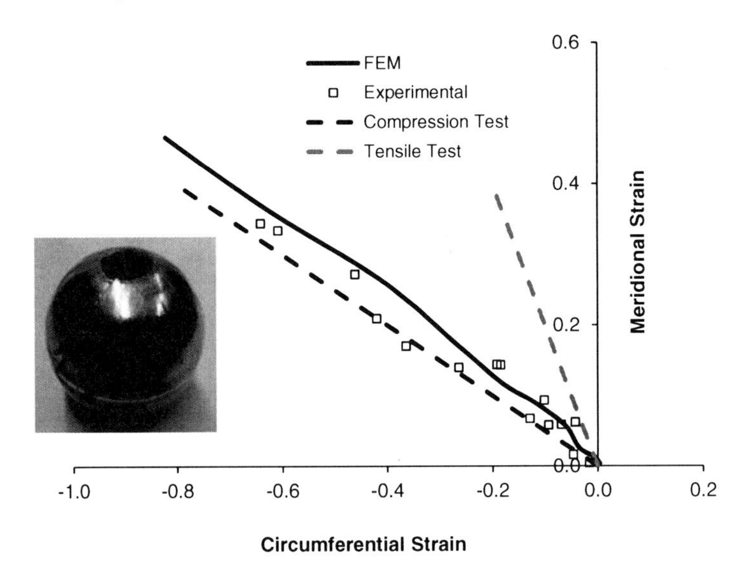

Fig. 7.12. Experimental and finite element predicted strain loading path in the principal strain space resulting from forming a tubular preform into a spherical reservoir.

7.5.3 *Forming Load*

Figure 7.13 shows the experimental and finite element predicted evolution of the load-displacement curve for a test case in which fabrication was successfully accomplished by means of the proposed forming process using a mandrel.

As seen, there is a steady monotonic increase of the forming load with the displacement of the upper die. Because the ultimate forming load for producing cylindrical reservoirs with semi-ellipsoidal ends made from AA6063T0 with 60 mm diameter is below 350 kN, the process can be industrialized in a low-cost, small capacity, press. The overall ability of the finite element computer model to predict, not only the ultimate forming load but also the general shape of the load-displacement curve, is very good.

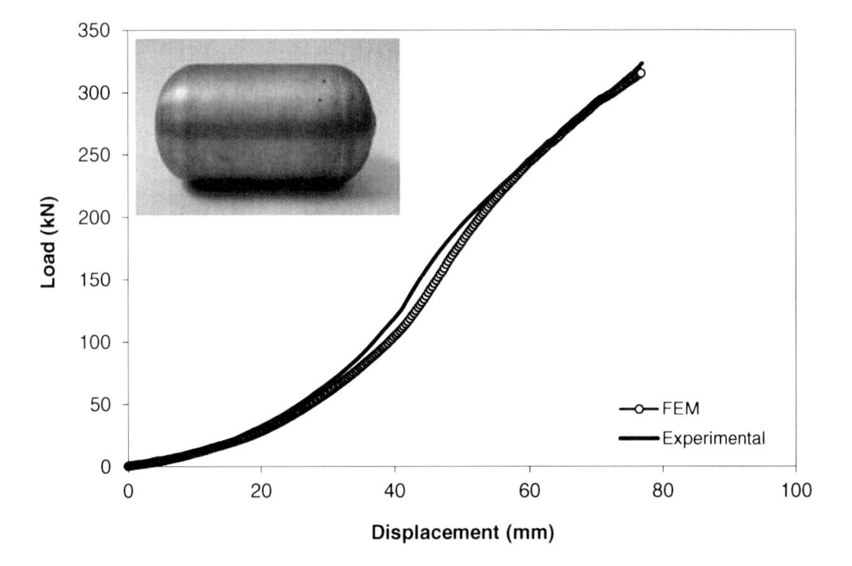

Fig. 7.13. Experimental and finite element predicted evolution of the load-displacement curve.

7.6 Applications

High pressure reservoirs are fundamental for several industries. For terrestrial applications their role is important in markets such as transportation where they are employed for the storage of compressed natural gas (over 11 million vehicles worldwide) [17] and hydrogen (hailed as the 'fuel of the future', which according to the DoE should translate to nearly 150 million in-circulation vehicles by 2050) [18]. Besides transportation, high pressure reservoirs are used for scuba diving applications, professional paint ball nitrogen high pressure bottles, etc.

7.6.1 Performance and Feasibility of the Process

Figure 7.14 shows the initial finite element discretization of the tubular preforms together with the computed predicted geometry of the reservoirs at the end of the process. The reservoirs have 60 mm diameter and are made from AA6063T0 at the end of the process.

As seen, forming with an internal mandrel allows fabricating sound reservoirs whereas forming without a mandrel will inevitably lead to a disastrous result similar to that previously obtained when attempting to fabricate spherical reservoirs without a mandrel (refer to Fig. 7.10). This result further confirms the key role played by internal mandrels in the overall success of the proposed manufacturing process.

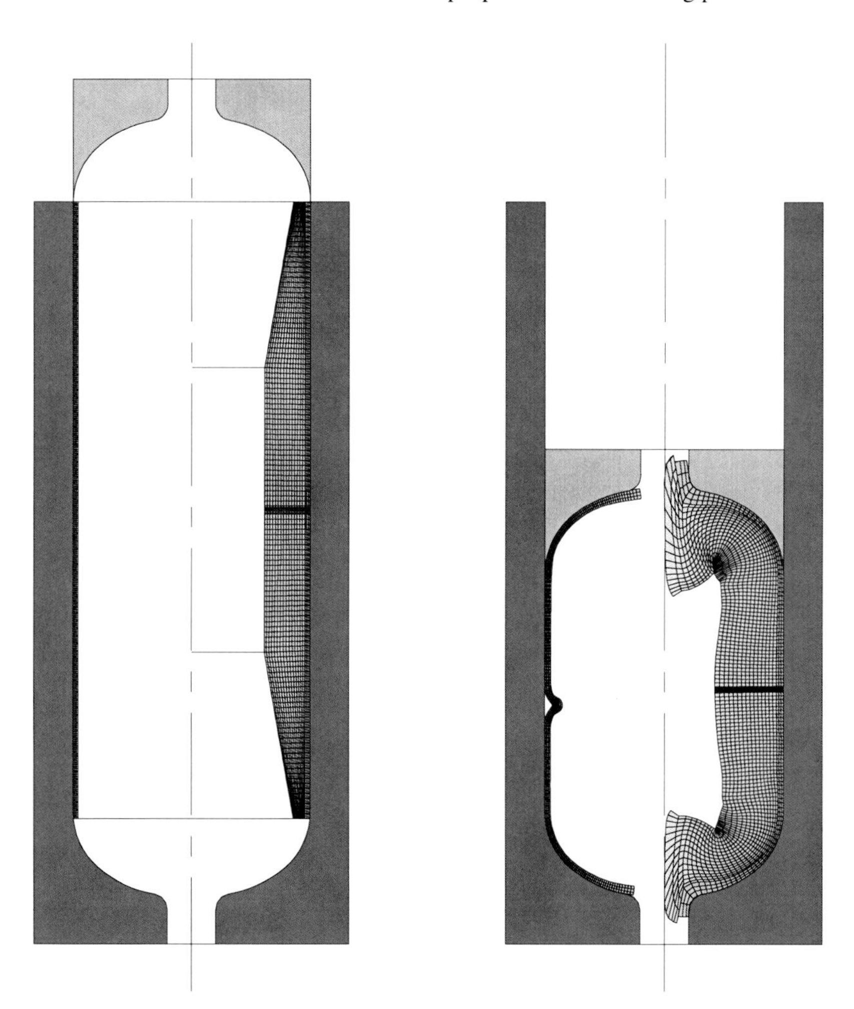

Fig. 7.14. Forming tubular preforms into cylindrical reservoirs with semi-ellipsoidal ends. Discretization of the tubular preform and mandrel (if exists) by finite elements and computed predicted geometry at the end of the process.

The finite element predicted distribution of effective stress at intermediate and final stages of the forming process with an internal mandrel is shown in Fig. 7.15.

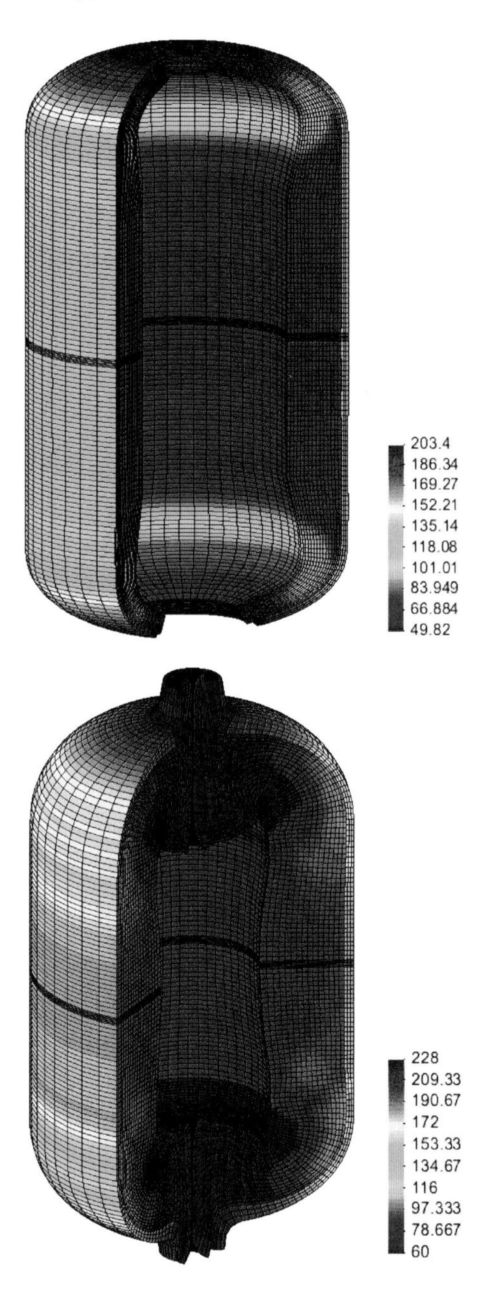

Fig. 7.15. Forming a tubular preform into a cylindrical reservoir with semi-ellipsoidal ends. Finite element predicted distribution of effective stress (MPa) after 45 mm and 90 mm displacement of the upper profile shaped die.

Figure 7.16 shows the finite element predicted distribution of average stress σ_m at the end of the forming process. As seen larger, compression values of the average stress are found at the semi-ellipsoidal tubular ends and justify the need to employ internal mandrels for avoiding collapse by wrinkling along the circumferential direction.

On the contrary, the polar openings of the reservoir and the opposite regions located at the internal mandrel show evidence of tensile average stresses.

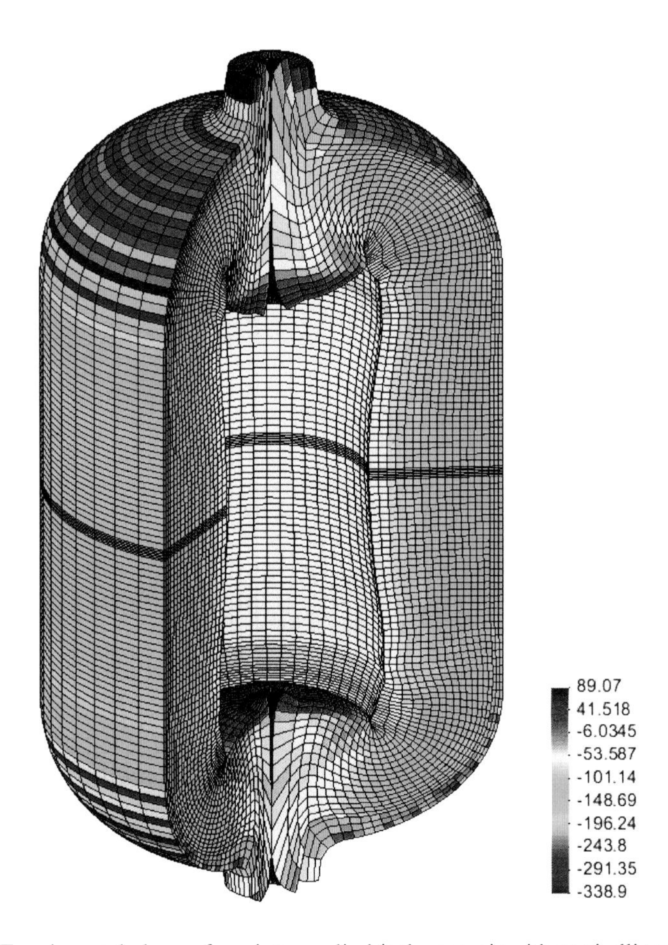

Fig. 7.16. Forming a tubular preform into a cylindrical reservoir with semi-ellipsoidal ends. Finite element predicted distribution of average stress (MPa) at the end of the forming process.

The feasibility of the proposed manufacturing process was further investigated by analysing the experimental and numerical predicted variation of thickness in the cross-section of a cylindrical reservoir with semi-ellipsoidal ends (Fig. 7.17). The meridional distance is measured from the mid-point of the finished reservoir and, as seen in the figure, thickness variation along the cross-section of the reservoirs shows a significant growth as the circumferential perimeter decreases with values above 150% at the open poles.

The initial flat region of the graphic corresponds to nearly unstrained material placed in the cylindrical region of the reservoir. The final thickness in this region of the reservoir remains practically identical to the initial thickness of the preform. The subsequent slight decrease in the variation of thickness is related to the portion of the tubular preform that starts to bend in order to match the contour of the die. Measurements and numerical predictions can even yield negative values, resulting in local thicknesses smaller than that of the original preform, as can be observed at 40 mm distance from the equatorial region.

The last part of the graphic (say, above 45 mm distance from the equator) shows a significant growth rate in thickness variation. This is due to compression in the circumferential direction and the significant increase of thickness at the polar openings of the reservoir being very advantageous for subsequent installation of valves and end caps by mechanical fixing or welding.

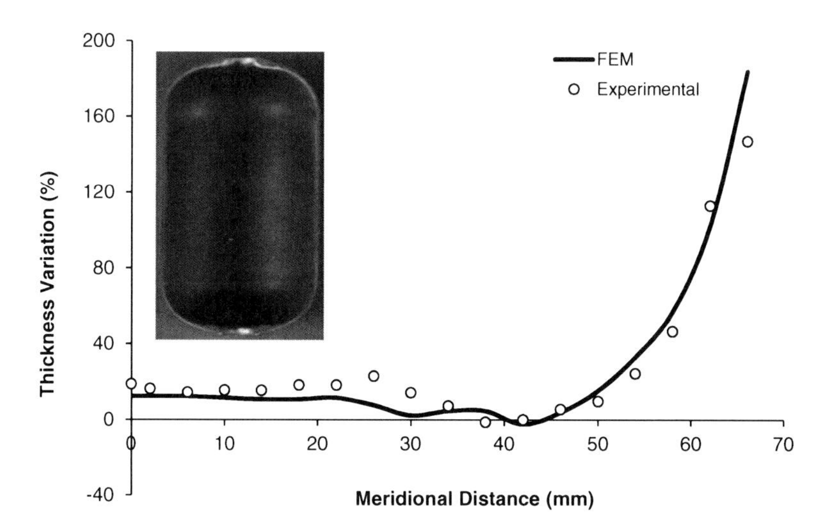

Fig. 7.17. Experimental and finite element predicted variation of thickness in the cross-section of a cylindrical reservoir with semi-ellipsoidal ends.

The initial flat region of the graphic corresponds to nearly unstrained material placed in the cylindrical region of the reservoir. The final thickness in this region of the reservoir remains practically identical to the initial thickness of the preform. The subsequent slight decrease in the variation of thickness is related to the portion of the tubular preform that starts to bend in order to match the contour of the die. Measurements and numerical predictions can even yield negative values, resulting in local thicknesses smaller than that of the original preform, as can be observed at 40 mm distance from the equatorial region.

The last part of the graphic (say, above 45 mm distance from the equator) shows a significant growth rate in thickness variation. This is due to compression in the circumferential direction and the significant increase of thickness at the polar openings of the reservoir being very advantageous for subsequent installation of valves and end caps by mechanical fixing or welding.

7.6.2 Requirements for Aerospace Applications

In the aerospace industry, high pressure reservoirs are required for the storage of gases, specifically noble gases (such as helium or argon) which, because they form no compounds and are difficult to store in compound form and, because they are light tend to have very low boiling points and severe boil-off losses.

Hence, these gases are usually stored in high pressure form to achieve significant volumetric density. The relevant application, as required by Omnidea, adds the fact that pressure might be required for something more than volumetric density, as pressure is required so that the stored gas can act as pressurant (helium and nitrogen) in chemical propulsion or directly as propellant (xenon) in electric-ionic space propulsion (Fig. 7.18). While xenon requires pressures of the order of 150-200 bar, helium and nitrogen require 280-350 bar.

To withstand these pressures while still delivering good performance, a high pressure vessel performance index I_p must be achieved,

$$I_p = \frac{pV}{m} \tag{7.20}$$

where p is pressure, V is volume and m is the tank's mass. The performance index is usually presented in J/kg^{-1} reflecting the fact that an energy density is stored.

The typical construction type of storage vessels with high values of the performance index is a hybrid solution in which a metallic liner is wrapped around with epoxy-embedded composite fibre in an arrangement usually known as 'composite overwrapped pressure vessels' (COPVs). The metallic liner provides mainly the shape, gas tightness and the tank's toughness, while the composite overwrapping provides the strength required to withstand the tank's internal pressure.

The application here under discussion, xenon storage for electric propulsion in satellites, requires the production of a metallic liner for a COPV, matching the performance of current state-of-the-art COPVs with significant (over 50%) cost and manufacturing time reductions versus conventional manufacturing technologies.

For this particular application, aluminium alloys best fit the requirements of fabricating seamless high pressure reservoirs with a high strength-to-weight ratio, adequate toughness, low cost and considerable availability in seamless extruded tube form, including a multiplicity of diameters and thicknesses. It is important to mention that this manufacturing technique cannot be employed without access to seamless extruded tube. Also important, from a manufacturing point of view, is

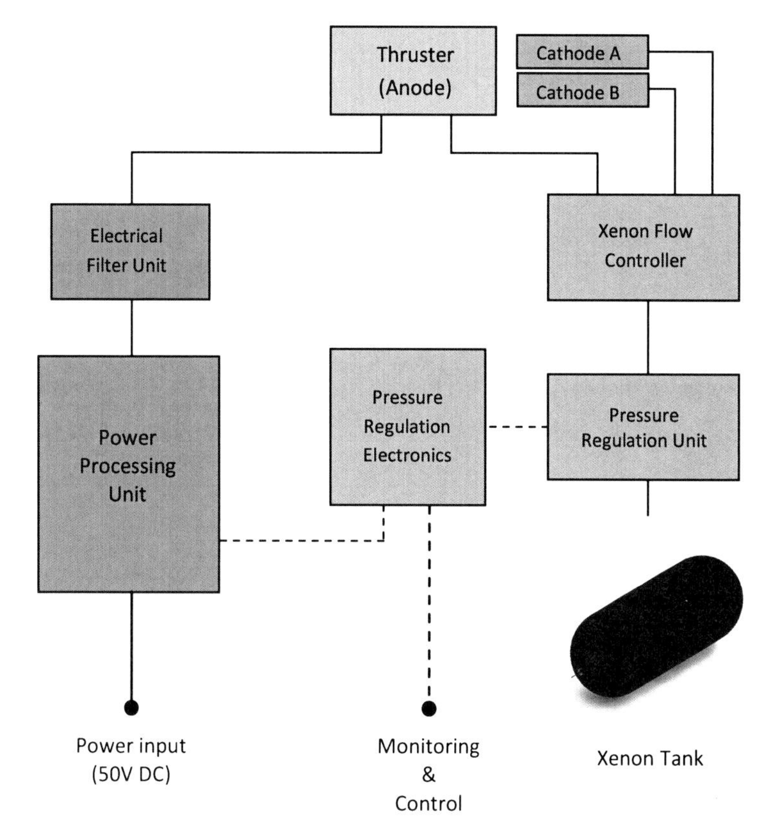

Fig. 7.18. Electric–ionic propulsion system layout in ESA's Smart-1 probe. The xenon tank is of US manufacture and a candidate for European replacement.

the very high heat treated-to-annealed strength ratio (see Fig. 7.19), allowing the forming process to be performed in annealed conditions, which in turn provides reduce polar apertures, good surface roughness characteristics (important for sealing) while maintaining a relatively small press force. This is not always easy to achieve with cold forming processes while the subsequent heat treatment process, applied to the already formed spherical pressure vessels, can avoid the residual stress remaining from the cold forming operation.

As it is easy to understand from Fig. 7.19 only aluminium alloys are capable of multiplying the yield strength by a factor of 3 to 4 between annealed and heat-treated conditions, especially when compared to alternative metals for this application: Titanium alloys (mainly the α-β Ti-6Al-4V alloy, the current industry benchmark), Inconel 718 and AISI 316 stainless steel.

Figure 7.20 shows a cylindrical reservoir with hemispherical ends made of AISI 316 stainless steel that was successfully formed with an internal mandrel together with a cylindrical reservoir with hemispherical ends also made of AISI 316 that was not successfully formed with an internal mandrel made of MCP137.

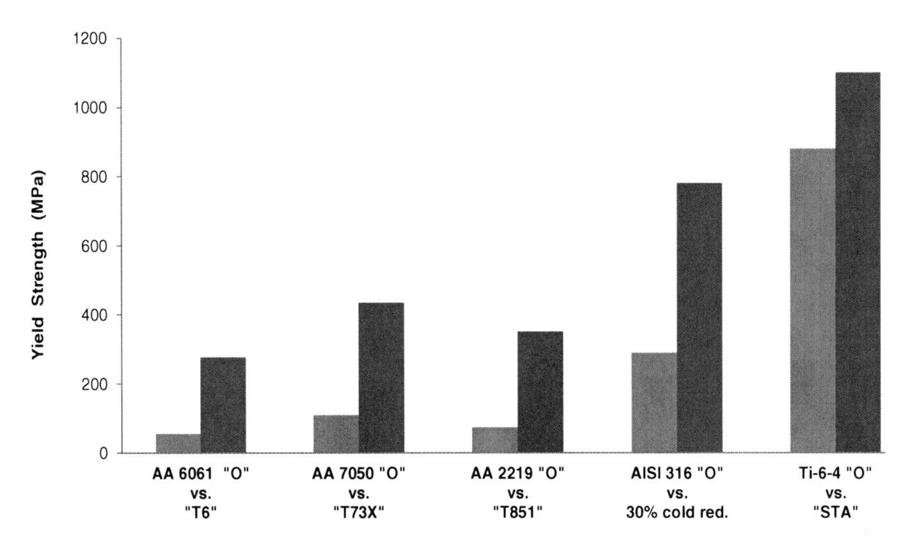

Fig. 7.19. Comparison, for different metallic alloys, of yield strength in annealed vs. treated conditions. Aluminium alloys exhibit the smaller annealed strengths and greatest response to treatment.

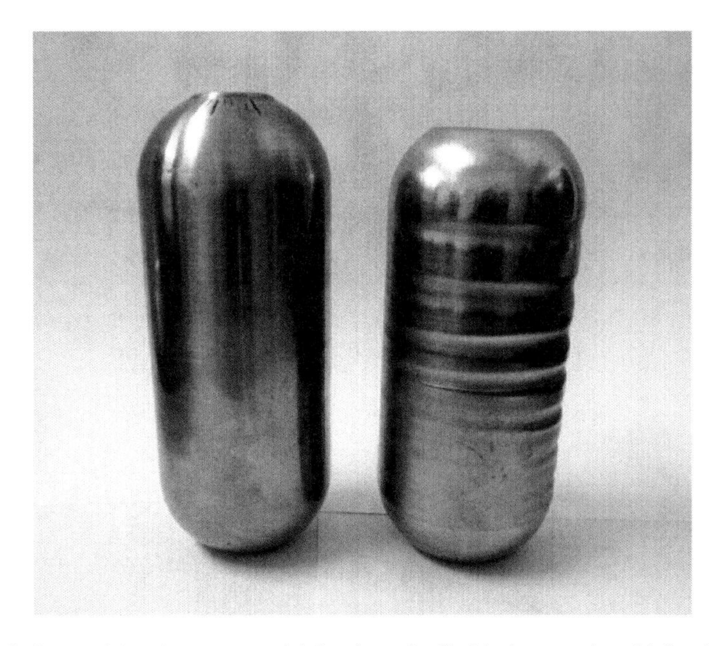

Fig. 7.20. Successful and non-successful forming of cylindrical reservoirs with hemispherical ends made from AISI 316 stainless steel.

7.7 Conclusions

The proposed manufacturing process was developed to fit the specific requirements of InnovGas project, performed between the Portuguese SME Omnidea and the European Space Agency, which required substantial development of manufacturing technologies capable of producing seamless high pressure reservoirs. The process extends the tools and techniques commonly utilized in tube forming in order to include two innovative features related to the utilization of sharp edge dies and internal, recyclable, mandrels made from low melting point alloys.

Sharp edge dies and internal mandrels proved crucial to fabricate spherical and cylindrical reservoirs from both aluminium and stainless steel, in a single forming operation without the risk of collapse by local buckling and/or wrinkling. Local buckling and wrinkling are associated with compressive instability in the axial and circumferential directions during the forming process. Wrinkling can also be attributed to the strain loading paths being close to uniaxial compression as has been experimentally observed and numerically predicted by means of finite element analysis.

The increase in thickness at the poles is useful for installing devices, fixing the outlet ports; also because overwrapping is much more complex near the polar regions, the metallic liner alone might be required to withstand the internal pressure, thus requiring higher thicknesses. In addition, the ultimate forming load for producing seamless reservoirs is small enough for enabling the process to be industrialized in a low-cost, small capacity, press.

References

[1] Hübner, A., Teng, J.G., Saal, H.: Buckling behaviour of large steel cylinders with patterned welds. International Journal of Pressure Vessels and Piping 83, 13–26 (2006)

[2] Kawahara, G., McCleskey, S.F.: Titanium lined, carbon composite overwrapped pressure vessel. In: 32nd AIAA/ASME/SAE/ASEE Joint Propulsion Conference, Lake Buena Vista, FL, USA (1996)

[3] Teng, J.G., Lin, X.: Fabrication of small models of large cylinders with extensive welding for buckling experiments. Thin-Walled Structures 43, 1091–1114 (2005)

[4] Lee, H.S., Yoon, H.S., Yoon, J.H., Park, J.S., Yi, Y.M.: A study on failure characteristic of spherical pressure vessel. Journal of Materials Processing Technology, 164–165, 882–888 (2005)

[5] Ostwald, P., Muñoz, J.: Manufacturing processes and systems. John Wiley & Sons, New York (1997)

[6] Fengman, H., Zheng, T., Ning, W., Zhiyong, H.: Explosive forming of thin-wall semi-spherical parts. Materials Letters 45, 133–137 (2000)

[7] Rosa, P.A.R., Alves, L.M., Martins, P.A.F.: Experimental and numerical modelling of tube end forming processes. In: Davim, J.P. (ed.) Finite Element Methods in Manufacturing Processes, pp. 93–136. ISTE –Wiley (2010)

[8] Alves, M.L., Rodrigues, J.M.C., Martins, P.A.F.: Simulation of three-dimensional bulk forming processes by the finite element flow formulation. Modelling and Simulation in Materials Science and Engineering – Institute of Physics 11, 803–821 (2003)

[9] Alves, L.M., Martins, P.A.F., Pardal, T.C., Almeida, P.J., Valverde, N.M.: Plastic deformation technological process for production of thin-wall revolution shells from tubular billets. Patent request no. PCT/PT2009/000007, European Patent Office (2009)
[10] Everhart, M.C., Stahl, J.: Reusable shape memory polymer mandrels. In: Proceedings of the SPIE - The International Society for Optics and Photonics, vol. 5762, pp. 27–34 (2005)
[11] Alves, L.M., Pardal, T.C.D., Martins, P.A.F.: Nosing thin-walled tubes into axisymmetric seamless reservoirs using recyclable mandrels. Journal of Cleaner Production 18, 1740–1749 (2010)
[12] Alves, L.M., Nielsen, C.V., Martins, P.A.F.: Revisiting the Fundamentals and Capabilities of the Stack Compression Test. Experimental Mechanics (in press, 2011)
[13] Alexander, J.M.: An approximate analysis of the collapse of thin cylindrical shells under axial loading. Quarterly Journal of Mechanics and Applied Mathematics 13, 10–15 (1960)
[14] Allan, T.: Investigation of the behaviour of cylindrical tubes subject to axial compressive forces. Journal of Mechanical Engineering Science 10, 182–197 (1968)
[15] Rosa, P.A.R., Rodrigues, J.M.C., Martins, P.A.F.: External inversion of thin-walled tubes using a die: experimental and theoretical investigation. International Journal of Machine Tools and Manufacture 43, 787–796 (2003)
[16] Kobayashi, S., Oh, S.I., Altan, T.: Metal forming and the finite element method. Oxford University Press, New York (1989)
[17] IANGV International Association for Natural Gas Vehicles, http://www.iangv.org
[18] Report to Congress, Effects of a Transition to a Hydrogen Economy on Employment in the United States, Department of Energy, US (2008)

Author Index

Subject Index

Printed by Publishers' Graphics LLC
CIMO20121002.13.05.21